Energy and U.S. Foreign Policy

A Report to the Energy Policy Project of the Ford Foundation

Energy and U.S. Foreign Policy

Joseph A. Yager
Eleanor B. Steinberg

with
Barry M. Blechman
Robert M. Dunn, Jr.
Edward R. Fried
Jerome H. Kahan
Arnold M. Kuzmack
Robert A. Mertz
Philip Musgrove
Phillip H. Trezise

Ballinger Publishing Company ● Cambridge, Mass.
A Subsidiary of J.B. Lippincott Company

Published in the United States of America by Ballinger Publishing Company,
Cambridge, Mass.

First Printing, 1974

Library of Congress Catalog Card Number: 74-19193

International Standard Book Number: 0-88410-027-8 H.B.
                                    0-88410-028-6 Pbk.

Printed in the United States of America

Library of Congress Cataloging in Publication Data

Yager, Joseph A
    Energy and U.S. foreign policy.

    "A report to the Energy Policy Project of the Ford Foundation."
    Includes bibliographical references.
    1. Petroleum industry and trade. 2. Power resources. 3. Energy policy—
United States. I. Steinberg, Eleanor B., joint author. II. Ford Foundation.
Energy Policy Project. III. Title.
HD9566.Y34   333.7   74-19193
    ISBN 0-88410-027-8
    ISBN 0-88410-028-6 (pbk.)

Contents

List of Tables

xi

List of Maps

Foreword

In December 1971 the Trustees of the Ford Foundation authorized the organization of the Energy Policy Project. In subsequent decisions the Trustees have approved supporting appropriations to a total of $4 million, which is being spent over a three-year period for a series of studies and reports by responsible authorities in a wide range of fields. The Project Director is S. David Freeman, and the Project has had the continuing advice of a distinguished Advisory Board chaired by Gilbert White.

This analysis of some of the international aspects of energy policy, entitled "Energy and U.S. Foreign Policy," is one of the results of the Project. As Mr. Freeman explains in his Preface, neither the Foundation nor the Project presumes to judge the specific conclusions and recommendations of the authors who prepared this volume. We do commend this report to the public as a serious and responsible analysis which has been subjected to review by a number of qualified readers.

This study, like many others in the Project, deals with sensitive and difficult questions of public policy. Not all of it is easy reading, and not all those we have consulted have agreed with all of it. Nor does it exhaust a subject which is complex, controversial, and subject to sudden changes; in fact, the study has had to be revised during preparation to take account of recent events which are still unfolding.

The study reflects many of the tensions which have been intrinsic to the whole of the Energy Policy Project—tensions between one set of objectives and another. As the worldwide energy crisis has become evident to us all, we have had many graphic illustrations of such tensions, and there are more ahead. This is what usually happens when societies face hard choices, all of them carrying costs that are both human and material.

But it is important to understand that there is a fundamental difference between present tension and permanent conflict. The thesis accepted by our Board of Trustees when it authorized the Energy Policy Project was that the very existence of tension, along with the inescapable necessity for hard choices, argued in favor of studies which would be, as far as possible, fair, responsible, carefully reviewed, and dedicated only to the public interest. We do not suppose that we can evoke universal and instantaneous agreement, and still less do we presume that this Project can find all the answers. We do believe that it can make a useful contribution to a reasonable and democratic resolution of these great public questions, one which will serve the general interest of all.

The current study is a clear example of what we aim at. It draws on a large body of political, economic and technical information and professional judgment to reach conclusions about likely developments in world energy markets and to suggest policy approaches for dealing with these developments. A wide range of outside experts and organizations reviewed the book; and, although not all of them agreed with all of the analysis and conclusions, we do believe that the authors have treated their hard subject with the respect it deserves. I commend their analysis to the attention of the American public.

McGeorge Bundy
President, Ford Foundation

Preface

The Energy Policy Project was initiated by the Ford Foundation in 1971 to explore alternative national energy policies. This book, *Energy and U.S. Foreign Policy*, is one of the series of studies commissioned by the Project. It is presented here as a carefully prepared contribution by the authors to today's public discussion about the international aspects of energy policy. It is our hope that each of these special reports will stimulate further thinking and questioning in the specific areas it addresses. At the very most, however, each special report deals with only a part of the energy puzzle; the Energy Policy Project's final report, *A Time To Choose*, which was published in October 1974, attempts to integrate these parts into a comprehensible whole, setting forth the energy policy options available to the nation as we see them.

This book, like the others in the series, has been reviewed by scholars and experts in the field not otherwise associated with the Project in order to be sure that differing points of view were considered. With each book in the series, we offer reviewers the opportunity of having their comments published in an appendix to the volume; none have chosen to do so for this volume.

Energy and U.S. Foreign Policy is the authors' report to the Ford Foundation's Energy Policy Project and neither the Foundation, its Energy Policy Project or the Project's Advisory Board have assumed the role of passing judgment on its contents or conclusions. We have expressed our views in *A Time To Choose*.

S. David Freeman
Director
Energy Policy Project

Authors' Note

Because of its size and scope, this study of the international aspects of energy policy required the specialized knowledge and skills of a large number of people. Persons who contributed directly to the study are listed on the title page, and the nature of their contribution is noted at appropriate points in the text. All of the authors, except Robert M. Dunn, Jr., who is associate professor of economics at The George Washington University, were members of the staff of the Brookings Institution at the time of their participation in the study.

Expert consultants were also employed in several fields. The consultants listed below prepared background papers on the areas indicated:

> Herbert Block—Soviet Union
> John C. Campbell—the Arab countries
> Robert M. Dunn, Jr.—Canada
> Jerome F. Fried—Iran, the less developed countries, and
>   the impact of oil income on development.
> Francis W. Herron—Latin America
> Thomas G. Rawski—China

In addition, the European Community Institute for University Studies prepared a background paper on Western Europe, and the Japan Economic Research Center and the Institute of Energy Economics of Japan jointly produced a background paper on Japan. The background papers were valuable contributions to the study. The authors of these papers are not responsible, however, for the way in which their material has been used.

The authors also acknowledge their debts to a number of other experts on various aspects of this study who gave generously of their time. A full list of persons in this category would be very long and would include numerous government officials, business leaders, and scholars both in this country and abroad. Special mention must, however, be made of Thomas Barger, Joel Darmstadter, Paul Frankel, John Gray, James McKie, and Milton Searl. Within Brookings, Henry D. Owen, director of the foreign policy studies program provided helpful comments on all parts of the study; A. Doak Barnett and Ralph N. Clough offered useful comments on the chapters dealing with China and Japan. None of these persons is responsible for the way in which his advice has been used.

Special acknowledgement must also be made of the contributions of Jean M. Shirhall, who edited the study and worked closely with the authors during the entire process of preparing the study for publication; Bonnie S. Wilson, who made major contributions during the research phase of the study; and Richard Gross, who collected and organized much of the statistical information.

The authors were fortunate in having the assistance of two unusually able secretaries, Aeran Lee and Annette M. Solomon. With commendable good nature and efficiency, they tolerated the uneven dispositions and penmanship of the authors and worked industriously over a period of many months to bring the study to completion.

The views expressed in the book are those of the authors and should not be attributed to the trustees, officers, or other staff members of the Brookings Institution or the Ford Foundation.

Joseph A. Yager
Eleanor B. Steinberg

The Brookings Institution
Washington, D. C.
July 1974

Energy and
U.S. Foreign
Policy

Part One

Introduction

Chapter One

The World Energy Problem

The world is not about to run out of energy. The known and probable deposits of oil, coal, gas, and uranium are still enormous, and there are no insurmountable physical barriers to the expansion of energy production from these sources in order to meet the world's needs through this century and beyond. The cost of finding and exploiting additional supplies of these exhaustible mineral resources can, however, be expected to rise as the more accessible and higher quality deposits are exhausted, and the world must be prepared to turn to other forms of energy, such as solar and geothermal energy and controlled nuclear fusion. But for the period relevant to most public policy decisions—especially in the field of international relations—the world energy problem is not one of absolute physical shortage. The difficulties lie instead in the shadowy borderland where politics, technology, and economics overlap. Rapid changes in all of these areas are having a strong impact on production and trade in energy materials. And the need to adjust to these changes has created the world energy problem.

The force that dominates all others in the energy field is the continuous, rapid rise in energy requirements in all parts of the world. By 1985, under current projections, the non-Communist nations may be consuming almost twice as much energy as they did in 1970.* (See Table 13-4.) Two fundamental energy problems facing the world are finding the means of meeting these rising

*Much of the analysis in this study focuses on the non-Communist nations. This approach reflects both the relative lack of reliable energy statistics for the Communist nations and the fact that those nations are not fully involved in the international markets for energy materials.

requirements efficiently and avoiding economically damaging inter-ruptions of energy supplies. Other, more specific, problems are created by changes in the world's reliance on various sources of energy.

During the 1960s, coal, which had powered the industrial revolution, lost its place as the dominant source of energy to oil. This change was not caused by the using up of the world's coal resources, which are still very large, but by the lower cost and greater convenience of oil. Natural gas also became a major source of energy during the middle decades of this century.

By 1970, oil was providing a little more than half of the total energy consumed by the non-Communist nations.* Coal accounted for nearly a quarter of the total, and gas for about a fifth. The remainder was mostly supplied by hydropower; nuclear power supplied less than half of one percent of all energy used.

A change in the pattern of energy supply is currently under way. By 1985, the share of oil may fall to about 40 percent of the total and that of coal to about 15 percent. Gas and hydropower should roughly hold their own. The big increase—if current technical and political problems are overcome—will be in nuclear power, which may supply some 15 percent of total consumption by 1985.

Since most of the non-Communist nations lack adequate oil resources of their own, their increased reliance on oil has made them increasingly dependent on imported energy. The United States was long an important exception. Since World War II, however, oil consumption in the United States has outrun, first, domestic oil-production capabilities and, more recently, all available sources of supply in the Western Hemisphere. In 1973, U.S. dependence on Eastern Hemisphere oil reached 36 percent of total imports, and, at least for the next few years, that dependence will almost certainly continue to grow.

As oil production increased to meet rising world require-ments, the dominant position of the states bordering on the Persian Gulf became more and more evident. In 1973, those states possessed approximately two-thirds of the proved oil reserves of the non-Communist world and accounted for 64 percent of non-Communist oil production. This geographic concentration of oil production has had important economic and political consequences. It has facilitated action by the major oil-producing countries to take advantage of the

*Appendix A describes world patterns of energy production, consumption, trade, and reserves. Regional and major country breakdowns are presented in tables accompanying the appendix. The data indicate the overriding importance of liquid fossil fuels in meeting the world's current energy needs.

sellers' market that arose in the early 1970s. And it enabled the Arab oil-exporting countries to use oil as an effective political weapon during and after the Arab-Israeli war of October 1973.

High oil prices and the danger of renewed manipulation of oil supplies for political purposes are two of the major foreign policy problems in the energy field that face the United States and other oil-importing countries today. These problems are also at the root of other serious problems.

The huge increase in price imposed by the oil-exporting countries at the end of 1973 also created new balance-of-payments difficulties for many industrialized and developing nations. For the industrialized nations, the problem, although serious, is essentially one of relieving sudden and unexpected strains on the international financial system. For the hardest hit developing countries, however, the problem is the more fundamental one of finding the means of paying for vital imports of food, fuel, and fertilizer.

The Arab oil supply restrictions and selective embargo of 1973-74 had a temporarily divisive effect on relations between the United States and some of its major allies and trading partners. This experience underlined the need to keep differences over energy policy from inflicting more lasting damage to these essential relations. It also demonstrated the need to consider more urgently precautions against possible future interruptions of oil supplies.

The United States, more than any other oil-importing country, must also grapple with serious political problems in the Middle East that could not only disrupt an important part of the world's oil supplies but could also lead to an exceedingly dangerous confrontation with the Soviet Union. The central issue is finding means of reconciling the conflict between the close U.S. ties with Israel and the need to ensure an uninterrupted and steadily increasing volume of oil exports from the Arab oil-producing nations.

The problems created by the impending rapid spread of nuclear energy are quite different from those relating to oil. The two sets of problems are, however, interrelated in that the fact that the oil market is beset by serious problems makes it all the more important that the problems surrounding a new source of energy be dealt with successfully.

Nuclear energy owes its origin and early development to governmental initiatives, and major aspects of this new industry are still in the hands of governments. It is therefore not surprising that nuclear energy is a subject of public policy in the United States and elsewhere. At the same time, a private nuclear power industry of global scope has emerged that is sustained by a large and expanding trade in fuel, reactors, and technology. This new international industry has injected a dynamism into the spread of commercial nuclear power and must be taken into account by governments in dealing with the problems associated with nuclear energy. These problems all proceed from the unique nature of the fuel (natural or enriched uranium) used by present-day nuclear power plants and of the by-products resulting from operation of such plants. The fuel itself or one of the by-products, plutonium, can be used directly or processed into the material for nuclear bombs or explosive devices. Moreover, the by-products are dangerously radioactive and, in some cases, retain their toxicity for thousands of years. And finally, the operation of nuclear power plants creates problems of thermal pollution and some danger, however slight, of accidental release of lethal radioactive materials into the atmosphere.

The environmental and safety problems of nuclear power are primarily matters of domestic concern for countries adopting this source of energy. Preventing the diversion of nuclear materials for military purposes or the theft of such materials by terrorist or other criminal groups, however, requires international action. Establishing effective safeguards against these contingencies is made difficult by the very rapidity with which commercial nuclear power is spreading in all parts of the world and by the development of new nuclear technologies, most notably the fast breeder reactors and the centrifuge method of enriching uranium.

This study is concerned with these and other related international problems arising from the production and trade in energy. The focus is on what the United States can do about these problems, both in its own interest and as a responsible member of the world community. Primary attention is paid to oil and nuclear energy. International trade in coal is relatively less important and involves few foreign policy problems. Trade in natural gas is increasing, but the problems associated with that trade are largely bilateral in nature and do not approach those of oil or nuclear energy in either scope or complexity. Coal and gas are treated when relevant to the energy positions of individual countries or to significant bilateral trade relationships. The analysis of U.S. policy problems, however, focuses on oil and nuclear energy.

The chapters that follow explore in some depth the various problems that relate to U.S. energy policy internationally before analyzing a range of U.S. policy alternatives. The remaining chapter in Part One gives an overview of the factors influencing world energy production and trade. Part Two is devoted to to an examination of the outlook in the major oil-exporting countries, and Part Three to the outlook in the oil-importing countries. Part Four considers the possible future roles of the Soviet Union and China in world energy markets. Part Five analyzes the possible future world oil balance, including market trends, the financial implications of high oil prices, and the relationship between oil and national security. Part Six deals with the commercial and security problems associated with the worldwide growth of nuclear power.

The concluding part of the study assesses U.S. policy options with respect to nuclear energy and oil. A brief final chapter highlights the difficult choices confronting the United States in the field of international energy policy in the years ahead.

Chapter Two

Structure of World Energy Markets

Before examining the complex factors that will determine energy requirements and production in the future, it is useful to have an overview of the present structure of world energy markets. Emphasis will be placed principally on the markets for oil and nuclear power, which are the most important sources of energy from the standpoint of U.S. foreign policy. Both markets are global in scope and both have developed under conditions that have resulted in production, price, and trade patterns different from those in more competitive markets.

THE WORLD MARKET FOR PETROLEUM

The world market for petroleum comprises three major entities: the international oil companies, the governments of the oil-exporting countries, and the consuming countries. Until the early 1970s, the production and marketing of petroleum were largely controlled by the first of these groups.

Structure of the Market

The key structural feature of the international oil industry has been vertical integration: eight major international companies have dominated every step of the oil process from exploration and production to transport, refining, and marketing. As the major companies extended their operations to more and more oil-rich countries around the world, horizontal integration became a characteristic of the industry as well. There has always been some diversity in the industry, however, and smaller, non-integrated companies have managed to survive and even thrive.

Of the eight major oil companies, all but three are U.S. owned. The Royal Dutch Shell Group (Royal Dutch Petroleum Company and Shell Transport and Trading Company, Ltd.) is owned by British and Dutch interests. The British Petroleum Company is British owned and the majority of shares belong to the government. The Compagnie Francaise des Pétroles (CFP) is partly owned by the French government. The five U.S. companies are EXXON, Standard Oil of California, Mobil, Gulf Oil, and Texas Company. By 1950, the U.S. firms had captured at least 64 percent of the proved reserves of the Middle East and controlled roughly 70 percent of production in both Canada and Venezuela. Wherever access was available and oil was to be found, the major corporations extended their empires and perfected their role as international traders.[1]

The interests of the second group—the oil-producing countries—are represented by the Organization of the Petroleum Exporting Countries (OPEC).* Initially formed in 1960 by the governments of Iran, Iraq, Kuwait, Saudi Arabia, and Venezuela, OPEC has expanded its membership in recent years to include Algeria, Indonesia, Abu Dhabi, Libya, Qatar, Nigeria, Ecuador, and Trinidad. Although the original motive for its formation was the stabilization of oil prices, OPEC has evolved into an instrument through which the oil-producing countries act collectively in an effort to maximize their receipts from oil production and to gain increasing control over oil operations.[2] Starting in the early 1970s, the producing countries began to gain for themselves some of the market power once exercised by the major oil companies and have acquired far greater control over their oil resources.

The third group with a major interest in the supply and price of oil—the governments of the oil-importing countries—has as yet not been able to exert much collective influence on the industry. In their relations with the oil-exporting countries, the governments of the importing countries have usually acted separately.

The domestic oil policies of the importing countries have varied. Some (e.g.,France and Japan) have imposed detailed controls over the industry in order to ensure orderly marketing and to limit the role of foreign companies. Others (e.g., West Germany) have tried to maximize competition in the interest of keeping the prices of petroleum products at desired levels. A number of oil-importing countries have also sought to protect domestic energy producers

*OPEC is not to be confused with OAPEC, the Organization of Arab Petroleum Exporting Countries. Founded in 1968, this body includes Saudi Arabia, Kuwait, Libya, Algeria, Egypt, Syria, Iraq, Abu Dhabi, Bahrain, Dubai, and Qatar.

from the competition of cheap foreign oil. Thus, several European governments have in the past kept oil prices high enough to preserve domestic markets for European coal. Similarly, between 1959 and 1973, the U.S. government imposed rigorous quotas on oil imports to protect the domestic oil industry, whose exploration, development, and operating costs were then above the international price of oil.

Over the years, a desire to secure independent sources of oil for their domestic markets has led a number of importing governments to sponsor the establishment of national oil companies and to subsidize their foreign exploration and production activities. In Western Europe, government ownership or control of oil companies is not new. The British government acquired majority ownership of British Petroleum during World War I, and the French government has long exercised control over French oil companies.

The national oil companies formed more recently in Japan, West Germany, Italy, and elsewhere have had little success in gaining secure sources of oil. They have entered the field too late to obtain concessions of the type that for many years provided the basis for the operations of the major companies. These newer national companies have therefore sought to operate through joint ventures with the national companies of the oil-exporting countries, but results thus far have not been impressive in quantitative terms.

Past efforts of some oil-importing countries to negotiate special deals with the exporting countries have usually yielded only higher prices and greater frustration, rather than secure sources of oil at stable prices. The experience of France in Iraq (where a "privileged position" meant that France was obliged to purchase oil at regular prices and extend $80 million in long-term credits) and in Algeria (where a "new type" of relationship resulted in confiscation of two-thirds of France's interests) are illustrative of what has happened.[3]

The search for security of oil supplies has led some observers to propose that the importing countries form a group to confront the exporters.[4] Others have suggested that the proper forum for resolving the problems of the international oil industry is a comprehensive conference or organization embracing all interested parties, both exporters and importers.[5] Either approach would of course have to contend with difficult political and economic problems, including the differing relations of various governments with the major oil companies. These and other possible policies will be considered in some detail later in this study.

Capsule History of Oil Negotiations
And Arrangements Prior to 1970

The period between the end of World War II and 1970 was marked by a steady erosion of the concessionary terms obtained by the oil companies during the era of their predominance and can be regarded as a prologue to the more recent demands of OPEC members for a greater role in the oil industry.

The original concessions were granted during the colonial era. In the Middle East, such agreements, signed at various times between 1901 and 1935, not only granted the companies control over huge areas (for periods from 60 to 99 years), but gave them complete power in determining production and pricing policy and exempted them from all taxes and customs duties. In return, the companies were obliged to pay the host governments a fixed royalty of from 20 to 25 cents per barrel of oil produced. Although this arrangement may have been warranted during a period when reserves were unknown, prices were falling, and huge capital investments were necessary before oil could be extracted, it became increasingly unacceptable to the oil-producing states in the years following World War II. Not only had inflation reduced sharply the purchasing power of the royalties paid to the producing countries, but the oil concessions were obvious targets for rising nationalist and anti-colonialist sentiments.

Discontent over royalty payments first erupted in Venezuela in 1943, as a result of which taxes were imposed that provided the government with 50 percent of the net income earned by the concessionaires from their Venezuelan operations. An agreement between Saudi Arabia and the Arabian American Oil Company (ARAMCO) extended similar terms to the Persian Gulf area in 1950, and by 1952 this 50-50 profit sharing arrangement was in effect in all oil-producing countries in the Middle East. In Iran, the International Consortium of oil companies, which was formed to operate the oil industry after the nationalization of foreign oil interests in 1951, also shared profits on a 50-50 basis with the government.

The profit-sharing agreements achieved the desired effect as far as producing-country earnings were concerned. Receipts rose by approximately 60 cents per barrel and government revenues increased nearly tenfold in the years before 1960. The new arrangements made the producing countries even more keenly aware of the price of oil, however, since price was the major variable in determining net profits.

Prices had undergone a steady increase during World War II and in the immediate postwar years and declined sharply in the

1947-49 period. However, during most of the 1950s, the major oil companies were relatively successful in avoiding sharp changes in world petroleum prices.* At that time, however, several factors combined to put pressure on the industry and on the existing price structure. Greater output and sales by independent producers, rising exports from the Soviet Union, and the rapid development of Libyan oil caused a decline in the market price of oil. The companies reacted by lowering the posted prices on which taxes and royalties paid to the exporting countries were based.

This action led the major oil exporters to form OPEC in an effort to resist the decline in their revenues. Although OPEC was unable to restore the posted prices, it was successful in preventing further reductions in producing-country revenues. Thereafter, posted prices were held constant and became a base for calculating the tax liabilities of the oil companies. Instead of being levied on income, the taxes were, in effect, a fixed charge per barrel of oil produced and a fixed cost to the oil companies.[6]

The period between 1958 and 1967 was characterized by conditions of excess supply and declining market prices for oil. In Iran and Kuwait, for example, oil revenues per barrel were higher in 1958 than they were at any time over the next ten years. After its formation in 1960, OPEC attempted to adopt a prorationing scheme whereby producers would agree to limit production because of the excess supply conditions, but the attempt failed because Iran refused to hold back production.

The relative calm that characterized the international oil market in the early 1960s was harshly disrupted by the events of 1967. The war between Israel and the Arab states in June of that

*Edith Penrose, in *The Large International Firm in Developing Countries: the International Petroleum Industry*, describes well the power the oil companies once exercised in the sphere of price determination. "As a result of their dominant position," she observed, "the major companies were able not only to regulate the rate of development of supply, but also to exercise a strong control over prices, so long as they refrained from price competition among themselves in the 'free' crude-oil markets . . . and in product markets. Except for the so-called 'price wars', they seem to have observed a kind of 'basing point system', remnants of which persisted well into the 1950s for crude oil, and for products into the 1960s . . . The ability of the Companies to regulate the rate of supply in spite of the discovery of vast new reserves made it possible for them to maintain a very high cash flow from their operations, and thus obtain from their earnings the funds required for the enormous investments needed to transport, process, and sell a rapidly increasing output. The combination of the vertical integration of these Companies with oligopoly in product markets provides the single most important key to the understanding of the economic policies of the industry, and of the past and present relations between oil Companies and the governments of both the countries exporting and those importing crude oil or its products."[7]

year resulted in the closing of the Suez Canal and an embargo on petroleum exports from several Arab oil-producing countries (Saudi Arabia, Kuwait, and Libya) to Great Britain and the United States.[8] * Although the availability of huge tankers that could transport oil around the Cape of Good Hope and the increased supply of oil from North Africa enabled Western Europe to escape the potentially disastrous effects of these events, the June war nonetheless served to underline the extent of Europe's vulnerability to events in the Middle East and simultaneously increased significantly European reliance on North African, particularly Libyan, oil supplies.

Indeed, it was the rapidly increasing West European dependence on Libyan oil that triggered the crisis of 1970-71. The failure of the United Arab Republic to reopen the Suez Canal, together with a mounting shortage of tanker facilities in the late 1960s, stimulated the oil companies operating in Libya to increase production from 1.5 million barrels a day in 1966 to over 3.1 million in 1969, about 30 percent of European requirements. In early 1970, this dependence on Libyan supplies was further increased by three developments—an unexpectedly rapid rise in demand, particularly in Europe, and a decrease of supplies from Nigeria (because of civil war) and from Saudi Arabia (because of the disruption of the Trans-Arabian Pipeline by Syria).[9]

In the spring of 1970 the new Libyan government, formed as a result of a left-wing military coup against the monarchy led by Colonel Qaddafi, saw its opportunity to act. Accusing some of the oil companies of overproducing, Libya cut back production in the name of conservation. This move was followed by a series of escalating demands for higher posted prices and tax rates, which were accompanied by threats of an embargo if the demands were not met. Occidental Petroleum, an independent U.S. company whose long-run strategy was to enlarge its share of Libyan output, took advantage of the Libyan government's order prohibiting production by some of the major companies and was the first to negotiate a new settlement giving Libya substantially improved financial terms. In September, the other concessionaires followed suit and agreed to an increase in posted prices of 30 cents per barrel and an increase in the tax rate to as much as 58 percent.[10] These price and tax increases triggered demands for new agreements on the part of other oil-producing states.

*Iraq banned exports to Great Britain, West Germany, Rhodesia, and South Africa.

The Tehran and Tripoli Agreements

Recognizing the importance of presenting a united front in their negotiations with the oil companies, OPEC members adopted a joint resolution at their December 1970 conference in Caracas calling for the elimination of "existing disparities" in crude oil taxes and prices and the acceptance of a "uniform general increase in oil prices reflecting the improvement of the international petroleum market." If these demands were not met within a short period of time, the oil companies were threatened with "concerted and simultaneous action"—an embargo on oil imports.[11] Negotiations between the oil producers and oil companies were set for February 1971 in Tehran.

In anticipation of the negotiations, the oil companies attempted to consolidate their forces by arranging a series of meetings with government officials of their own countries.[12] Efforts were also made among the consuming countries to arrive at a common approach to the oil producers; the United States led the way by convening a meeting of the OECD Oil Committee in Paris in January 1971. Following this gathering, U.S. Undersecretary of State John N. Irwin went to Tehran, the site of the forthcoming negotiations, to meet with the heads of Persian Gulf states and attempt to convince them of the importance of "stable and predictable" prices for petroleum.[13]

Representatives of national oil companies, such as those of Japan, France, and Italy, were conspicuously absent from the 1971 Tehran negotiations. Nonetheless, representatives of the major Anglo-American oil companies sought to speak with a common voice for the industry and attempted to arrange a uniform agreement with all OPEC members on questions of taxes and prices. This approach was rejected by the oil-producing states on the grounds that varying conditions in, and different transport costs from, various producing areas necessitated separate negotiations on a regional basis. Thus, the Tehran agreement applied only to Persian Gulf oil. Further negotiations with Libya took place in Tripoli in the spring of 1971, and at approximately the same time Algeria began to put pressure on French companies to increase government revenues by 42 cents per barrel. As for Venezuela, the government ignored existing concessionary agreements and legislated major tax and price increases as of March 1971.

The Tehran agreement, which was concluded on February 14, 1971, increased posted prices for Persian Gulf oil by approximately 35 cents per barrel (to be raised by about 5 percent a year until 1975) and expanded the producing countries' share of oil profits to 55 percent. The Tripoli agreement of April 1971, on the other hand, granted Libya a corresponding profit share of 55 percent, but provided for a 90-cent increase in posted prices to take account of the low-sulphur content of Libyan oil and the shorter transport distance to Europe.

The market stability that the oil companies hoped had been achieved through the agreements proved short lived. In late 1971, the oil countries, having achieved both increases in posted prices and expansion of their share of the profits, began to press for higher payments to compensate them for the losses they claimed they sustained as a result of the devaluation of the U.S. dollar in August 1971. The Tehran and Tripoli agreements, in addition to providing for an annual increase in the posted price of oil to compensate for inflation, had also included within their broad outlines provisions for subsequent negotiations to take account of changes in the value of a major world currency. On this basis, the OPEC countries were able to obtain, in what came to be known as the Geneva agreements, an additional 8.5 percent increase in posted prices in January 1972.[14]

The Aftermath of the Agreements

Within a few months of the Tehran and Tripoli agreements, Algeria confiscated properties responsible for two-thirds of the output of French oil companies, and Libya seized the properties of British Petroleum—ostensibly in retaliation for Britain's failure to prevent Iran's seizure of two islands at the mouth of the Persian Gulf. In 1972, Iraq, which had nationalized the southern fields of Rumaila in 1967, demanded that its concessionaire, Iraq Petroleum Company (IPC), make a long-term commitment to expand output by 10 percent annually. When IPC made a counter-proposal, Iraq responded by seizing the Kirkuk field in June of the same year.[15] In early 1973, Venezuela, again dispensing with the formality of negotiated agreements, legislated another tax increase, nationalized exploration and production of natural gas, and required that concessionaires deposit increasing sums of money to prevent their properties from deteriorating before the national takeover, which is expected to take place sometime during the next few years and no later than 1983, when existing concessions are due to expire.

Those countries that had not nationalized the physical assets of the foreign oil companies began in 1972 to press for further progress toward their long-standing goal of control over their oil resources. To this end, they turned to the device of "participation," whereby over a period of some years they would assume majority control of the oil companies' holdings in their respective countries. In the meantime, the oil companies were to continue to produce and market the oil and to act as "tax collectors" for the producing countries, thus giving the oil states the satisfaction of ownership (and higher revenues) without the responsibility of marketing the oil.[16] As an alternative to nationalization, participation thus provided a device by which the producing countries could acquire a dominant position in the world oil market and at the same time retain for themselves the services and technological skills of the major oil companies.

Negotiations between various countries and the oil companies were carried on throughout the 1971-73 period. During that time, participation accords were signed by the oil companies involved and the governments of Saudi Arabia, Abu Dhabi, Qatar, and Kuwait; participation negotiations were also begun by Iraq, Libya, and Nigeria.

In the midst of this period of continuing pressures by producer governments for higher revenues and greater national control of the oil industry came the Arab-Israeli war of October 1973, the embargo and production cutbacks by the Arab oil states, and the ensuing disruption of the international oil industry. As of mid-1974, the future structure of the world oil market is still uncertain. It is apparent, however, that the participation agreements will be short lived, and that producers will assume majority ownership of the oil industry in their respective countries much sooner than anticipated. Their already strong control over production levels will thereby be solidified and the roles of their national oil companies in both the production and marketing of oil increased. Opportunities for bilateral arrangements (service contracts, joint ventures, and purchase agreements) between those companies and both the independent companies and the national companies of the oil-importing countries should as a consequence also increase. These structural changes in the international oil industry will inevitably decrease the role of the major companies; what effect, if any, they will have on overall production levels and prices is much less clear.

NATURAL GAS

Natural gas is frequently found in conjunction with crude oil, so that countries with large oil reserves generally have significant gas resources as well. The structure of the gas industry differs from that for oil, however; in most gas-rich countries (except the United

States), state-owned companies produce and sell the gas and, there-fore, trade agreements have to be arranged on an intergovernmental basis. Prices for gas exports are determined by bilateral arrangements between governments, rather than by the operation of a complex international market, as is the case with oil.

Moving natural gas from producing to consuming areas is often quite expensive and requires heavy investment in pipelines or in special processing and transportation facilities. Thus, unlike the situation in the international oil market, the governments controlling large gas resources are not able to exact large economic rents from consumers. Instead, they operate in a buyers' market and are eager to enter into bilateral agreements with importing countries able and willing to make the necessary large investment.

Increased international trade in natural gas will probably be largely by pipeline—e.g., from the Soviet Union to Western Europe, from Canada to the United States, and possibly by underwater pipeline from North Africa to Western Europe. Where movement by pipeline is not feasible, natural gas must be converted into liquefied natural gas (LNG) or methane before it can be transported by ship; upon delivery it must then be reconverted to gas.

Japan, which has only limited domestic gas resources, has concluded a number of long-term LNG contracts and is the most likely of the major energy-consuming countries to increase LNG imports significantly in the future. The United States is not a natural market for LNG, even though it has entered into a long-term contract to import LNG from Algeria. The potential for increased gas production in the United States is very large, and the price of Algerian LNG is much higher than the U.S. price for domestic gas would be even if it was determined by market forces rather than being artificially depressed by government regulation. Western Europe also does not appear to be likely to import much LNG. The gas resources under the North Sea are probably large, although their full extent has not yet been determined. If additional gas imports become necessary, they might possibly be obtained by pipeline from the Soviet Union, North Africa, or the Middle East.

COAL

Although coal has lost its former dominant position in energy consumption and trade, it is possible that the decline will be arrested or decelerated if there is some shift away from oil con-sumption as a result either of higher prices or insufficient production. Much of any future increase in coal consumption would take place in those countries or regions where it is produced (e.g., the United States, India, Europe, and Africa), but there may also be a relative increase in the share of coal moving in world energy trade. In some countries, notably the United States and Japan, increased coal usage would entail a relaxation of environmental standards until a satisfactory technology for controlling stack emissions is adopted.

It is primarily in Asia that the major increases in coal trade are likely to occur. Japan could decide to increase coal imports as a means of lessening its dependence on Middle Eastern oil. Its sources of supply would probably be the United States, Australia, Indonesia, China, and the USSR. With respect to the last three potential sources, Japan would probably have to enter into bilateral deals to help finance the production and transport of increased supplies of coal. Some less developed countries may also seek to increase their coal imports if coal prices compare favorably with the cost of an equivalent amount of oil. Such an increase in coal trade would most likely occur in Asia also, particularly if Japanese capital was available to develop new sources of supply. It is also conceivable that the United States would increase coal exports to developing countries.

There is no reason to expect the emergence of an international cartel in coal. Coal in the United States is owned and produced by a substantial number of independent private companies, none of which dominates the U.S. market. In Western Europe, the coal industries have been nationalized; and in most other coal-producing countries, state-owned companies control production. Trade in coal, therefore, can be expected to continue on a bilateral basis involving private firms and government entities.

THE EMERGING NUCLEAR POWER MARKET

Nuclear energy is widely regarded as the principal energy source to supplement and ultimately succeed conventional fossil fuels, but it is not the only alternative to those fuels. Solar energy and fusion, for example, are also potential alternatives. The reason the use of nuclear energy is far more advanced than other new energy sources and is seen as the next fuel after fossil fuels is of course that the governments of several major countries developed nuclear power for military purposes. In the United States, the existence of an extensive research and development effort, a scientific community, a number of industries, and a bureaucracy concerned with nuclear development provided the base for widespread interest in and commitment to the development of commercial nuclear power. In

Western Europe, Japan, and in a few less developed countries—notably India—the interest in developing nuclear power was stimulated at least partly by a desire to develop domestic energy security.

Military security, the complexity of the technology, the extremely high developmental costs, and the long lead times required before nuclear power could become commercially viable, all dictated that nuclear energy initially be developed by governments. In the non-Communist world, the United States has been the leader, not only because of its early start through its weapons program, but also because of the heavy investments it was able to commit to the project and because nuclear power, which is presently economical for only large power plants, appeared suitable for the U.S. electric power market.

Two trends are discernible in the development of the international nuclear power market. One is the turning over from government to private industry of various phases of the nuclear fuel cycle, so that, in the United States, for example, the uranium enrichment process is the only phase still handled by the government. (The U.S. Atomic Energy Commission has expressed its interest in having the next U.S. enrichment facility built and operated by private industry.) In Western Europe, Canada, and Japan, private firms are also engaged in several phases of the nuclear fuel cycle. Interestingly, the major oil companies have begun to enter the nuclear fuel industry, which is indicative of the fact that nuclear power is showing signs of becoming a large-scale international energy business.

The second major trend is the increased interest in Western Europe and Japan in developing enrichment facilities and in competing with U.S. manufacturers in the world reactor market. At present, the United States has a near monopoly over the provision of enrichment services and dominates the markets for power reactors in the non-Communist world.

The light water reactor (LWR), which was developed in the United States and is the principal type of reactor in use, under construction, or on order both in the United States and abroad, requires uranium fuel that must undergo a highly complex enrichment process before it can be fabricated into fuel elements and used in the reactor. This is the only phase of the nuclear fuel cycle for which foreign countries are almost entirely dependent on the United States. Stimulated by a strong desire to break this dependence and by a growing fear that the United States might not be able to supply their future enrichment requirements, several countries in Western Europe have mounted a major effort to develop an enrichment capability using the new gas centrifuge technology.

Small, but significant inroads into the world reactor market are being made by the heavy water reactor (HWR), which was initially developed primarily in Canada and in West Germany. Perhaps the major reason that this type of reactor has had some success in the world export market is the fact that it is fueled by natural uranium, thereby enabling the user to avoid the enrichment process. This has strong appeal for a country, like India, that is seeking nuclear independence.

New developments in reactor technology, including the breeder reactors and high temperature gas reactors, as well as changes in international marketing patterns, are discussed in detail in Chapter 16. Here it will suffice to note that the reactor market is expected to expand greatly over the next fifteen years and that several factors will work to diminish dramatically the once dominant market position of the United States. Licensing agreements entered into by U.S. companies with foreign firms, for example, will be expiring in the 1980s, after which the foreign firms will be able to produce reactors independently. In addition, barriers to imports of U.S. power plants and equipment have been imposed in a number of industrialized countries that export reactors.

Tables A-9 through A-11 in Appendix A indicate that the vast majority of the world's uranium resources are concentrated in a few countries and that the principal holders of reserves, with the exception of the United States, are not the leading consumers of uranium. Thus, the world uranium market would appear to bear some similarities to the oil market in that it is characterized by a few "haves" and a large number of "have-nots."

At the same time, there are at present several important differences between the oil and uranium markets. First, while the number of large uranium producers (or potentially important producers) is small, the uranium-rich countries are widely dispersed geographically and are politically diverse. Second, there is no concern about a uranium shortage in the foreseeable future; present known reserves of uranium will be sufficient for world needs until sometime in the 1990s. (See Chapter 16.) Indeed, until 1973 the world uranium market was very depressed because nuclear power had not grown nearly as rapidly as had been widely forecast. The market was characterized for almost ten years by falling prices, both in the United States and abroad.* The uranium-producing countries felt compelled to take a number of steps in the face of market conditions. The United States, for example, imposed an embargo on imports of natural uranium for domestic use in 1966 in order to

*Foreign prices for uranium began to rise in mid-1973.

maintain a floor under faltering domestic prices; the Canadian government shored up the uranium-mining communities by purchasing uranium ore concentrates it could neither use nor sell; Australia stopped production in 1970; and South Africa is known to be stockpiling substantial quantities of uranium.* A further distinction between the uranium and oil markets is that uranium exploration, mining, and milling are done by numerous small independent companies in most of the producing countries.

So strong is the desire for national nuclear security and independence that several "have-not" countries, including Japan and West Germany, have entered into joint-venture arrangements in several producing countries in order to assure supply continuity and sufficiency. In addition, France has made direct investments in uranium mines in several former African colonies for the same purpose.

While the specter of four or five major uranium producers one day forming a cartel and choking every last dollar out of the "have-nots" or threatening an embargo may haunt some consumer governments, the threat at this point does not appear serious. World uranium resources have not been extensively explored; as nuclear power expands, exploration and discoveries are bound to increase markedly.

*   *   *   *   *   *

The above review of the structure of world energy markets indicates that trade in coal and natural gas will continue to be conducted on a bilateral basis and probably without important implications for U.S. foreign policy. The markets for oil and nuclear power, however, are undergoing rapid, major changes that create both problems and opportunities for U.S. foreign policy. The fact that the markets for these two major sources of energy are in a state of flux—and hence may be unusually susceptible to influence—is a circumstance of considerable potential importance.

In the case of oil, control over production levels and prices is passing from the once-dominant major oil companies to the governments of the oil-exporting countries. As will be brought out later in this study, the challenge for the international community will be to find the means of stabilizing the production and marketing of

---

*In South Africa, uranium is mined as a by-product of gold mining so that uranium production is not dependent on market conditions.

oil in a way that takes into account both the needs of consumers for this vital commodity and the desires of producers to derive maximum benefit from the ownership of a nonrenewable resource.

In the case of nuclear power, the United States is losing its dominant market position as supply centers arise in other parts of the world. At the same time, control over the market is passing from government to private hands in the United States and elsewhere. The major problem for all governments concerned, which will also be discussed later in this study, is how to shape the development of the nuclear power market in a way that will facilitate expanded use of this important new source of energy without increasing unduly the dangers of nuclear weapons proliferation or nuclear theft.

NOTES FOR CHAPTER 2

1. Robert Engler, *Politics of Oil: A Study of Private Power and Democratic Institutions* (Ill.: University of Chicago Press, 1967), p. 66.

2. Walter J. Levy, "Oil Power," *Foreign Affairs*, July 1971, p. 652.

3. M. A. Adelman, "Is the Oil Shortage Real? Oil Companies as OPEC Tax-Collectors," *Foreign Policy*, Winter 1972-73, pp. 96-101.

4. James E. Akins, "The Oil Crisis," *Foreign Affairs*, April 1973, p. 488.

5. See for example, "International Organ for Energy Resources Needed," *Nihon Keizai* (Tokyo), December 22, 1971.

6. Sam H. Schurr, Paul T. Homan, and associates, *Middle Eastern Oil and the Western World: Prospects and Problems*, Resources for the Future report (New York: American Elsevier, 1971), p. 15.

7. London: Allen and Unwin, Ltd., 1968, pp. 61-62.

8. George W. Stocking, *Middle East Oil: a Study in Political and Economic Controversy* (Nashville: Vanderbilt University Press, 1970), pp. 458-59.

9. Schurr, *Middle Eastern Oil and the Western World*, p. 16.

10. Levy, "Oil Power," p. 654.

11. Cited in Mostafa Elm, "Oil Negotiations: A View from Iran," *Columbia Journal of World Business*, November-December 1971, p. 89.

12. Ibid.

13. Adelman, "Is the Oil Shortage Real?" p. 81.

14. Schurr, *Middle Eastern Oil and the Western World*, p. 16.

15. Adelman, "Is the Oil Shortage Real?" pp. 97-99.

16. Ibid., p. 70.

Part Two

The Outlook in the Oil-Exporting Countries

Chapter Three

The Arab Countries

Nearly 70 percent of the world's proved oil reserves outside Communist countries are located in the Arab countries of the Middle East and North Africa. This geographical concentration of a vital resource has made it possible for the Arab oil states to exercise an enormous influence on the international oil market for both political and economic purposes. Before turning to an analysis of the links between Arab oil supplies and world politics, this chapter provides a brief sketch of the oil industry in the Arab world.

PATTERN OF OIL RESERVES, PRODUCTION, AND TRADE

Prior to the Arab-Israeli war of October 1973, the Arab countries of the Middle East and North Africa produced four of every ten barrels of oil consumed daily in the non-Communist world. In 1973, Arab oil accounted for an estimated 43 percent (2.3 million barrels a day) of Japan's oil imports and about 70 percent (11.3 million barrels a day) of Western Europe's oil imports. In the United States, which is much less dependent on oil imports than Japan and Western Europe, Arab oil supplied slightly less than 10 percent (1.6 million barrels a day) of total U.S. oil consumption in 1973.[1] In the near future, Arab oil may supply much of the incremental growth in world demand for petroleum imports.

Oil resources were not bestowed on the Arab states on any principle of equality. Some have been highly favored, others have as

Note: This chapter is based in large part on research and analysis conducted by Robert A. Mertz.

yet found little or no oil (see Table 3-1). Only a few will be of significance as major producers in the future.

Table 3-1.   Arab Oil Reserves, Recent Production, and Production Capabilities 1980 and 1985

Countries	Proved Reserves (billion barrels)	Production (million b/d)		Possible Production (million b/d)	
	Jan. 1974	1972	1973	1980	1985
Middle East					
Iraq	31.5	1.5	2.0	5.0	6.0
Kuwait[a]	64.0	3.3	3.0	3.0	3.0
Saudi Arabia[a]	132.0	6.0	7.6	15.0	20.0
United Arab Emirates	25.9	1.2	1.5	4.0	6.0
Others[b]	36.3	1.0	1.0	1.5	2.0
Total	289.7	13.0	15.1	28.5	37.0
North Africa:					
Algeria	11.8	1.1	1.2	1.5	2.0
Libya	30.4	2.2	2.2	3.0	3.0
Others[c]	6.2	0.3	0.3	0.8	1.0
Total	48.4	3.6	3.7	5.3	6.0
Total Arab	338.1	16.6	18.8	33.8	43.0

Sources: Estimates for reserves are from *Oil and Gas Journal,* December 31, 1973, pp. 86-87; 1972 and 1973 production figures are from U.S. Department of Interior, Bureau of Mines, *Mineral Industry Surveys,* June 24, 1974, pp. 2-3; potential production figures for 1980 and 1985 are from Table 13-7 in Chapter 13 of this study.

Note: The figures for potential production are not predictions of output but estimates of production capabilities, given known geological and technical constraints. Actual future production will also depend on world demand and the marketing strategies of the various oil-exporting countries.

[a]Includes one-half of Neutral Zone.

[b]Includes Bahrain, Oman, Qatar, and Syria.

[c]Includes Egypt and Tunisia.

Of the North African producers, Egypt, whose present output of about 200,000 barrels a day meets its own domestic requirements, has a modest potential for export and hopes to raise production to about 1 million barrels a day by 1982.[2] Libya and Algeria have been important producers and exporters since the mid-1960s. The relatively small size of Libya's proved reserves, however, and the fact that the production of some existing fields has already begun to decline indicate that there may be physical limitations to expanding Libyan production much above 3 million barrels daily—unless new reserves are discovered. Algeria, which began producing about 1 million barrels a day in 1970, also seems to have reached the limit of existing fields, and estimates for production do not exceed 1.5 million barrels daily by 1980.

North African oil is likely to play a relatively minor part in meeting the energy demands of Western Europe and the United States. Despite their rapid growth as suppliers of crude oil to Western Europe in the 1960s, the North African states have been providing a declining share of total Arab crude oil production; between 1970 and 1972 their share slipped from 33 percent to 20 percent.[3] This reduction was attributable in part to Libya's decision to decrease production by 1 million barrels a day and to Algeria's embargo on exports to France in 1971, while at the same time the Arab states of the Persian Gulf were expanding production to meet increased world demand.

The Arab states of the Persian Gulf region present a different picture. Saudi Arabia has the most extensive proved reserves of any country in the world—an estimated 132 billion barrels, or almost one-fourth of the total non-Communist world reserves.[4] The other major Arab producers, Kuwait, Iraq, and Abu Dhabi, hold proved reserves of another 117 billion barrels. From the standpoint of physical capacity, Saudi Arabia and the other gulf oil states could produce almost 30 million barrels a day in 1980 and about 40 million barrels a day in 1985—amounts that would exceed the estimated import requirements of the United States, Western Europe, and Japan in those years. How much Arab oil will in fact be made available and on what terms will depend on the resolution of difficult economic and political questions.

THE TREND TOWARD NATIONAL CONTROL

Arab nationalism, as it has affected the oil industry in the Middle East and North Africa, has been a major source of friction between the Arab states and the West. More and more producing states have demanded control over their natural resources. The sovereignty of the producing nations and their right to nationalize the local oil industry in return for fair and prompt compensation were never in question. Much of the controversy about nationalization has stemmed from differences over whether the compensation offered was fair. The more important issue for the consumers of Arab oil, however, is whether and for how long nationalization and the controversy over compensation will adversely affect the continued supply of oil.

The chief examples of nationalizations, as of mid-1974, were those carried out by Algeria, Libya, and Iraq. Algeria took over the non-French oil companies in 1967-68, and in 1971 demanded and obtained a controlling interest in the French oil companies operating in the country. Libya nationalized the holdings of the British Petroleum (BP) Company in late 1971, and in 1973, it nationalized BP's U.S. partner, Bunker Hunt, and took over a 51 percent interest in all the foreign companies operating in the country.[5] In 1972, Iraq nationalized the Western-owned Iraq Petroleum Company (IPC) concession, which included the major oil fields in the northern part of the country. In the wake of the 1973 Middle East war, Iraq nationalized the U.S. and Dutch shares of the Basrah Petroleum Company, the last Iraqi oil operation owned by foreign companies. In each case, the national oil companies of the producing countries took control of the properties and operations of the nationalized companies.

What effect has nationalization had on the continued supply of oil to consuming countries? In most cases, the expropriated company attempted to cut off the nationalizing country's access to the market pending satisfactory settlement of its claims to adequate compensation, but because of the lack of solidarity on the part of oil companies and consumers, the attempts were only partially successful. Non-French companies, for example, did not heed French pleas to boycott Algerian oil in 1971, and major U.S. companies rushed to tie up long-term supplies of Algeria's low-sulphur crude. In Iraq, the French component of IPC chose to work out a separate deal with Baghdad for long-term purchases of oil from the nationalized fields. Generally, exports from the nationalized fields initially declined, partly as a result of legal pressure brought by the nationalized companies and partly from delays while the new national owners lined up alternative buyers.

In Algeria and Iraq, however, agreements on compensation were successfully negotiated within less than a year, largely on the terms desired by the nationalizing country, but not without compensation for the nationalized assets. After settlement of the Algerian dispute, production and exports climbed back to normal levels, while in Iraq exports rapidly exceeded pre-nationalization levels. Both countries, which are heavily dependent on oil revenues, restored good relations with the oil companies. Once compensation had been agreed upon, the foreign companies were willing to resume operations, although on a revised contractual basis. On the other hand, Libya, which was having no problems selling its oil at high prices, was in no hurry to settle its nationalization dispute with the majors, and the independent companies quickly agreed to the compensation terms offered by Libya.

In the early 1970s, as was noted in Chapter 2, an alternative to nationalization, known as "participation," emerged. In principle, participation need not adversely affect production, since it shifts ownership gradually and maintains the essential technical, managerial, and marketing services of the international oil companies.

All things considered, the key issues regarding the volume and security of oil supply are not nationalization, participation, or the disappearance of the oil concessions, but a producing country's decisions on whether its own political and economic interests are served by slowing down or stepping up production and exports. That it can do at will, regardless of the status of ownership of the oil or the facilities of production and transport.

OIL AND POLITICS IN THE ARAB WORLD

From the beginning, the formation of the state of Israel- -on land the Arabs consider a part of their national patrimony—has troubled the relations of the West with the Arab world. The U.S. role in the creation and support of Israel, in particular, made it very difficult for the United States to establish a basis for solid, cooperative relations with those Arab states directly involved in the conflict with Israel.

For a long time, however, Western oil interests in the Middle East were virtually immune from any side effects of the Arab-Israeli dispute. The rulers of Saudi Arabia, Kuwait, and Libya, though avowed enemies of Israel, were not inclined to disrupt their profitable arrangements with Western oil companies or their friendly relations with Western governments as a means of demonstrating their concern for the Arab cause. Thus, although the U.S. standing with some Arab states, notably Egypt and Syria, was disastrously affected by the continuing conflict with Israel, U.S. relations with the major oil-producing states remained good.

The immediate events attending the six-day war of 1967 seemed to leave this pattern intact. At the time of the war, several oil states tried without much success to impose an embargo directed primarily against the United States and Great Britain.* Nevertheless, the 1967 war ultimately brought changes. The magnitude of Israel's victory and the continued Israeli occupation of large sections of Arab territory stimulated Arab solidarity on the Israeli issue and drew the

*The 1967 attempt to impose an embargo failed because of the speed of the Israeli victory and the unwillingness of several key oil states to sacrifice their flourishing economic and political relations with the United States.

oil-producing states more closely into general Arab efforts to find effective policies. (Saudi Arabia, Kuwait, and Libya, for example, began to provide regular subsidies to Egypt and Jordan to compensate them for their losses and for the burden of being on the front line.) In addition, the Israeli annexation of the Arab sector of Jerusalem was bitterly resented by King Faisal and introduced a discordant note in the previously harmonious Saudi-American relationship.

As various U.S. and UN efforts to promote an Arab-Israeli settlement following the 1967 war failed to show progress, it was almost inevitable that Arab restiveness and resentment would spill over and affect relations with the Western oil companies and with the United States. Although the successive waves of OPEC pressure on the oil companies for higher revenues and the nationalizations that marked the early 1970s can be explained by a number of factors other than the Arab-Israeli conflict, there is no doubt that the entire oil picture was heavily colored by the question of Israel and U.S. support of Israel. The policies of the major Arab oil countries were bound to reflect this, whether or not the fourth Arab-Israeli war had come in October 1973.

A few days after the outbreak of hostilities in October 1973, the Arab states led by Saudi Arabia, at the urging primarily of Egypt and Syria, announced a policy of graduated reductions in oil production and a selective embargo on exports to those countries supporting Israel. Shortly after their announcement of initial production cutbacks of 5 percent per month, the Arab oil producers (including Saudi Arabia, Algeria, Abu Dhabi, Qatar, and other small Persian Gulf states) announced that they would initiate an immediate production cutback of 25 percent.

Based on the diplomatic position taken on the Middle East dispute by various foreign countries, the Arab states drew up a list of "friendly," "neutral," and "hostile" countries. The hostile countries were to receive no oil, the friendly countries would receive their normal level of imports, and the neutral countries would have to divide up the remaining oil produced. In an effort to assure that friendly countries did not suffer shortages, the participating nations periodically adjusted their production levels.

Because oil marketing and distribution operations are controlled by a few major oil companies, which also have access to much non-Arab oil, the oil-importing countries did not necessarily receive oil supplies in accordance with their status on the Arabs' list. The Netherlands and the United States, for example, which were considered hostile countries, felt the impact no more severely than most nations on the friendly list. In addition, the policy of unified production cutbacks was followed more faithfully by some of the participating states than others. Iraq followed a policy of non-compliance with the cutbacks and expanded output. Nevertheless, there is no doubt about the efficacy of the "oil weapon" in forcing many nations to abandon or modify their support of Israel. The oil weapon probably also added urgency to U.S. efforts to end the hostilities and devise a settlement, although even if the weapon had not been used the United States would presumably have done what it could to defuse a serious threat to world peace.

ARABISM AS A POLITICAL FORCE

"Arab Unity," whether viewed in terms of a united position on a particular issue, such as Israel, or as a more far-reaching concept of brotherhood and cooperation for the benefit of the Arab world, is an ideal that inspired Nasser and various other Arab leaders over the years. Until the recent coordinated bargaining on oil matters and the Middle East war of 1973, this goal in modern times always lay beyond the grasp of Arab leaders.

Although they were always agreed in their opposition to Israel, the Arab nations during the 1950s and 1960s showed a marked lack of unity on other matters of common interest. They could come together periodically for wars with Israel, but they moved apart once the fighting was over. Lack of unity sprang from their diverse historical, political, economic, religious, and social development. During the 1950s, a pronounced split developed between the radical regimes (Egypt, Syria, Iraq, and Algeria) and the conservatives (Saudi Arabia, Jordan, Kuwait, Morocco, and—until the 1969 coup—Libya). The radicals were in the militant forefront of the Arab struggle against Western imperialism and Israel, and they looked to Moscow for aid. The conservative states, fearing both Arab radicalism and communism, looked to the West for support.

Arab radicalism for most Arabs was personified by Egypt's Gamal Abdel Nasser. Nasser was the one Arab leader who had successfully defied the Western powers and who had won a following in every Arab state. The failure of the radicals to dominate the Arab world is traceable, however, largely to the failures of Nasser himself. The breakup of the Egyptian-dominated union with Syria in 1961 tarnished the fundamental ideal of Arab unity and Nasser's position as its champion. Nasser's unsuccessful attempts to assert control over Yemen and to overthrow the government of Saudi Arabia and his devastating defeat by Israel in 1967 irreparably damaged his prestige.

Nasser died in 1970 and Nasserism as an effective force in Arab politics died with him. His successor, Anwar al-Sadat, is more conservative politically and more domestically oriented. As for the other "radical" states, Iraq and Syria are ruled by weak military governments beset with internal problems and are unable to influence other Arab states by their example. Qaddafi's Koran-thumping mixture of militant Islamic traditionalism and radical nationalism, though a force for other leaders to contend with, has tended to isolate Libya from other Arab states.

The ineffectiveness of the radical Arab regimes, contrasted with the suddenly augmented importance of the oil-rich conservative states, combined in the early 1970s to produce a new political situation in the Arab world. No one figure and no single state today dominates Arab politics. Egypt may still be the Arabs' political capital, but Saudi Arabia, Kuwait, and even tiny Abu Dhabi speak with new authority in Arab circles. Instead of being Nasser's victim, King Faisal of Saudi Arabia has become a financial prop of the Egyptian economy.

Despite the decreasing polarization among Arab states, the crucial problem of Israel remained. In the years following the 1967 war, little international pressure was exerted on Israel to return the occupied lands, and it became apparent to the Arabs that those lands might well be permanently lost. The humiliation of their defeat in 1967, bitterness over the occupied territories, and the plight of the Palestinian refugees continued to dominate the policies and politics of many of the Arab governments—particularly Egypt, Syria, and Jordan, which had suffered the territorial losses, but also others, notably Saudi Arabia.

During this period, increasing pressures, both external and domestic, were being exerted on the Arab oil producers, particularly Saudi Arabia, to use the oil weapon in their struggle against Israel. During the spring and summer of 1973, King Faisal warned the United States that he would not expand production so long as the United States did nothing to press Israel to return the occupied lands, especially the Muslim holy places in Jerusalem. Yet he took no action until the October war was under way. When he and the leaders of other Arab states did decide to use their oil for political purposes, unity was important, but Saudi Arabia's participation was decisive.

At the Algiers summit conference after the war, the Arab leaders stressed unity on such issues as the oil embargo and the Palestinian problem. They agreed to manipulate Arab oil exports according to the attitude of each country toward the Arab cause, and they recognized the Palestinian Liberation Organization (PLO) as the sole legitimate representative of the Palestinian people.

Whether Arab solidarity can be maintained if the pressures of the conflict with Israel are substantially lessened is not clear. In particular, it is not clear whether solidarity can be retained on oil market matters, or whether national self-interest will cause the Arabs to go their old, disparate ways. In order to answer this question it is necessary to look at the policies of the individual Arab states.

OIL IN THE PRODUCING STATES: THE POLICIES OF INDIVIDUAL GOVERNMENTS

Saudi Arabia

Before the production cutbacks instigated in connection with the Middle East war of 1973, Saudi Arabia was the Middle East's leading oil producer. From a physical and technical standpoint, Saudi Arabia's output could be as high as 20 million barrels a day by 1985 or earlier.

Political life in Saudi Arabia centers on the king. His authority, which is great but not absolute, reflects the allegiance and support he has been able to command from his main political constituency—the extensive royal family intermarried with prominent and influential families and tribes throughout the country. In recent years, the machinery of government has been modernized by the creation of ministries and a bureaucracy that includes many well-educated officials.

Two salient problems loom ahead. One is that political instability and perhaps the seizure of power by a new ruling group could attend the succession to the king. The other is that changes in Saudi society, which are already creating a new class of educated professionals, technicians, and entrepreneurs, could drastically modify or fracture the existing political structure.

King Faisal is now in his late sixties. One of his half-brothers has been designated to succeed him, but he probably does not have comparable leadership qualities, and another half-brother is expected to carry the real responsibility of governing. There could be troubling rivalry, but the royal family has weathered difficult succession problems in the past and probably could do so again. The military is a possible source of revolt, but the army (about 40,000 strong) is kept scattered in border districts, has no outstanding leaders, and is counterbalanced by another military force, the tribally based National Guard, which is loyal to the royal family and has the task of preserving internal security. A coup by dissident officers is not an impossibility, but the odds are against it. At present, no nonmilitary group is capable of seizing power.

Saudi Arabia has long been a conservative, isolated society, suspicious of foreigners, proud, and independent. The combination of Bedouin independence and religious conservatism has served to reinforce loyalty to the monarchy in the past and helped to shield the country from the external sedition attempted by Egypt under Nasser and by Iraq. New forces in the society, however, are dissatisfied with the slow pace of reform and change. In time, they will want to open up the political system, but the country's inherent conservatism may grant a period of ten years or so during which good jobs and the government's allocations of oil money to economic and social development may deflect most criticism. The regime has also shown itself capable of taking drastic police action when it deems necessary. Yet the incentive to revolution, with control of immense wealth as the prize, is not to be underestimated.

A change in the political structure, either by evolution or by revolution, could affect oil production policies, depending on the orientation of the government. It is likely, however, that any future government would want to maintain the oil revenues upon which its development programs and foreign policies would depend, although the level of oil exports might be well below what consuming countries would like.

Before 1970, the growth of budget expenditures, development investment, and the demand for imports paralleled the rise of oil income. Beginning in 1971, the prospect of sharply higher oil revenues enabled Saudi Arabia to accelerate its spending on many areas of human and physical development—education, health care, social welfare, infrastructure, economic diversification. Total expenditures for the first five-year development plan (1971-75) were set at $10 billion. The total general budget in 1972-73 was expanded to $3.3 billion, including development expenditures of $1.6 billion. Before the 1973 war broke out, the budget for the 1973-74 fiscal year had been almost doubled to $6.2 billion (including outlays for development).[6] Actual expenditures almost certainly exceeded that level substantially, in light of Saudi Arabia's role as principal financier of Arab military imports from the USSR and elsewhere for the war with Israel.

Over the next decade, developmental requirements, defense programs, and the furthering of foreign policy objectives will place large financial demands on the government. National development alone will require substantial sums. The new development plan for 1976-80 is projected to cost about $50 billion over the five-year period, or an average annual expenditure of roughly $10 billion. The new plan envisages capital outlays five times greater than the outlays budgeted for the 1971-75 plan. During 1971-73, planned development targets of between $1 and $2 billion were unfulfilled by a considerable margin due to the limited absorptive capacity of the Saudi economy. Thus, there appears to be a real question as to whether the level of development expenditures called for in the new plan can be realized. The budget surplus that would result from oil revenues in future years cannot be predicted with a high degree of confidence, given uncertainties concerning future oil prices and production levels. Nevertheless, it seems clear that Saudi Arabia will continue to be a financial surplus country.*

Money not needed at home can of course be used abroad. It can be invested in development projects elsewhere in the Arab world or sent directly to Arab governments, as subsidies now go to Egypt and Jordan. And it can be used to continue to finance Arab arms if that is deemed necessary. The key question, then, is not whether Saudi Arabia can make good use of its oil revenues; that is not a real problem. The important decision for the Saudi leaders is choosing the marketing strategy that will maximize oil revenues over a period of years. This problem will be analyzed in Chapter 13. At this point, it need only be noted that the fact that prospective revenues greatly exceed likely domestic requirements provides the Saudis with a range of choices not open to many of the major oil-exporting countries.

Kuwait

Long a trading nation, Kuwait is commercially oriented and outward looking. Political power in Kuwait remains in the hands of the royal as-Sabah family, which governs by what has been called inspired traditionalism. Basically a conservative society, Kuwait has evolved its own political system to meet the demands of modernization and to cope with its oil wealth. Kuwait has a parliamentary structure and a measure of representative government, a legal opposition though no formal political parties, and a politically vocal public that includes many non-Kuwaitis (more than half of the total population of about 850,000). Reforms are forthcoming, but whether rapidly enough to satisfy some of the more radical and

*See Chapter 14 for a more detailed analysis of the future financial positions of Saudi Arabia and the other oil-exporting countries.

nationalist Arab elements is a question. Oil money has made it possible to establish a thoroughgoing welfare state and to buy off unrest at home. One method used by the government to redistribute wealth between 1952 and 1971 involved disbursement of almost $2 billion to Kuwaiti citizens in a unique property-acquisition scheme.[7]

The non-Kuwaiti segment of the population, which does not enjoy political or economic equality with Kuwaiti citizens, includes many Palestinians (estimated at about 150,000), Iraqis, and other Arabs, plus Iranians, Pakistanis, Indians, and others. Many of the foreigners are well educated and hold important positions in business and the government. Thus, Kuwait has generally been more sensitive to developments elsewhere in the Arab world, especially in relation to Palestine, than has Saudi Arabia.

Kuwait has generally remained neutral amid competing Arab states and ideologies and has served often as a mediator. Its greatest insurance for survival, however, has been its generosity. In the early 1960s, the government established the Kuwait Fund for Arab Economic Development (KFAED) to create a mutuality of interest between Kuwait's continued prosperity and that of the recipient countries. The fund has extended 39 loans totaling over $200 million for development projects in other Arab countries of all political persuasions. In addition, the government has made direct grants to the Arab states of the Persian Gulf and Arabian Peninsula. Following the six-day war in 1967, Kuwait provided subsidies of $98 million annually to Egypt and $56 million to Jordan (Jordan's subsidy was cut off in 1970 but restored in October 1973 when it entered the war), and unrevealed sums to Palestinian resistance organizations. Kuwait also helped arm Arab forces for the 1973 war. As oil revenues grow, Kuwait is likely to increase its direct grants and subsidies, as well as development loans through KFAED.

Generosity alone may not be enough to guarantee Kuwait's survival, however. Iraq maintains a claim to sovereignty over all of Kuwait and eagerly desires to improve its own position as a Persian Gulf power by gaining usable coastline and ports at Kuwait's territorial expense. Iraq's 1973 invasion of Kuwaiti border territory and the subsequent demand that Kuwait cede Iraq two strategically located islands underlined Kuwait's vulnerability. As a result, Kuwait has begun to build up its armed forces and is planning large-scale arms purchases. Moreover, Kuwait does not stand alone. Other Arab states, especially Saudi Arabia and Egypt, are determined to deny Iraq control over Kuwait's oil and have opposed Iraqi aggression against Kuwait in the past. However, the most effective deterrent is the power of Iran. The Shah has made it clear that he will protect Kuwait if the latter requests it, and he might do so without an invitation if necessary.

Kuwait has become used to the income it has been receiving from oil over the past decade and it is counting on at least that much in the future. Imports have continued to grow annually but lag far behind exports. With a chronically large trade surplus and limited local investment opportunities, Kuwait has built a sound portfolio of international investments. Private and public foreign holdings at the end of 1973 were approximately $6 billion.

In 1970, the government stabilized oil production at about 3 million barrels a day in order to conserve its resources. When Saudi Arabia and the small gulf states decided to wield the oil weapon in October 1973, Kuwait participated in both the production cutbacks and in the embargo. With an impressive annual trade surplus and official foreign exchange reserves of $357 million at the end of 1973, Kuwait would be able to limit oil production further and maintain its comfortable patterns of consumption without serious economic hardship.[8]

Iraq

Iraq has been governed by a succession of juntas since the monarchy was overthrown in 1958. Successive ruling groups have lacked the stability and solid support required for consistent policy making. In a poisonous political atmosphere of plot and counterplot, the nation has been unable to overcome its internal divisions and problems. Relations with the United States have been generally bad, as governments in Baghdad have taken a strong line on the issues of Israel and Western imperialism.

Under its current Ba'thist regime, Iraq is also at odds with many of its neighbors. It has continued the long-standing disputes with Iran over the border between them and the treatment of their respective minority populations. It has threatened and clashed with Kuwait. It regards Saudi Arabia as a reactionary country tied to the United States. It does not even get along well with Syria, which is ruled by another faction of the Ba'th movement. As both cause and effect of these relationships, Iraq has preached radical revolution in the gulf area, has attempted to lead popular opposition to existing conservative regimes, and has supplied arms and money to various subversive movements. The regime has also developed close ties with the USSR, although that relationship has had its ups and downs. Iraq has continued to receive Soviet military and economic aid, and in 1972, the two countries concluded a fifteen-year treaty of friendship and security.

Beginning in the early 1960s, the general atmosphere in Iraq was unfavorable for the oil industry. Successive governments conducted a running struggle with the Iraq Petroleum Company (IPC) over prices, production levels, and the area and terms of the IPC concession. The quarrel with IPC led to repeated curtailments of production and reduced new exploration and the development of new reserves and production facilities to a minimum, but Iraq never shut down IPC production totally. During this period, oil production increased very slowly. From 1 million barrels a day in 1961, output had climbed to only 1.7 million barrels a day in 1971.

The dispute between Iraq and the IPC culminated in the nationalization of the company in 1972. In that year, production fell to 1.5 million barrels a day. Once the controversy over nationalization and compensation payments to IPC was settled (February 1973), however, production increased rapidly, reaching 2.1 million barrels a day only four months later. Oil began to flow again to Western markets. Previously, exports had gone primarily to the Soviet Union and to Eastern Europe as part of barter agreements. A portion of future oil production will continue to be exported to the USSR and Eastern Europe as called for by these agreements, but most of the anticipated production increases should be available to non-Communist countries because of Iraq's need for freely convertible foreign exchange for its own development.

Thus, a substantial increase in total production—3 million barrels a day by 1976, and 5 million barrels a day by the early 1980s —is possible.[9] With additional discoveries, these production levels could be surpassed. On the basis of already proved and probable reserves and the promising geological structures known to exist, Iraq's oil potential is considered exciting enough by Western companies to cause them to take the political risks of operating in Iraq and negotiate new service contracts with the Iraqi government for exploration. For its part, Iraq is counting on the technical expertise of the big Western oil companies to help it expand production quickly. To emphasize its commitment to expand production and to shelter oil exports from the political whims of Syrian governments, which in the past have shut down the portion of the IPC pipeline that passes through Syria, Iraq is planning a new pipeline to carry all production from its northern fields to a new deep water port to be built on the gulf.

From the Western standpoint, however, Iraq may be anything but a reliable source of oil. Estimating the future depends partly on interpreting the past. Was the 10-year struggle over nationalization an inevitable and decisive act of emancipation that, with the assertion of Iraq's sovereignty and its acceptance by the company and the Western governments, clears the way to new relationships under which oil will flow without any political barriers or grievances about exploitation and imperialism? Or were these events but the most recent manifestation of profound political and social forces that will continue to disrupt Iraq's oil industry?

The answer probably lies somewhere between these alternative hypotheses. Iraq's politics will continue to reflect the instability of its internal political and social divisions, but government oil policy seems now to be based on a more pragmatic, commerical attitude toward relations with Western oil companies and oil-consuming nations. The reason apparently is a mixture of politics and economics. Iraq seeks a policy of lessened dependence on the Soviet Union, internal political stability, and broad economic development. Its plans are ambitious and they depend on oil. This pragmatic approach was very much in evidence in the fall of 1973, when Iraq refused to join in the Arab production cutbacks on grounds that they would harm only Western Europe and Japan and not the United States. Iraq participated in the embargo, but its production and exports to countries not under the embargo were increased rapidly in this period.

With a mud-flat coastline of less than 40 miles on the Persian Gulf, Iraq still has ambitions to be a "gulf power." It is trying to counter the growing influence of Iran, to develop military forces that can operate in gulf waters and in the general region, to get a larger share of offshore oil resources in the upper gulf than others are willing to concede it, and to assume a role of leadership or sponsorship of revolutionary forces seeking to gain power in a number of gulf states.

No one can predict whether the government in Baghdad will be successful in so extending its influence and in establishing revolutionary forces in positions of power in other countries. Formidable obstacles to Iraq's ambitions exist within the gulf region. Iraqi support or prompting of any attempted revolution in a small gulf state would prejudice the interests of Saudi Arabia and Iran, both of which would be alert to prevent or to reverse it. Moreover, there is no indication that Iraq is the chosen instrument of the Soviet Union in furthering Soviet interests or ambitions in the gulf, or that the Soviets would support adventures undertaken by the Iraqis on their own initiative. The Soviet Union would not be likely to jeopardize its long-term relations with other Arab states and with Iran by committing itself solely to the policies and interests of one country in the region, particularly an unstable one.

The main problem posed by Iraqi radicalism is that it can lead indirectly to instability within the various states of the region and in their relations with each other. The buildup of local armed forces throughout the gulf, in part to counter the perceived Iraqi threat, will certainly enhance the ability of the gulf states to ward off aggression, but it could also have the effect of encouraging military adventures. The revolutionary overthrow of one sheikdom could pose a serious threat to the stability of other conservative regimes in the gulf.

Such events are not the inevitable outcome of Iraqi policies. Indeed, it is not out of the question that in the future Iraqi radicalism may mellow, or that the present regime may give way to a more moderate one. There are already indications that a certain moderation may be creeping into Iraqi policies and that the regime may be looking more seriously to the country's internal problems and development.

The Small Gulf States

Some of the small gulf states now produce oil, and some may develop into major producers in the future, but the only one likely to export a substantial amount of oil to the United States and other large consuming countries is Abu Dhabi, the leading state in the United Arab Emirates (UAE). A relatively new producer, Abu Dhabi's output rose from about 282,000 barrels a day in 1965 to about 1.3 million barrels a day in mid-1973. Dubai is the only other member of the UAE with significant oil production at present. Output had reached 222,000 barrels a day in mid-1973 and could perhaps be double that amount in the next decade. Sharja, the newest gulf producer, hopes to be producing about 200,000 barrels a day by late 1974 from its recent discoveries off Abu Musa Island. Total oil production in the other small gulf states not belonging to the UAE was slightly less than 1 million barrels a day in mid-1973. Qatar was producing about 600,000 barrels a day; Oman, which will be unable to increase production much without further discoveries, reached 300,000 barrels a day; and Bahrain, whose output has long been declining, produced only 68,000 barrels a day.[10] Total production from the UAE and the other small gulf states could be as high as 8.0 million barrels a day by 1985 with Abu Dhabi producing over half of the total—an impressive figure but one tinged with optimistic speculation.

Abu Dhabi and the other oil sheikdoms have no other significant resources. They are tribal societies with absolute rulers and few modern institutions. With the growth of the oil industry, these societies have begun to change. New people have immigrated (the indigenous population in Abu Dhabi is now a minority), and new ideas and demands are developing among the people. Political changes are being made slowly; in Abu Dhabi they involve the limited diffusion of power from within the ruling family to a group of important tribal leaders and, in the execution of policy, to a growing corps of foreign technocrats.

Traditional power and oil wealth in Abu Dhabi and the other Arab sheikdoms may make the government difficult to overthrow, but the system is subject to sudden upheavals at the top because of the personalization of power. Family and tribal rivalries, palace revolutions, and coups d'etat are commonplace. Tribal and territorial disputes between the Arab gulf states also cause tensions.* Numerous lesser disputes, although they seldom erupt into fighting, undermine efforts at cooperation. Moreover, Abu Dhabi and other small gulf states do not live in a vacuum. Small revolutionary movements supported by South Yemen and Iraq have already tried unsuccessfully to infiltrate the armed forces of these small states. If a revolt or coup was to be inspired or carried out by radical forces based in South Yemen, Iraq, or possibly the Soviet Union, then Iran or Saudi Arabia would intervene to make sure that it did not succeed. If Iran intervened militarily, however, the general Arab reaction would be one of anger and resistance, and Saudi Arabia might not be able to stand aside from it. Either eventuality would impair efforts at political and economic cooperation in the gulf and might well affect the continuing advance of oil production in Abu Dhabi and several of the other gulf states.

The smaller gulf states differ in their ability to absorb their oil revenues domestically. Oman has enough oil to initiate the development process, though perhaps not enough to carry it through. Dubai's canny merchants will easily turn its oil revenues into advantageous channels. But this is not the case in Qatar and Abu Dhabi. Qatar, with less than 50,000 citizens, earned an estimated $250 million from oil in 1972 but budgeted expenditures of only $52.5 million. With 90 percent of its population resident in the capital area and with limited development potential, Qatar could not easily absorb more than $100 million annually. Abu Dhabi, which has a native population of about 30,000, spent only one-half of its total oil revenues between 1968 and 1972 at home and within the

---

*A compromise settlement of Saudi Arabia's long-standing claim to much of Abu Dhabi's territory was signed on August 19, 1974.

other six emirates of the UAE. Taking a lesson from Kuwait, it has formed the Abu Dhabi Fund for Arab Economic Development (ADFAED). In 1972, Abu Dhabi distributed an estimated $75 million abroad, and the available funds will increase spectacularly. If the whole of the UAE, with its native population of approximately 120,000 is considered as a potential consumer, the level of affluence would still be exceptional.

Abu Dhabi and the other small gulf producers followed Saudi Arabia's lead in the fall of 1973 and participated in the production cutbacks and the embargo. With this experience as precedent, and given their strong financial position, these states might be willing to join Saudi Arabia in limiting the rate of growth of their production in the interest of maintaining prices.

Algeria

Algeria, which joined the camp of radical Arab states through an anti-colonial war rather than by a coup d'etat in the Iraqi style, came out in roughly the same place on oil policy. By taking over a majority interest in the French companies in 1971, Algeria was finally in the position of controlling its oil resources instead of collecting revenues and royalties from foreign companies.

A charter member of the radical Arab group since it won its independence from France in 1962, Algeria has taken consistently strong positions on the Arab-Israeli conflict. However, in recent years Algeria has shown surprising economic pragmatism despite its outspoken radical political positions. In part, this pragmatism results from its close economic relations with the European Community. In fact, its economic relations with Europe are so close and so important that it is difficult to see Algeria's disrupting them, now that the questions of ownership and control of the oil industry have been resolved.

Algeria's pragmatism of the past few years extends even to the United States. Despite the rupture of diplomatic relations in 1967 in retaliation for the alleged participation of U.S. planes in the six day war and despite strong antipathy to the continued U.S. support of Israel, the Algerian government has separated business from politics to a large degree. The "neutralizing" of commercial dealings was not politically possible when the Arab-Israeli war broke out in the fall of 1973, however, and Algeria joined in both the oil production cutbacks and the embargo. Nevertheless, the government did not cut back on natural gas production and announced its intention of honoring its long-term gas supply contracts with Western customers.

Although, as noted earlier, Algeria is not a major source of oil supplies for the United States and Western Europe, the country could become a major supplier of natural gas. Algeria's natural gas reserves are estimated at 3 trillion cubic meters, the fourth largest in the world, and the country is already placing increasing reliance on gas as the foundation of its economic development program. West European and U.S. firms have signed long-term supply contracts for LNG. Included in the deals are provisions for the supply by the Western firms of technical assistance, management training, and equipment. Negotiations have been under way for some time to supply Western Europe with Algerian gas by pipeline, but difficult technical problems involved in laying the pipe across the Mediterranean must be overcome.

Future oil and, increasingly, gas revenues will finance a large part of Algeria's planned investment of $12 billion between 1974 and 1977.[11] The remainder will be financed largely by foreign borrowing. To guarantee the success of its social and economic development goals, Algeria must maintain its credit rating and access to Western capital markets. Thus, the prospects for further exploration by foreign companies in partnership with Sonatrach (the national oil and gas company) for the development of Algerian gas for export appear good.

Libya

Under King Idris, oil concessions were given out to most major international companies and a large number of independent firms, and the royal coffers were soon bulging. Production tripled from 867,000 barrels a day to 2.6 million barrels a day between 1964 and 1968. The closing of the Suez Canal in 1967 and the resulting tanker shortage put a high premium on North African oil, just across the Mediterranean Sea from Europe.

Following Iraq and Algeria, Libya was the third oil-producing country to join the radical camp. When Colonel Qaddafi and several other young military officers took over the government in 1969, they were in an excellent position to put pressure on the foreign companies for higher revenues. In 1970, Libya took the lead in forcing the companies to pay more in taxes, and after the companies made similar concessions to the other Persian Gulf oil-producing states in the Tehran agreement of 1971, the Libyans came right back to demand, and get, an even higher tax rate because of the difference in shipping costs from North Africa and the gulf area.

Libyan production peaked in 1970 at 3.3 million barrels a day and had leveled off at about 2.2 million barrels a day prior to October 1973. The production cutback was part of a deliberate government "conservation" policy, which was motivated by several considerations, including continuing controversies with the oil companies, an interest in extending the lifespan of the country's limited oil resources, and a desire to pressure the Western oil-consuming nations, especially the United States, into a more pro-Arab policy on the dispute with Israel. Libya has learned since 1971 that it can export smaller quantities of its low-sulphur, easily accessible oil and still earn greater revenues by raising prices. In 1971, while production dropped 17 percent below that of 1970, earnings rose 23 percent.[12]

When the Middle East war broke out in October 1973, Libya, which had taken over additional investments of foreign oil companies and which had been thwarted in a move to form a political union with Egypt, participated in the Arab oil production cutback and embargo. A few of the major oil companies continued to produce and ship oil from their expropriated properties, but the Libyan government on its own sold crude oil in bilateral deals at high prices.

As for the future, Libya is not under a great deal of economic pressure to increase oil earnings. Nevertheless, the government clearly wants to maximize per barrel revenues as long as selling oil is in harmony with its political goals. With more money than the country needs for domestic development, the Qaddafi government has used oil revenues to play the game of international politics in Malta, Sudan, Uganda, Northern Ireland—almost anywhere it has seen a revolutionary cause to support—and has embarked on a "cultural revolution" at home. Indeed, the regime places such a high priority on its ideological and political ends that it would not shrink from sacrificing economic interests for their sake. Thus, at some time in the future, if Libya saw a political advantage in reducing or stopping exports to Western Europe or other customers, it probably would not hesitate to do so. The dispute with Israel is the most likely cause of some kind of punitive action on Libya's part, but other situations deemed to have been brought on by Western imperialism could have the same result.

ARAB OIL AND THE SOVIET UNION

Thus far, the close relations between the USSR and several Arab states have had little impact on the oil industry or Western access to

Middle Eastern oil. But this could change—particularly during the next ten or so years of heavy Western and Japanese dependence on Arab oil.

Arab Oil in Soviet Strategy Against the West

The growth of Soviet influence in the Arab world has been based primarily on political support for the Arab side in the conflict with Israel and in disputes with Western powers, on military and economic aid to some Arab states, and on a general solidarity with governments and political movements professing radical nationalism or some form of socialism. The Soviet Union's principal friends have been Egypt, Syria, and Iraq (plus Algeria for a while). Saudi Arabia, Kuwait, and Libya (until 1969) were among the "reactionary" states whose regimes were the foremost targets of the Soviet-supported radical states (although the campaign had little success). Of the Arab oil states in the Persian Gulf, only Iraq is subject to Soviet influence today, and only Kuwait and Iraq have diplomatic relations with the Soviet Union. King Faisal of Saudi Arabia is fiercely anti-Communist and anti-Soviet.

Revolution remains a possibility in every Arab state, of course, and the Soviet Union can be expected to try to exploit any instability that develops. But the record of relations between the Soviet Union and the radical regimes that have come to power in oil-producing states suggests that Soviet influence over Arab politics and over Arab oil is likely to remain circumscribed. After Libya had its revolution in 1969, the new regime made a limited arrangement with the USSR to market nationalized oil, but it accepts no outside directions whatever, and it makes its decision on oil policy for its own reasons, not those of the Kremlin, of which it is highly distrustful. Algeria accepts Soviet technical aid, in its oil industry as in others, and has certain ideological affinities with the Soviet Union, but it is independent in economic policy and is turning increasingly toward economic cooperation with the West. Iraq is the one Arab country where political solidarity and military alignment with the USSR are matched by extensive collaboration on oil matters. But Iraq is no different from Libya and Algeria in keeping policy decisions in its own hands, and it is presently engaged in expanding its oil exports to hard-currency markets.

Soviet propaganda has consistently applauded moves by Arab governments against the "imperialist oil monopolies," with the obvious purpose of gaining political credit with the Arabs. Actually, however, the Soviet Union is not in a position to gain much by ambitious attempts either to play a major role in Arab oil affairs or to disrupt oil relations between the Arab states and the West. The theory that the Soviets can gain control of Arab oil through political means, and thus have the economy of the West at their mercy through their ability to turn the tap on and off at will, is a hobgoblin. To control the flow of oil, the Soviets would have to impose puppet governments on the producing countries, a task which seems beyond their capabilities and their calculus of acceptable risks. Or they would have to establish a pattern of cooperation, display a competitive level of expertise in oil matters, and provide a market for Arab oil--all of which would have to strike the governments of producing countries as an acceptable alternative to their oil trade with the West. Thus far, despite arms deliveries, large-scale trade, and general political support, the Soviets have not been able to provide such an alternative.

For the present, the Soviets can probably gain more by cautiously exploiting the troubles that inevitably will enmesh the Western countries if the Arabs, for their own reasons, restrict the flow of oil to traditional markets. In that event, the Soviets stand to gain from the weakening of the Western economies and the strains on the Western alliance. In the Middle East area itself, they can profit politically by maintaining a public posture as friend and benefactor of the Arabs in their struggle with the private oil "monopolies" and the "imperialist" governments, and they can lose by trying to do much more than that.

At the same time, the deals that the Kremlin is trying to negotiate with U.S. and Japanese companies for the large-scale exploitation of energy resources in the Soviet Union are indirectly related to Soviet policy in the Middle East. The Soviets are not concerned over the rising prices of Arab oil on the world market, as the price of their own oil can be raised correspondingly. The higher the price of Middle Eastern oil for the West and Japan, the more interest the richer capitalist nations may have in helping to open up the energy resources of Siberia.

The Soviet leadership is aware that an active policy directed at disrupting the oil supply of the major consuming countries would introduce great tension into its relations with those countries. Nevertheless, the general Soviet policy of détente with the United States and Western Europe is not irreversible. Nor is it certain that the degree of détente achieved on European questions will be evident in the Middle East, where the Soviets may see opportunities for political gains without undue military risk. The Soviet Union's conduct in the Arab-Israeli war of October 1973 showed it to be ready to move for such gains, but in working with the United States to bring about the cease-fire it demonstrated the importance it attaches to the bilateral relationship with the United States and its desire to avoid a confrontation with the United States. However, in light of Soviet displeasure with Sadat's policy of working for a settlement through U.S. auspices, the USSR may weigh how much support it will give to the settlement effort. The USSR's policy is likely to reflect its double and sometimes conflicting aim of wanting to go along with the idea of a settlement (although preferably not one masterminded by the United States) in the general interest of peace, and wanting to keep the pot boiling in the Middle East in order to play a diplomatic and political role in the area.

The Soviet Factor in Arab Policy

There is no doubt that certain Arab states have found the connection with the Soviet Union to be a useful element in the process of working out their oil relations with the Western oil companies and the major consuming countries. On the general political front, the heavy support given to the Arabs by Soviet propaganda throughout the world in all their controversies with Western governments and oil companies has provided welcomed encouragement. More concretely, the Soviets could and did furnish technical aid to the Arab national oil companies in Algeria, Iraq, and Syria as they took over unexploited concessions or producing fields from Western companies. The ability to turn to Moscow for such help provided Arab governments with at least a partial alternative to reliance on the West. Nevertheless, the fact that the Soviet Union, itself a net exporter of oil, cannot offer an economic alternative to the Western and Japanese markets sets limits on the bargaining value of the Soviet card. The real basis of Arab bargaining power is not Soviet support but the West's demand for oil.

There is always the possibility, if there is no progress toward an Arab-Israeli settlement, that the Soviets and the Arab states might join in a concerted effort to force the United States to end its support of Israel or to insist on Israel's unconditional withdrawal from occupied Arab territories. They might undertake a diplomatic offensive in the United Nations, backed up by Soviet political pressure and an Arab slowdown or selective embargo of oil exports. Such an offensive would place the United States in a very uncomfortable position. But the basic factors in such a situation would be not the Arab-Soviet "alliance" but the relations of the United States with each of the parties. The effectiveness of such an offensive would depend on the ability of the Arab states to take the necessary measures and incur the consequences. The Soviet Union could not do much to make the action more effective or ease the Arabs' burdens and risks, even if it was prepared to engage in economic warfare with the United States and revive the cold war.

FUTURE ARAB OIL POLICIES AND THE WEST

Whether the Arab countries will be a reliable source of oil for the industrialized countries and for the United States in particular depends on both political and economic factors. Most important will be the course of the Arab-Israeli dispute and perceptions of the U.S. attitude toward Arab aims and interests. The key countries whose views and actions are crucial to the United States are Egypt and Saudi Arabi. A mutually satisfactory settlement between Israel and Egypt, combined with future Egyptian and Israeli moderation, should keep Saudi Arabia's oil flowing to the West.

But if war breaks out again, or if little or no progress is made on substantive issues, there is every reason to expect that Saudi Arabia, along with some of the other producers, would resume use of the oil weapon. Clearly, the Arabs regard oil as an effective weapon, and they would continue to use it whenever they thought the occasion demanded. Having invoked it on the issue of Israeli withdrawal from Arab territory, they will not easily forego the use of the oil weapon if progress is lacking on that issue.

The economic factors center on how the oil-producing countries, both Arab and non-Arab, will tailor their policies on production and export of oil to fit their interests in the use and development of their own resources. The key issue is whether the oil states will agree to hold down production in order to maintain or raise prices. The economic factors influencing the availability of Arab oil are therefore essentially a matter of marketing strategy. These factors will be analyzed in detail in Chapter 13.

NOTES FOR CHAPTER 3

1. Department of State estimates cited in U.S. House of Representatives, Subcommittee on the Near East and South Asia, Committee on Foreign Affairs, *The Impact of the October Middle East War*, Hearings, 93rd Cong., 1st sess. (October-November 1973), Table VIII, p. 159. (Note: The figure for U.S. imports from Arab countries includes both direct crude oil imports plus estimates of Arab crude oil that was sent to refineries overseas and then shipped to the United States.)

2. *Oil and Gas Journal*, June 5, 1972, p. 22.

3. *Petroleum Press Service*, June 1973, p. 217; and United Nations,

Department of Economic and Social Affairs, *World Energy Supplies*, Statistical Papers, Series J (New York), various years.

4. For world "proved" reserves see appendix Table A-6.

5. *Petroleum Intelligence Weekly*, August 13 (p. 1), 20 (p. 8), and 27 (p. 1), 1973; and *New York Times*, September 2, 1973, p. 1.

6. *Umm al-Qura* (Saudi newspaper, Riyad), July 30, 1973.

7. Kuwait Planning Board, *The Kuwait Economy, 1969-1970* (1970), p. 119.

8. International Monetary Fund, *International Financial Statistics*, March 1974, p. 226.

9. *Oil and Gas Journal*, August 6, 1973, pp. 23-24.

10. The mid-1973 production figures given here for the small gulf states are from *Oil and Gas Journal*, December 31, 1973, pp. 108-35.

11. *Arab Oil and Gas*, January 1973, p. 17.

12. Organization of the Petroleum Exporting Countries, Statistics Unit, *Annual* Statistical *Bulletin*, 1971 (Vienna, 1972), pp. 28 and 128.

THE MIDDLE EAST

Chapter Four

Iran

DEVELOPMENT OF THE OIL INDUSTRY
IN IRAN

Iran has played a central role in the development of the Middle Eastern petroleum industry.[1] The first oil concession in the Persian Gulf was granted by Iran in 1872, and in 1908 the country was the scene of the first commercial oil strike in the Middle East. In 1913, the first Middle Eastern oil exports—at an initial rate of 5,000 barrels a day—were loaded at Abadan. Iran remained the region's leading oil producer until 1951. In that year, under the leadership of Prime Minister Muhammad Mossadegh, Iran became the first country in the Middle East to nationalize a foreign oil company, the British-owned Anglo-Iranian Oil Company, whose properties were turned over to the newly established National Iranian Oil Company (NIOC). In retaliation, the Anglo-Iranian Oil Company withdrew its operating personnel, and Iranian oil was boycotted by the major oil companies. As a result, Iran quickly slipped from first place (664,300 barrels a day in 1950) to last (61,400 barrels a day in 1954) among Middle Eastern oil producers, and the Arab producers began to pick up Iran's share of world markets.[2]

Events in Iran between 1951 and 1953 were dominated by a power struggle between Mossadegh and the Shah. At one point during this struggle, the Shah fled the country. In mid-1953,

Note: This chapter is based in large part on research and analysis conducted by Robert A. Mertz.

53

however, the army forced Mossadegh out of office and restored the Shah to power. Negotiations with foreign oil interests were then resumed. In 1954, an agreement was forged between Iran and a consortium of foreign companies.[3] Iran was recognized as the legal owner of its oil resources, and terms were set for the consortium's operations. Provisions were also made for the division of oil revenues and for the allocation to NIOC of a share of the crude oil produced.

In the years after the 1954 agreement, Iran evolved several other forms of contractual arrangements with foreign oil companies, including service contracts and joint ventures by NIOC and foreign companies. These arrangements, which applied to areas outside the area covered by the consortium agreement, gave Iran several advantages over the terms of that agreement: a higher percentage of net profits; acknowledgment of NIOC's right to at least 50 percent of the oil produced, which has enabled it to enter foreign markets on its own or through the market facilities of its partners and contractors; exploration offshore and onshore in areas relinquished by the consortium; and shorter contract length.

Between 1956 and 1972, Iran's oil production increased tenfold—from approximately 500,000 to over 5 million barrels a day—and proved reserves roughly doubled—from 32 billion to 60 billion barrels.[4] The major share of the new reserves is attributable to a revision of the estimated reserves of fields already discovered by the Anglo-Iranian Oil Company prior to 1951 and to the consortium's discovery of eleven new fields between 1959 and 1972. About 25 percent of these new reserves—7 billion barrels by the end of 1972—were discovered offshore, at very high cost, by four joint-venture operations. The output from these offshore discoveries, which in 1972 amounted to some 10 percent of total crude oil production in the country, accounts for all of the oil currently produced by the joint-venture operations. (See Table 4-1.)

While the amount of proved reserves has continued to rise each year, the rate of increase has not been dramatic and vast new onshore discoveries are not expected. Iran's onshore areas have been

Table 4- 1.   Iran's Crude Oil Production (1,000 b/d)

Producer	1966	1967	1968	1969	1970	1971	1972
Consortium	2,017.1	2,466.6	2,704.0	3,100.5	3,496.1	4,144.4	4,541.7
Joint Ventures	87.6	121.1	134.2	265.1	322.3	383.1	467.7
NIOC	8.6	8.9	9.4	9.3	10.2	12.1	14.7
TOTAL	2,113.3	2,596.6	2,847.6	3,374.9	3,828.6	4,539.6	5,024.1

Source: National Iranian Oil Company communication.

explored thoroughly, and the general prognosis is that onshore production will not rise more than about 1 million barrels above the target of 8 million barrels a day that the consortium expects to meet in the late 1970s. The outlook for additional offshore discoveries is uncertain.

In 1973, Iran negotiated a new twenty-year service contract, which supersedes the old consortium agreement and which reflects a basic change in the relationship between NIOC and the oil companies. Under the 1954 agreement, the companies basically determined production levels, which the Shah from time to time tried to alter by private exhortation or by publicly requesting increases. The 1973 agreement places NIOC in the driver's seat by giving it the power to set production targets and other oil policies. The role of the companies is now that of performer of services under contract to NIOC and purchaser of crude oil.

Although NIOC has not yet become a significant producer of crude oil through its own exploration and production activities, since 1966 it has used its crude oil from joint-venture operations and the nearly 150 million barrels of oil it received from the consortium between 1967 and 1971 to extend its operations abroad. Limited to selling its oil from the consortium in noncompetitive markets, NIOC has signed barter agreements with several East European countries to exchange Iranian oil for industrial plant and equipment.

A desire to earn convertible currency and an interest in gaining equity participation in downstream facilities, however, have also prompted NIOC to seek markets outside Eastern Europe for its share of crude oil produced by joint ventures. In 1971, NIOC won a contract over Soviet and Afghan companies to serve Afghanistan's internal market and to provide refueling services at the Kabul and Kandahar airports. In its first equity investment overseas, NIOC and AMOCO (a subsidiary of Standard Oil of Indiana) each purchased a 13 percent interest in a refinery in Madras, India (the remaining 74 percent is owned by the Indian government). In the confusion prevailing in world oil markets in the winter of 1973-74, the requirement that NIOC export only to markets in which the consortium was not in business vanished. NIOC has since engaged in direct sales of crude oil throughout the world.

For the future, NIOC hopes to encourage foreign investment in refineries to to be built in Iran, in return for which refined products would be shipped to the home markets of the foreign partners. A desire to enter into various phases of the industry makes such an arrangement attractive to NIOC, and an important advantage to the potential foreign partners is that locating the refinery in Iran would eliminate the difficult task of persuading their pollution-conscious governments to permit the construction of new refineries at home. Moreover, NIOC would guarantee to supply the refineries with agreed amounts of crude oil.

In addition to producing and marketing oil, NIOC has begun to form subsidiaries to diversify its operations. Most important was the utilization of Iran's immense natural gas reserves. These reserves, estimated at 270,000 billion cubic feet at the end of 1973, are the second largest in the world and comprise more than an eighth of the world's proved reserves of natural gas.[5] In 1966, at a time when over 90 percent of the gas produced in conjunction with oil production was being flared, Iran signed an agreement with the Soviet Union for the exchange of natural gas for a large Soviet steel mill, a machine-tool factory, and other industrial plants. The USSR also agreed to take 60 percent of the initial throughput of a 1,200-mile Iranian gas trunkline, which was to be built with Soviet technical assistance for the National Iranian Gas Company, and to build its own pipeline south to the Iranian border to link up with it. Gas deliveries to the Soviet Union began in 1970, and by 1973 nearly 40 percent of the gas produced in Iran was being used. As for the future, large liquefied natural gas (LNG) projects involving investments by industrialized countries in exchange for long-term deliveries of LNG appear to offer distinct possibilities.

Since 1964, NIOC's National Petrochemical Company has established three petrochemical plants in Iran in partnership with leading U.S. chemical companies and one chemical fertilizer complex. Another subsidiary, the National Iranian Tanker Company, owns several small tankers and is expanding its operations. In early 1972, NIOC announced that it had formed a subsidiary, the Iranian Oil Company (UK), which in equal partnership with the British Petroleum Company had won two concessions in the North Sea to explore for oil. This step marked the first time that a national oil company from an OPEC country had become involved in exploration outside its own territory, and it symbolizes the expanded role NIOC hopes to play in the world petroleum market.

The destination of Iran's oil exports shifted dramatically in the 1960s. (See Table 4-2.) Japan, which took only 3 percent of Iran's exports in 1960, received 37 percent by the end of the decade, while Western Europe's share dropped from 47 percent to 31 percent over the same period. In 1972, 39 percent of Japan's total crude oil imports came from Iran, primarily in the form of heavy, low-sulphur crudes.

Table 4- 2.   Destination of Iranian
Oil Exports (percentage distribu -
tion)

Destination	1960	1965	1972
Western Europe	47	41	31
Japan	3	17	37
Other	50	42	32
TOTAL	100	100	100

Source: Organization of the Petroleum Exporting Countries, Statistics Unit, *Annual Statistical
Bulletin, 1972* (Vienna, 1973), p. 57.
Note: The United States, which in 1970 purchased about 4 percent of Iran's total oil exports,
is included in "Other."

OIL AND ECONOMIC DEVELOPMENT

The restoration of oil production in late 1954 following the change
in domestic political leadership that returned the Shah to power and
the subsequent consortium agreement launched Iran on a path of
accelerating economic development. Between 1965 and 1971, Iran
attained an annual growth rate of over 10 percent.[6] Not surprisingly,
economic progress has been interdependent with the internal
political stability that has attended the Shah's dominance over the
political system during the past two decades.

An expanding oil sector has directly contributed an
increasing share to the gross domestic product, has made possible the
development of a petrochemical industry, and has provided the
energy needed for general economic growth. In addition, Iran's
growing oil revenues have facilitated borrowing abroad for development. Tables 4-3 and 4-4 indicate the impact of oil on the growth of
the Iranian economy between 1954 and 1972.

During the 1960s, the share of gross domestic product
contributed by the oil sector almost doubled, from about 17 percent
in 1960 to nearly 32 percent in 1971. The rapid growth of the
economy in the late 1960s was largely due to the increased oil
revenues pumped into the expanding activities of the public sector.
Government revenues from oil accounted for 45 percent of total
revenues during 1965 and climbed to nearly 60 percent of total
revenues by 1971. Due to the general economic upswing, government
non-oil revenues, primarily from indirect taxes, almost tripled from
$660 million in 1965 to $1.8 billion in 1972. As a result of heavy

Table 4-3.   Iran's Government Budget and Gross Domestic Product
(Solar Years ending March 20) (in millions of current US dollars)

	1960/61	1965/66	1970/71	1971/72	1972/73
Total Government Revenues:	n.a.	1,192	2,265	3,453	4,431
Non-oil revenues	n.a.	660	1,136	1,399	1,813
Oil revenues	295	532	1,129	2,054	2,618
Oil Revenues as % of Total Govt. Revenues	n.a.	45	50	59	59
Total Government Expenditures:	731	1,297	3,050	3,883	5,254
Operating expenditures	490	809	1,782	2,331	3,167
Development expenditures	241	488	1,268	1,552	2,087
Budget Balance	n.a.	-105	-785	-430	-823
Gross Domestic Product (GDP)	4,050	6,040	10,380	12,750	
Value Added by Oil Sector	680	1,090	2,295	4,050	
Oil Sector as % of GDP	16.8	18	22.1	31.8	

Sources: Iran Central Bank, *Annual Report and Balance Sheet*, various years;
International Financial Statistics, various issues.

public sector investments, facilitated by rising government revenues, the investment objectives of the third and fourth development plans (1962 to 1972) were surpassed.

Revenues, however, did not keep pace with the rising rate of government spending. Because the terms of trade were unfavorable to oil producers throughout the 1960s, the increase in oil income resulting from expanding output did not keep up with the rise in the cost of imported goods and services on which much of Iran's economic growth depended. Whereas import prices increased 12 percent between 1963 and 1970, oil export prices rose only 1 percent in the same period.[7] Moreover, the total volume of imports rose dramatically in the 1960s due to vigorous demand in the private sector for capital goods and raw materials, a sharp increase in public sector development expenditures, and growing purchases of military hardware.

In the last half of the 1960s, sharply higher expenditures for development and security consistently outpaced the growth in government revenues; as a result Iran experienced substantial budget deficits, a deteriorating net trade deficit, and a shortage of foreign exchange. Net foreign exchange reserves fell steadily from $326 million in 1965 to only $76 million in 1970, and the trade deficit rose to $437 million. This imbalance was financed by heavy foreign borrowing. Between 1967 and 1971, excluding purchases of military hardware on credit, Iran borrowed nearly $2.4 billion abroad, which, minus repayments, increased its net foreign indebtedness by about $1.6 billion.[8] In 1971 alone, Iran borrowed $710 million in

Table 4-4. Iran's Balance of Payments (Solar Years Ending March 20) (in millions of current US dollars)

	1954/55	1960/61	1965/66	1970/71	1971/72	1972/73
Total Exports (Goods & Services):	217	887	1,466	2,895	4,327	4,997
Oil exports (f.o.b.)	89	683	1,183	2,426	3,669	4,145
Non-oil exports (f.o.b.)	120	142	181	287	379	509
Other (services)	8	62	102	182	279	343
Oil Exports as % of Total Exports	41%	77%	81%	84%	85%	83%
Total Imports:	298	1,023	1,644	3,332	4,303	5,317
Imports (c.i.f.)	239	623	857	1,917	2,485	3,164
Investment income paid to oil companies	44	285	512	906	1,267	1,507
Other (services)	15	115	275	509	551	646
Net Trade Balance	-81	-136	-178	-437	24	-320
Oil Sector Retained Value[a]	30	356	624	1,296	2,231	2,598
Gross Foreign Exchange Earnings[b]	158	560	907	1,765	2,889	3,450
Oil Sector Retained Value as % of Gross Foreign Exchange Earnings	19%	64%	69%	73%	77%	75%
Net Official Borrowing[c]	21	39	35	376	631	570

Source: International Monetary Fund, Bureau of Statistics, *Balance of Payments Yearbook* and *International Financial Statistics*, various years; and Iran Central Bank, private communication to authors.

a. Government oil revenues plus local currency expenditures by oil companies.

b. Non-oil exports (goods and services) plus oil sector's retained value.

c. Public sector foreign loans and suppliers' credit minus loan repayments including interest payments.

long-term loans and over $250 million to cover short-term foreign exchange shortages—nearly one-half of its total oil earnings in that year.[9] From 1967 to 1971, the annual interest on long-term foreign debts alone rose from $11 million to $109 million.[10]

By the end of the 1960s, Iran's current and development expenditures were approaching the limit that its oil revenues (supplemented by foreign borrowing) could sustain. Part of its future export earning capacity was mortgaged to service the growing foreign debt. Drastic action was required if Iran was to maintain its growth momentum. In November 1970, the consortium agreed to an increase of the tax rate on oil from 50 to 55 percent of posted prices. The agreement signed shortly thereafter in Tehran in February 1971 between the gulf oil-producing countries and the oil companies operating there increased the posted prices of Persian Gulf crude oils and thereby further increased Iran's oil revenues.

The various price increases since 1971 have opened up the prospect of greatly increased oil revenues for Iran, and the 1973 agreement with the consortium guarantees Iran a continued flow of oil exports and a growing amount of oil for direct export by NIOC. Iran played a leading role in bringing about the unprecedented price increases that went into effect in the aftermath of the October 1973 Middle East war. Higher oil revenues should facilitate the fulfillment of Iran's ambitious development goals and can also be expected to finance a growing level of defense expenditures.

As in the past, Iran should have no difficulty in finding uses for its oil revenues. The ambitious development program is projected to cost on the order of $4 billion annually during the present five-year plan (1973-78).[11] Defense expenditure can also be expected to grow.

FOREIGN POLICY AND OIL

Since 1946, Iran has moved, with growing self-assurance, from a position of dependence on the United States to one of close, yet independent, friendship. In 1967, U.S. economic aid was terminated, but credits for military purchases have continued. Iran's shift toward a somewhat more independent international position has given it added stature in the councils of the Third World and has helped to deflect domestic and foreign criticism of the regime.

Iran has approached the problem of relations with the Soviet Union cautiously. Impelled by a long-standing fear of traditional Russian territorial designs on Iran (which pre-date 1917 by centuries), the Shah began, in the early 1960s, to edge toward a détente with the Soviet Union based on growing economic cooperation.[12]

The Shah hoped that closer political and economic relations with the Soviet Union would help relieve the tensions in their relations engendered by the cold war and also temper the Soviet commitment to Iraq, Iran's principal foe in the gulf area.* In particular, Iran considers Iraq, backed by the USSR, as the only state within the region interested in, and capable of, supporting subversive movements within the gulf states, as well as of overt aggression against them. It was in part to check Iraq that Iran initiated an expensive rearmament and expansion of its military forces.

Despite Iranian displeasure, however, the Soviets have continued to provide economic and military assistance to Iraq, including helping it to develop the North Rumaila oil field and dispose of the crude oil from the nationalized Iraq Petroleum Company (IPC) field at Kirkuk, and building a modern Iraqi naval base at Umm Qasr near the Persian Gulf. In August 1971, Iran extended diplomatic recognition to Peking, in part to bring further pressure to bear on the Soviets, but the USSR signed a treaty of friendship with Iraq in early 1972, which formalized their already-close relations.

Iran has traditionally seen the Persian Gulf as an area of strategic, economic, political, and cultural importance, and its natural sphere of influence. The Shah's personal and dynastic ambition to fulfill Iran's "imperial destiny" in the gulf is an important determinant of Iran's foreign policy.[13] Iran strongly favored Britain's announcement in 1968 that it would withdraw its military forces from the gulf region by 1971. Iran viewed the decision as an opportunity to establish hegemony in the gulf, a goal that had been frustrated by 150 years of British dominance. Once it became clear that the British decision was irrevocable, Iran moved swiftly to elaborate a policy to meet the problems likely to be created by the British withdrawal and to assure that Iran, and not an Arab gulf state or an outside power, would be in a position to determine the tempo and direction of subsequent development in the gulf region.

As the self-appointed arbiter of economic and political stability in the gulf, Iran intends to protect the smaller gulf states, like Kuwait and the United Arab Emirates, from external aggression or internal subversion. In keeping with its regional defense posture,

---

*The points at issue between Iran and Iraq include the demarcation of borders; navigation rights; clandestine support for subversive nationalists within each country; ideological antagonisms; historically competitive religious, ethnic, and cultural traditions; and rivalry for influence in the gulf area.

Iran is developing a military force capable of reacting quickly and effectively throughout the gulf area and beyond into the Indian Ocean. Since late 1972, the government has ordered over $3 billion of sophisticated military hardware from the United States, Britain, and France.[14] The impressive arsenal of weapons already on order guarantees Iran overwhelming military superiority to deter overt aggression by one state against another within the gulf region. However, these forces would be inadequate to counter direct Soviet intervention and are unsuited to coping with the more subtle problem of political subversion from within the Arab gulf states. Moreover, there is a very real danger that Iran's arms buildup will have a severe negative impact on the Arab gulf states. Although the conservative Arab regimes, which fear the threat of further Iraqi aggression against Kuwait or subversive efforts in Bahrain or in the United Arab Emirates, are reassured by Iran's military power, they are also alarmed at the scale of the buildup and can be expected to take steps to build up their own military capabilities.

Iran has flexed its military muscle on several occasions in the name of overall stability in the gulf area and to protect its own interests. In late 1972, Iran extended military assistance (a company of air-mobile marines) to the Sultanate of Oman, which since 1965 had been unsuccessfully trying to eradicate a Marxist-led insurrection in the southern province of Dhufar that threatened to spread to the other gulf states. Of the Arab states, only Jordan had come to the aid of the Sultan. Arab reaction in the gulf to Iran's involvement was mixed. Fearful of the spread of the Dhufari rebellion to the gulf area, the Arab gulf states seemed relieved by Iran's decisive action, yet reluctant to say so publicly. Although unhappy about the intervention of an outside power, especially Iran, in an Arab problem, that involvement won the grudging acquiescence of Saudi Arabia and Abu Dhabi, both of which later extended financial and limited military assistance to Dhufar.[15]

Iran's military occupation of three small islands (Abu Musa and two Tunb islands) guarding the gulf approaches to the Strait of Hormuz was an indication that Iran will risk Arab displeasure to take actions it considers in its national interest. Ownership of the islands had been recognized by the British as belonging to two states in the United Arab Emirates. Although the islands are not of vital strategic value, since Iran can defend the entrance to the gulf from onshore air and naval bases that are situated much closer than the islands to the Strait of Hormuz, the government apparently decided that Iranian control of the islands would prevent them from being taken over by any other power or used as bases from which guerrillas, of any nationality or political persuasion, might launch an attack on tankers carrying petroleum through the strait.

The issue of sovereignty over the islands goes far beyond the immediate question of the three small islands. Geologically, the offshore areas of the lower gulf offer exciting petroleum potential, which was confirmed in early 1973 by the discovery of oil nine miles off Abu Musa. The possibility of additional oil strikes makes the future delimitation of territorial boundaries in and around the islands in the lower gulf a potentially explosive issue.

Iran's ability to influence developments on the Arab side of the gulf is circumscribed by the fact that the norm of past Arab-Iranian relations has more often been conflict and hostility than peace and cooperation. A further issue that divides the Arab states and Iran is the latter's relatively close ties with Israel. Iran did not participate in the Arab oil boycott after the 1967 war or again in 1973. Indeed, some of Iran's oil goes to Israel by way of the Red Sea. At the same time, Iran has tried to soft-pedal its relations with Israel and has generally supported the Arab view on Israeli withdrawal from occupied territories.

Despite differences in policy toward Israel and an undercurrent of national rivalries, a number of disputes have been resolved in the past few years, which could help pave the way for closer Arab-Iranian cooperation. Included on the list of settled issues are the satisfactory resolution of the long-standing problem of demarcation of the offshore boundaries between Iran and Saudi Arabia and Qatar, and Iraq's relinquishment in 1969 of its claim to Bahrain. The resumption of diplomatic relations between Iraq and Iran in 1973 and the resolution of a boundary dispute appeared at the time to be a further hopeful sign, but relations between Tehran and Baghdad deteriorated again in early 1974, when serious border clashes erupted.

Saudi Arabia and Iran recognize that cooperation between them is essential to the achievement of their common goals of maintaining the political status quo and regional security. Iraq's radical saber-rattling and King Faisal's religious hatred of the atheistic, Communist governments (the USSR and China) that support Iraq and South Yemen have pushed the conservative Arab states closer to Iran. How much cooperation will actually take place is impossible to predict in light of the highly volatile political and military situation in the area. With respect to oil, Iran and the Arab oil-exporting states share the objective of maximizing revenues, but they may not always agree on the means of achieving that end. In 1973-74, Saudi Arabia and Iran reportedly differed on price policy; the Shah favored even higher prices than those actually agreed upon by the Persian Gulf oil producers. Iran's unwillingness to join the Arabs in using oil as a weapon in their dispute with Israel has already been mentioned. A further test of the possibilities of Arab-Iranian cooperation will be faced if oil prices weaken. Iran might then be willing to join Saudi Arabia in limiting production, but it is also possible that Iran's greater need for oil revenues would cause it to pursue a policy of maximizing output.

PROSPECTS FOR POLITICAL STABILITY

Since the Shah returned to power following the overthrow of the Mossadegh government in 1953, he has sought to bring about political stability by winning over or neutralizing potential rivals and centralizing political power. Throughout the 1950s and 1960s, the Shah juggled policies of political repression and social reform in an attempt to solidify control. Dissident elements were suppressed and in 1963 the Shah embarked on a series of social and economic reforms that came to be known as the "White Revolution." Economic development plans were designed to coopt liberal elements and to provide the regime with a new basis of popular support.

Economic success brought to prominence a group of energetic, highly competent technocrats, who have benefited from economic development and have a strong personal interest in its continuance. A growing middle class, moderate-to-conservative in outlook and including most of the military and the civil service, also has a vested interest in continued economic progress. This middle class, which for the most part views the Shah with approval as a modernizer and reformer, has become the most important political force in the country and the major support of the monarchy and the political status quo. Almost all of those to whom the Shah appeals (except for the young university graduates) lived through Iran's political turmoil in the 1950s and view the Mossadegh era unfavorably.

A basic weakness in Iran's political system is the fact that the Shah has not established a structure of political institutions to guarantee continuity of leadership. The Shah's White Revolution and other reforms, plus the rapid rate of development, have enabled him to carry on successfully without meeting this problem, but it is questionable whether any successor regime would be able to resist as he has the demands of growing middle class elements in Tehran and elsewhere for a share of political power.

Iran's political system is bound to undergo change when the Shah's reign ends. It is unlikely that the same degree of stability could be maintained under a regency for his son, and a struggle for power among leading political figures is conceivable. Such a struggle would not necessarily disrupt the production and export of oil since there would be a national interest on the part of all to maximize the country's oil revenues. Widespread civil strife, on the other hand, with parts of the country possibly falling under the control of rival factions, could lead to sabotage, interruptions of production, or other actions that would decrease the availability of oil for export.

It is also conceivable that a group of military officers might seize power. While there is no way to predict the policies that such a group would adopt, they might well include radical internal change and general hostility to Western interests. While the national interest in maximizing oil revenues for development would certainly continue, there might be a period during which the availability of oil to the West would be subject to deliberate reduction or interruption.

IRAN'S FUTURE OIL POLICY

Iran's Fifth Development Plan can be taken as a good indication of what the Shah wishes to achieve in the economic area during the period covered by the plan (March 1973-March 1978). This plan is one of the most ambitious ever undertaken by a developing country. It envisions an accelerated rate of economic growth over and above the impressive record achieved in the past decade. It projects real growth in GNP at 11.4 percent annually; an increase in per capita GNP of just over 100 percent (from $418 to $851); creation of 1.4 million new jobs; and a continuation of major developmental programs to which the government is already committed: dams, electric power, gas distribution, the Isfahan steel mill, and the huge Sar Chesmeh copper mine.[16]

Even before the large price increases of late 1973, Iran's oil revenues were expected to cover at least 50 percent of total government expenditures during the period of the fifth plan.[17] Because of the prospect over the next several years of higher oil revenues than projected in the plan, estimated revenues and expenditures in the plan are already being revised upward. A large portion of the increased oil revenues, however, could be absorbed by Iran's soaring defense expenditures, which since 1972 have been running between $2 and $3 billion a year and could go higher.

To provide adequate government revenues and foreign exchange, Iran has sought to maximize its oil revenues. In the first instance, this can be accomplished by increasing crude oil production, which has long been a major tenet of Iran's oil policy. It seems probable that Iran will be able to increase crude oil production from the consortium area to about 9 million barrels a day by 1980. The consortium's major fields have been developed under a sensible, long-range plan of primary production; the introduction of secondary recovery and additions to proved reserves on the order of about 10 billion barrels should allow production to continue at that level during most of the 1980s. Iran can also earn higher net revenues per barrel by demanding higher prices and a greater share of profits. Whether oil prices will be sufficiently high over the long term to provide revenues needed to finance the development plan and a high level of military imports, thus releasing Iran from the necessity of extensive foreign borrowing, is not clear. It is conceivable that the hope of greatly increased revenues will stimulate a level of expenditures for development and imported military equipment that would force Iran to continue borrowing considerable amounts from abroad. What is abundantly clear, however, is that Iran, with its large population and dynamic economy, can probably absorb all of the oil revenues it is likely to earn, though for political and economic reasons it may wish to divert some of its revenues to foreign aid and overseas investment.*

Iran's choice among the various options to maximize oil revenues will be determined by conditions in the world petroleum market in general and the state of its relations with the major Arab oil-producing countries, especially Saudi Arabia. The ultimate criterion of Iranian action will obviously be self-interest. During the mid-1960s Iran followed a line independent of OPEC in pursuit of its own interests, and—along with Saudi Arabia and Libya—helped scuttle an earlier attempt at production controls. Although Iran is likely to support any OPEC policy to increase the economic value of its oil resources, it is unlikely to agree to any restrictions on its output if to do so might interfere with its economic growth.

Assuming no drastic change in political orientation, Iran's oil policy will for the foreseeable future involve as much cooperation as possible with other oil-producing countries and active cooperation with Western financial institutions, oil companies, and private

*In February 1974, after conferring with the presidents of the World Bank and the International Monetary Fund, the Shah proposed creation of a new development fund to ease problems caused by higher oil prices and pledged more than $1 billion to the fund.[18]

capital. At the same time Iran, with its limited reserves and the prospect of a leveling off of production sometime in the 1980s, could not by itself (or with other non-Arab countries now producing) make up the loss to world supplies that would result from an effective Arab embargo coupled with significant reductions in production.

NOTES FOR CHAPTER 4

1. Edith T. Penrose, *The Growth of Firms, Middle East Oil and Other Essays* (London: Frank Cass Ltd., 1971), p. 296; and Fuad Rouhani, *A History of O.P.E.C.* (New York: Praeger Publishers, Inc., 1971), p. 87. For an overall history of Iranian oil development see Stephen H. Longrigg, *Oil in the Middle East: Its Discovery and Development,* 3rd ed. (London: Oxford University Press, 1968); and George W. Stocking, *Middle East Oil* (Nashville: Vanderbilt University Press, 1970).

2. Organization of the Petroleum Exporting Countries, Statistics Unit, *Annual Statistical Bulletin, 1971* (Vienna, 1972), pp. 25-30.

3. For a concise account of the provisions of the consortium agreements, joint ventures, and service contracts see, Sam H. Schurr, Paul T. Homan, and associates, *Middle Eastern Oil and the Western World: Prospects and Problems,* Resources for the Future (New York: American Elsevier, 1971), pp. 114-15, 127-28, 130-41, and 134-36; Longrigg, *Oil in the Middle East,* pp. 276-79; and Stocking, *Middle East Oil,* pp. 158-80.

4. *Oil and Gas Journal,* December 31, 1973, pp. 86-87.

5. Ibid.

6. Agency for International Development, *Gross National Product: Growth Rates and Trend Data* (Washington, D.C., 1971), p. 4.

7. Charles Issawi, *Oil, the Middle East, and the World* (New York: Library Press, 1972), pp. 31-32.

8. International Monetary Fund, Bureau of Statistics, *Balance of Payments Yearbook* (Washington, D.C.) and *International Financial Statistics,* various years.

9. Iran Central Bank, *Bulletin,* September-October 1972, pp. 282-83.

10. *Balance of Payments Yearbook* and *International Financial Statistics,* various years.

11. Figures supplied by Iranian Economic Mission, Washington, D.C.

12. For a thoughtful analysis of Iran's approach to relations with the major powers see, Rouhollah K. Ramazani, "Iran's Changing Foreign Policy," *Middle East Journal,* Autumn 1970, pp. 421-37.

13. For the "Psycho-Cultural" background of Iran's goals in the Persian Gulf, see Rouhollah K. Ramazani, *The Persian Gulf: Iran's Role* (Charlottesville: University of Virginia Press, 1972), Chapter 1.

14. *Wall Street Journal,* December 19, 1972, p. 4; *Newsweek,* May 21, 1973, p. 40; and International Institute of Strategic Studies, *The Military Balance: 1972-1973* (London, 1973), p. 31.

15. *New Middle East,* December 1972, p. 35.

16. *The Economist,* December 2, 1972, p. 86 and "Iran's Economic Development and Prospects for Iran-United States Cooperation," speech delivered in New York in April 1973 by A. Mirbaha, member Iranian Economic Mission, Washington, D.C. See also *New York Times,* January 14, 1973, pp. 62-63.

17. Mirbaha, "Iran's Economic Development and Prospects for Iran-United States Cooperation." The figures used throughout Mirbaha's discussion of the plan do not take into account the 1973 devaluation of the U.S. dollar.

18. *New York Times,* February 22, 1974, p.1.

Chapter Five

Latin America

by Philip Musgrove

Oil has been produced in Latin America since the turn of the century and exported in large amounts for almost as long.[1] First Mexico and then Venezuela was, for a while, the world's leading oil exporter. Oil production grew faster than consumption in the region until the early 1960s, and net exports reached a peak of 2.7 million barrels a day in 1964. (See Table 5-1.) In the next decade, however, regional consumption increased by about 65 percent, but production rose by only 12 percent, and net exports declined by almost 400,000 barrels a day.

Table 5-1. Latin American Net Exports (thousand b/d)

	Late 1930s	1950-51	1962-64	1972
Production	752	2,030	4,241	4,884
Consumption	227	621	1,526	2,544
Net Exports	525	1,409	2,715	2,340

Sources: Joseph Grunwald and Philip Musgrove, *Natural Resources in Latin American Development*, Resources for the Future (Baltimore: The Johns Hopkins University Press, 1970), Tables 8-2, 8-3, 8-9, 8-10, and 8-11; Venezuela, Ministerio de Minas e Hidrocarburos, *Memoria y Cuenta, 1972* (Caracas, 1973); and Francis W. Herron, "Latin American Aspects of Energy Policy," an unpublished background paper prepared for The Brookings Institution (July 1973).

LATIN AMERICA'S PETROLEUM BALANCE

Until the early 1960s, Venezuela supplied the bulk of the crude and (directly or through the Netherlands West Indies) refined products imported by other Latin American countries. As crude from the

Middle East became available in the 1960s at prices considerably under what Venezuela charged, the oil-importing countries in the region turned increasingly to Middle East crude, and Venezuela's share of the Latin American market fell below 50 percent. Table 5-2 shows the amounts and sources of Latin American crude oil imports in 1971.

The current petroleum balance of the region is shown in more detail in Table 5-3. The countries listed can be divided into three groups: net exporters (Bolivia, Colombia, Ecuador, Trinidad-

Table 5-2. Latin American Imports of Crude Oil by Source, 1971 (thousand b/d)

Imports from by	Venezuela	Other Latin America[a]	Middle East	North Africa	Indo-nesia	Soviet Union	Total
Consumers:							
Argentina	8.2	15.1	22.3	4.8	...	...	50.4
Brazil	42.4	10.6	227.1	99.1	...	...	379.2
Uruguay	1.4	...	22.7	9.6	...	...	33.7
Other South America[b]	32.9	69.3	16.3	...	...	...	118.5
Cuba	...	...	...	...	...	92.5	92.5
Jamaica	31.8	...	...	...	...	...	31.8
Other Central America and Caribbean[c]	87.2	1.2	47.0	134.3	...	...	269.7
Total	202.9	96.2	335.4	247.8	...	92.5	975.8
Refining Centers:							
Netherlands West Indies	682.4	5.0	...	90.0	...	...	777.4
Trinidad-Tobago	70.4	1.8	106.5	104.3	11.9	...	294.9
Panama	78.4	...	...	...	...	...	78.4
Total[d]	831.2	6.8	106.5	194.3	11.9	...	1,150.7
TOTAL	1,034.1	103.0	441.9	442.1	11.9	92.5	2,126.5

Source: United Nations, Department of Economic and Social Affairs, *World Energy Supplies, 1968-1971*, Statistical Papers, Series J, No. 16 (New York, 1973); data converted at 1 m.t./year=0.0201 million b/d.

Notes: [a]Argentina, Bolivia, Brazil, Colombia, Ecuador, Mexico, Peru, and Trinidad-Tobago.

[b]Principally Chile, Ecuador, and Paraguay.

[c]Excludes Puerto Rico and the U.S. Virgin Islands.

[d]Sum of Netherlands West Indies, Trinidad-Tobago, and Panama: countries that reexport much of the crude as products.

Tobago, and Venezuela); producers, countries that import less than half of their requirements and are potentially self-sufficient or nearly so (Argentina, Mexico, and Peru); and importers, countries that must import the bulk of their needs (Brazil, Chile, Cuba, Dominican Republic, Haiti, Paraguay, and Uruguay). The judgment that Argentina and Mexico are potentially self-sufficient is based on their history of intermittent exports as well as imports and their reserve positions and exploration programs. Peru will probably become self-sufficient and in time a net exporter, as the resources of the upper Amazon Basin, now being actively explored, are developed. There is no prospect of self-sufficiency in the near future for any of the countries in the third group. From the U.S. viewpoint, countries in this group are competitors for oil supplies.

Table 5-3. Latin American Production, Consumption, and Net Exports of Petroleum, 1972 (thousand b/d)

	Production[a]	*Consumption[b]*	*Net Exports*
Argentina	455.7	495.2	−39.5
Bolivia	50.0	14.0	36.0
Brazil	162.0	620.0	−458.0
Chile	34.5	115.0	−80.5
Central America[c]	. . .	100.5	−100.5
Colombia	192.0	128.4	63.6
Cuba	2.5	115.0	−112.5
Ecuador[d]	59.4	22.0	37.4
Mexico	500.0	500.0	0.0
Paraguay	. . .	6.0	−6.0
Peru	65.0	100.0	−35.0
Trinidad-Tobago	143.0	66.0	77.0
Uruguay	. . .	35.0	−35.0
Venezuela	3,220.0	227.0	2,993.0
TOTAL	4,884.1	2,544.1	2,340.0

Source: Venezuela, Ministerio de Minas e Hidrocarburos, *Memoria y Cuenta, 1972* (Caracas, 1973).

[a]Includes natural gas liquids.

[b]Crude equivalent.

[c]Includes Dominican Republic and Haiti.

[d]Production in Ecuador rose rapidly during 1972. By the end of the year, it had reached 200,000 b/d and exports were up to 198,000 b/d.

As for the exporting group, Venezuela clearly dominates the scene. In 1972, 85 percent of the region's total gross exports originated in Venezuela. (See Table 5-4.) Ecuador, the region's next largest exporter, accounted for only 5 percent of total gross exports.

Table 5–4. Latin American Gross Exports of Crude and Products, by Destination — 1972 (thousand b/d)

Destination	Origin	Venezuela[a]	Colombia[b]	Bolivia[c]	Ecuador[d]	Mexico	Trinidad-Tobago[e]	Total
U.S.[f]	Crude	569	42		60	36	0	707
	Products	1,202	25		4	0	159	1,390
	Total	1,771	67		64	36	159	2,097
Canada	Crude	325			29		0	354
	Products	92			0		3	95
	Total	417			29		3	449
Europe	Crude	295						295
	Products	78						78
	Total	373						373
Asia	Crude	9						9
	Products	5						5
	Total	14						14
Africa	Crude	0						0
	Products	13						13
	Total	13						13
Trinidad-Tobago	Crude	37			51			88
	Products	3			0			3
	Total	40			51			91
Other Latin America[g]	Crude	272	30	42	49			393
	Products	135	3	0	0			138
	Total	407	33	42	49			531

								Total
Other Areas	Crude	0	14					14
	Products	37	1					38
	Total	37	15					52
Total Gross Exports[h]	Crude	1,507	86	189	42	36	0	1,860
	Products	1,565	29	4	0	0	162	1,760
	Total	3,072	115	193	42	36	162	3,620

Sources: U.S. Bureau of Mines, *Minerals Yearbook*, Vol. III, "Area Reports (International)" (Washington, D.C., 1973); Venezuela, Ministerio de Minas e Hidrocarburos, *Memoria y Cuenta, 1972* (Caracas, 1973); Ecuador, Ministerio de Recursos Naturales, private communication; and Francis W. Herron, "Latin American Aspects of Energy Policy," an unpublished background paper prepared for The Brookings Institution (July 1973).

[a]Includes exports from Netherlands West Indies (Aruba and Curaçao).

[b]Figures are for 1970.

[c]Year-end rate; includes 12,000 b/d to Brazil beginning in mid-year.

[d]Year-end rate; annual total divided by 132 exporting days.

[e]Includes exports of products refined from crude oil imported from outside Latin America.

[f]Includes Puerto Rico and Virgin Islands.

[g]Chile, Uruguay, Argentina, Brazil, Mexico, Central America, and the Caribbean.

[h]Estimated using year-end figures and 1972 estimate for Colombia. Excludes small amounts exported in return for other crudes by Argentina and Brazil.

Unlike most oil-producing regions, Latin America exports large amounts of refined products.* In 1972, products accounted for 50 percent of Venezuela's exports and 100 percent of Trinidad-Tobago's exports. (The latter imports crude from many sources and then reexports it to the North American market in the form of products.) The United States is by far the region's largest customer, taking over one-half of Venezuela's exports and about 60 percent (approximately 2 million barrels a day in 1972) of total exports from the region. The United States is a major market for residual fuel oil, which is the chief Latin American export product and which was less affected than crude oil by the U.S. import quota system in effect until 1973.

Venezuela can be expected to be the region's leading exporter at least through the 1980s, with exports unlikely to go much above 3 million barrels a day. Ecuador and Peru are the only other countries with significant export potential for the foreseeable future.

PRIVATE COMPANIES AND NATIONAL GOVERNMENTS

Private foreign enterprises (chiefly from the United States) have found and produced the bulk of Latin America's oil, and those companies currently produce about three-fourths of the region's crude output. If Venezuela is excluded, however, the share produced by private industry drops to 30 percent, which reflects the extent of state production elsewhere in the region. The private share for the entire region has been declining (it was about 84 percent in the early 1960s) as state-owned enterprises in several countries have taken over private concessions or expanded their operations into new producing areas. During the 1960s, the state share was stable in most countries, but it increased sharply in Peru with the expropriation of the International Petroleum Company in 1968, and it fluctuated in Argentina and Bolivia with changes in national policy.[2] The state share will increase dramatically in Venezuela as concessions begin to expire, and the state may take a share in Ecuador and a larger share in Peru.

Every oil-producing country except Trinidad-Tobago has a state oil company (the Ecuadorian company has not yet produced any oil). Currently, Venezuela, Peru, and Colombia still depend primarily on private companies, the state shares being about 2, 34,

*About three-fourths of world petroleum trade is in crude oil.

and 14 percent, respectively. The state companies of Brazil, Chile, and Mexico have monopolies on production, and two other countries have effective monopolies as a result of the annulment of contracts and concessions in 1963 (Argentina) and the expropriation of Gulf Oil in 1969 (Bolivia). New private concessions were offered, but not taken, in Argentina, and it seems likely that the government will retain full state control of production.[3] The Bolivian government is permitting the return of private companies under concession or joint-production arrangements, but because of Bolivia's general political instability, the extent of state versus private production cannot be predicted with any confidence.

Concessions of the traditional sort are still in force in Venezuela, Colombia, and Trinidad-Tobago. In those countries, the state collects a royalty at the wellhead plus a tax on the private company's profits. To encourage concessionaires to develop or return their holdings to the state, there are also taxes on acreage, and companies pay a variety of import and other minor duties. As a protection against price cutting, the states collect royalties and income taxes on the basis of posted or reference prices; if the oil is actually sold for less, the full impact falls on the companies. This system, elements of which are common in all exporting countries, was largely developed by Venezuela in the years after World War II.

Only in Ecuador have large new concessions been let in recent years. The Latin American governments have increasingly turned to service contracts to attract private foreign capital and technology. Venezuela has used such contracts to explore areas in which no concessions have been let; Bolivia has indicated it will use them instead of concessions in new areas; and Argentina turned to service contracts a decade ago to expand output and reserves without relinquishing state ownership of oil. The Peruvian government has chosen to rely entirely on contracts under a formula that imposes all risks and costs on the contractor in return for 46 percent of the oil produced. Under contractual arrangements, there are usually no royalties or taxes. Still another arrangement, under consideration by nearly all governments, is the joint-venture company comprised of the state oil company and a private firm.

The shift away from concessions is part of a worldwide movement to limit the power and profits of private oil companies and to reserve to governments all the monopoly rent and, increasingly, the power of decision. The result of this process may be to reduce the companies entirely to sellers of services. Such limitations do not necessarily put an end to private exploration and development; contractual arrangements can still be profitable for private companies as is attested to by the large number of companies active in Peru.[4]

Much more important than the concession-contract distinction or the exact share of profit a company can retain is the stability of government-company relations. The search for oil and the development of technology are inherently expensive and risky; increasing the risk politically does much more to retard investment than reducing the rate of profit. Setting aside the possibility of a change of policy in such countries as Argentina or Bolivia, and assuming a fairly stable arrangement of concessions or joint ventures in Colombia and Trinidad-Tobago, the question of stability or uncertainty is most serious in Venezuela and possibly also Ecuador. Given the considerable lead time required to exploit oil resources, and in some cases the necessity to develop the appropriate technology, uncertainty in government policies may affect the availability of resources many years later.

In Venezuela, no new concessions have been granted since 1956-57. The oil concessions were originally scheduled to begin to expire in 1983. The government, however, is speeding up the process of state control, and the state oil company (Corporación Venezolana de Petroleo, or CVP) will probably have absorbed the oil company holdings within the next few years. Because of the prospect of reversion, net private oil investment has been negative in most years since 1959, and as the time of reversion approaches the companies can only become more reluctant to invest unless they are offered a satisfactory alternative arrangement. Within a fairly short time, the state must decide how it is going to continue operations once the concessions expire or are nationalized.

In Ecuador, the government launched the oil industry in 1964 by offering traditional concessions; after substantial oil was found, however, these were revised. Acreage taxes were sharply raised; several concessions, including one for natural gas, were cancelled; and the Texaco-Gulf consortium, the major concessionaire and producer, was forced to return 60 percent of its holdings ahead of schedule. In mid-1973, the Texaco-Gulf concession was renegotiated as a 20-year agreement allowing the government to acquire up to 25 percent of the company in 1977 and to export up to 51 percent of the oil found. These exports would include, in addition to the royalty and participation share, purchases (at agreed prices) of up to 10 percent of output. Concessions let to other firms will presumably include similar conditions. This settlement removes some uncertainty as to the government's program; if there are no further changes, exploration may proceed rapidly.[5]

RESERVES AND RESOURCES

The proved reserves of conventional crude in Latin America are approximately 30 billion barrels, of which about 14 billion (45 percent) are in Venezuela and about 6 billion (20 percent) are in Ecuador. (See Table 5-5.) Most of the rest is distributed among Argentina, Bolivia, Brazil, Chile, Mexico, Peru, and Trinidad-Tobago. The principal changes over the last decade in the distribution of oil reserves are the steady decline in Venezuelan reserves and the emergence of Ecuador, which formerly had less than 1 percent of the region's reserves, as the possessor of the region's second largest reserves. The distribution of oil among the remaining countries has changed very little. If the current reserves of Argentina, Brazil, Chile, Mexico, and Peru are excluded, on the assumption that during the next decade they will be used largely or exclusively to meet domestic demand, there remain some 23 billion barrels in exporting countries (part of this will also go for domestic use).

Table 5-5.   Latin American Proved Reserves of Oil and Gas, January 1973 (million barrels of oil; billion cubic meters of gas)

Country	Oil	Gas
Argentina	2,800	249.3
Bolivia	200	141.6
Brazil	857	26.0
Chile	125	141.6
Colombia	1,500	68.0
Ecuador	5,750	85.0
Mexico	2,832[a]	325.8
Peru	750	85.0
Trinidad-Tobago	2,000	212.5
Venezuela	13,700[b]	980.2
Others	19[c]	. . .
TOTAL	30,514	2,315.0

Sources: *Oil and Gas Journal*, December 25, 1972; U.S. Geological Survey, *Summary Petroleum and Selected Mineral Statistics for 120 Countries, Including Offshore Areas*, John P. Albers, et al, Geological Professional Paper 817 (Washington, D.C.: Government Printing Office, 1973).

[a]Figures supplied by Petróleos Mexicanos; excludes reserves of condensate in gas.

[b]Excludes reserves of the heavy oil belt (*faja*), estimated at 70 billion barrels (at 10% recovery of oil in place).

[c]Cuba and Guatemala; in Guatemala oil has been discovered but not yet produced.

These figures are very much less than the amounts that may ultimately be recoverable; Venezuela's resources alone have been estimated as high as 125 billion barrels, excluding the *faja* or heavy-oil belt. Appraisals by the U.S. Geological Survey, which include onshore and offshore resources, place Mexico and Venezuela among the countries with ultimately recoverable resources of from 100 to 1,000 billion barrels; the estimates for Argentina, Bolivia, Colombia, Ecuador, and Peru range from 10 to 100 billion barrels; and for Brazil, Chile, and Trinidad-Tobago from 1 to 10 billion barrels.[6] Extensions or reevaluations of known resources probably account for most of the recent increases in the estimated reserves of nearly all Latin American countries. Large, rapid increases in reserves, however, have occurred in Ecuador as the result of the discovery of entirely new fields.

The search in Venezuela is directed to the southern part of Lake Maracaibo (the bulk of Venezuelan reserves and production are in the east-central part of the lake), the Gulf of Venezuela, and the Gulf of Vela off the Paraguaná Peninsula. Service contracts to develop southern Lake Maracaibo were signed in 1970. The Gulf of Venezuela has been extensively surveyed, and resources are thought to be as large as 25 billion barrels, but no reserves have been proved and development is frustrated by a boundary dispute between Colombia and Venezuela. Talks aimed at resolving the issue so that both countries could proceed with development of the gulf ended unsuccessfully.[7] Some oil has been found in the Gulf of Vela, and much of the Venezuelan continental shelf is regarded as likely to hold large resources. The long-run prospect is for substantial amounts of oil to be found in these and possibly other areas where exploration or development has only recently begun; it will be several years at best, however, before proved reserves are significantly increased.

Apart from developments in Venezuela, the area most likely to yield significant new supplies in the near future is the upper Amazon Basin of Peru and Ecuador, which is now being explored by a number of companies. In Ecuador, an intensive search for oil is continuing, and production of very large amounts—a million barrels a day or more—may well be feasible. The Peruvian Ministry of Mines announced in February 1973 that Peru has decided to build a pipeline across the Andes; initial capacity will be 211,000 barrels a day. It is hoped that the line will be completed in 1976, which would give Peru an export capacity of about 100,000 barrels a day in 1976 and double that amount as the line reaches full capacity. Present reserves may not quite justify the investment, but exploration is advancing so rapidly in the Peruvian Oriente that the required oil is expected to be found. A natural gas pipeline to serve domestic consumption is also under consideration, which would make somewhat more oil available for export. Before 1980, oil reserves of around 10 billion barrels may be proved in Ecuador and Peru.[8] If all the projects in Peru and Ecuador go forward, the west coast of South America might be exporting as much as 1 million barrels a day by 1980.

Elsewhere in Latin America there appears to be little prospect of major discoveries of conventional sources of oil. In Colombia, production from the Magdalena Valley fields is declining at about 10 percent annually, and the country's only hope of remaining an oil exporter lies in further discovery in the Putumayo-Orito area near the Ecuadorian border or in new areas, such as the Gulf of Venezuela or the eastern Llanos. The interior of Brazil was extensively explored between 1954 and 1964 at great cost and without success. Considerable exploration in Central America has disclosed several areas of geological promise but so far has not yielded discoveries of commercial amounts of oil. Bolivia's exports may increase appreciably if oil is found in the northern (Amazonian) region, which has just been opened to exploration. The amount is not likely soon to exceed 100,000 barrels a day and will be absorbed by neighboring importing countries. Discovery in Argentina and Mexico, while substantial, has in general only kept pace with domestic consumption.

Thus far, relatively little oil has been found on the continental shelf in Latin America, but offshore production is already important in Brazil and Trinidad-Tobago and is likely to increase relative to total output in most countries. Exploration is aimed at developing offshore areas in Peru, Mexico, Colombia, Venezuela, and Argentina, and may be so directed in Ecuador and Chile.[9]

Secondary and tertiary recovery as a means of increasing reserves is and will be undertaken whenever the cost is justified by the additional availability of oil. Such techniques as reinjection of gas, water flooding, steam injection, and in-place combustion of heavy oil are all in use, and in some basins, particularly in Venezuela, these methods have been used for so long that they no longer yield sufficient output.[10]

Latin America also has substantial reserves of natural gas—about 2.3 trillion cubic meters, roughly 16 billion barrels of oil equivalent, or one-half the region's proved crude oil reserves (see Table 5-5). Ultimately recoverable gas reserves are appraised at between 2.8 and 28.3 trillion cubic meters in Bolivia, Brazil, Mexico, and Venezuela, and at between 0.3 and 2.8 trillion cubic meters in Argentina, Chile, Colombia, Ecuador, Peru, and Trinidad-Tobago.[11] Thus far, the largest natural gas reserves have been developed in Argentina, Mexico, and Venezuela; in each country gas provides a significant share of domestic energy needs and is the principal non-vehicular fuel. At present, two-thirds of the gas produced in Venezuela is used, 44 percent for reinjection and 24 percent for consumption. (The remainder is flared.) Most of the gas produced in Chile is reinjected or liquefied for domestic consumption.

Natural gas has not yet been exported outside Latin America, but the existence in some countries of larger reserves than can be used domestically makes such exports attractive. The chief obstacles are the location of some gas fields (gas found in the Amazon Basin cannot be economically exported) and the possibility that taking off associated gas will decrease ultimate recovery of oil. Thus far, private ventures to export gas from Venezuela have been stopped by the state's reserving gas exports to itself, but the Venezuelan government is proceeding with plans to export gas in 1976.[12] A plant with a capacity of 19 million cubic meters per day is contemplated, which would be roughly equivalent to 135,000 barrels of oil per day. Only gas which would otherwise be flared is to be used.

Besides its conventional resources of crude oil and gas, Latin America has two very large unconventional sources of oil. Deposits of oil shale, similar to that found in the western United States, are estimated to hold some 380 million barrels of oil in Argentina, 125 million barrels in Chile, and 820 billion barrels in Brazil. Oil from shale was first produced in Brazil in 1972, but the plant is small (1,000 barrels a day) and is intended chiefly to help Brazil accumulate experience in this field; significant production increases are not expected for many years. The costs and uncertainties of this venture have held down its development despite Brazil's need for oil.[13]

Of much greater immediate interest is the *faja petrolifera*, the belt of tar sand or extremely heavy oil that runs for more than 500 kilometers north of the Orinoco River in Venezuela. (The *faja* is similar to the Canadian tar sands.) The oil in place is estimated at 700 billion barrels, of which about 10 percent is recoverable with present technology. Small amounts of the lightest oil are being produced in concessions bordering the tar belt. (The recovery rate drops sharply as the oil gets heavier.) Greater production in the near future appears necessary if Venezuela is to maintain its exports; however, the extreme density of much of the oil and its high content of sulfur (almost 4 percent) and metals require the development of production techniques that are still largely experimental. Including the currently recoverable reserves of oil, *faja* oil would approximately triple Latin America's proved oil reserves; the amount of oil available for export a decade from now may well depend on how rapidly that oil can be made accessible.

VENEZUELA'S OIL INDUSTRY

Clearly, Venezuela will continue to dominate Latin American reserves, output, and exports of oil. The country is also extremely dependent on oil, obtaining from it about 25 percent of gross national product, 90 percent of export revenues, and over 60 percent of government revenues. Despite a substantial program of "sowing the oil" to develop new industries (such as steel, aluminum, and chemicals), this dependence has not lessened. Only one other product, iron ore, is exported in quantity, and the newer domestic industries require oil-financed imports.

Venezuela's policy has followed two goals rather consistently for the last three decades: to maximize oil revenues and to minimize dependence on foreign oil companies. It is because these goals are at least partly in conflict that designing such a policy has been difficult. Venezuela's reserves are nowhere near so large as those of the Middle East; it has no transport advantage to Europe; and its traditional advantage in Western Hemisphere markets was reduced in the postwar decades by a decline in tanker costs. Moreover, the country does not have the financial resources to dispense with foreign participation in the industry.

Venezuela has for years followed a policy of maximizing revenues per barrel while letting the volume of exports stagnate or even decline. To this end, it has raised reference prices and begun to set them unilaterally; it has raised tax rates, based on those prices, and as noted earlier, it has instituted penalties for departures from the government's production goals.[14] Rather than compete in price with Middle Eastern and African oil exports, Venezuela has worked to limit production and raise prices in cooperation with other OPEC countries.

The average value of Venezuela's exports rose sharply between 1950 and 1957, reaching a peak of $2.71 a barrel in 1957 while the Suez Canal was closed. Thereafter, average value declined for 12 years to a low of $1.85 a barrel in 1969, despite increases in the reference prices. Only with the large price increases achieved by OPEC since 1970 have Venezuela's average revenues risen again.The total value of exports ($2.8 billion in 1972) has risen, despite some decline in volume, after almost a decade of very slow growth. Venezuela has been able to increase its share of these revenues only by increasing its participation in company profits from 50 to almost 90 percent.

The private companies responded to this squeeze by decapitalizing their operations by nearly $1.2 billion between 1960 and 1968. This disinvestment is of course due not only to the high tax rate, since taxes have been almost as high in other producing countries. It is also in response to the prohibition on new concessions, the fact that all concessions will expire by 1983 or soon thereafter, and the greater attractiveness of investment in other countries with lower production costs. In principle, Venezuela could replace the private companies entirely with the state oil company, the CVP, and finance the needed investment out of revenues. The capital needed to develop significant new resources could not come from current company profits, however, and the state company would either have to borrow abroad or absorb funds from the national budget. The first alternative requires significant new discoveries, and the second is politically difficult. Moreover, the interest on foreign borrowing might be as onerous as the profits retained by foreign enterprises.

The government seems to have settled on a policy of continued use of foreign companies under contractual arrangements that would permit them to keep 10 percent or less of the profits. There are several complicating factors in this policy, however. First, at most a decade and probably less remains in which to experiment and adopt a new policy; given the remaining lifetime of Venezuela's reserves, a leisurely approach, which might have been feasible in 1960, is no longer possible. Except for a slight increase in 1971-72, the reserve lifetime has been dropping and is now only about ten years, against seventeen in the late 1950s. Some 8,000 shut-in wells are maintained, but they represent only about a 10 percent potential increase in output.

As noted earlier, the largest known Venezuelan resources awaiting development are those of the *faja petrolifera*, and much of that oil will require the development of new techniques for production and refining. This raises investment requirements and makes lead times very critical. The most promising means of treating the *faja* oil involves generating large quantities of hydrogen to crack the heavy crude into a lighter product (the sulfur and metals must also be removed). Present indications from technological experiments are that at least some of the *faja* oil could be produced commercially at crude oil prices of roughly $5.00 per barrel (landed price at the U.S. East Coast.)[15] A CVP program of stratigraphic testing, which is expected to be completed in 1976, has been laid out for the *faja*, and a study of processing techniques is well advanced.[16]

The principal question concerns the terms on which foreign firms would participate in this development. The Venezuelan government wants to negotiate a bilateral agreement on special access to the U.S. market, not only for *faja* oil but for all the country's oil exports and later perhaps also for other Latin American producers. The government is unwilling to conclude contracts with U.S. oil companies while seeking an agreement on this point.

Given these uncertainties, projections of Venezuela's crude oil production must be made with caution. Output in the present concession areas could fall by 50 percent from 3.2 to 1.6 million barrels a day by 1985 (see Table 5-6). The decline might be less if very large investments are made in the concession areas, and output

Table 5-6. Projected Venezuelan Crude Oil Production (million b/d)

Year	*Concession Area*	*CVP*[a]	*Faja*[b]	*Total*
1973	3.20	.08	. . .	3.28
1974	3.20	.12	. . .	3.32
1975	3.20	.16	. . .	3.36
1976	3.00	.22	. . .	3.22
1977	2.80	.29[c]	.03	3.12
1978	2.60	.38	.04	3.02
1979	2.40	.50	.06	2.96
1980	2.20	.60	.07	2.87
1981	2.00	.70	.10	2.80
1982	1.90	.85	.12	2.87
1983	1.80[d]	1.00	.16	2.96
1984	1.70	1.00	.30	3.00
1985	1.60	1.00	.50[e]	3.10

Source: Francis Herron, "Latin American Aspects of Energy Policy," an unpublished background paper prepared for The Brookings Institution (July 1973).

[a]Outside present concession areas.

[b]Excludes about 0.06 million b/d of *faja* oil produced in present concessions by private companies.

[c]Government target for 1977 is 0.5 million b/d; projection assumes this level will not be reached until 1979.

[d]Concessions revert to state oil company starting in 1983.

[e]Production as high as 1 million b/d in 1985 is regarded as feasible.

could perhaps be in the range of 1.9 to 2.5 million barrels a day.* It is assumed in Table 5-6 that by 1985 the CVP (or private firms under contract to the CVP) will be producing 1 million barrels a day from new areas, such as the Gulf of Vela, the Gulf of Venezuela, or the Orinoco Delta. The government's plans actually call for more rapid development of these resources than is reflected in the table. The greatest proportional increase is expected to come from the *faja*, which by 1985 should be producing between 500,000 and 1 million barrels a day. Thereafter, the *faja's* importance would increase steadily, unless very large reserves were developed in the Gulf of Venezuela or elsewhere offshore. On these assumptions, total Venezuelan output will decline in the mid-1970s and then rise slowly in the early 1980s. Even given fairly optimistic assumptions—that there are no technological obstacles to exploiting the *faja* and no political delays in arriving at contracts or some other alternative— Venezuela's output a decade from now is not likely to exceed the current level by more than a few hundred thousand barrels a day.

LATIN AMERICAN OIL POLICIES

The emphasis on Venezuela in this chapter reflects the view that nowhere else in Latin America are comparable amounts of oil likely to be found or produced. Even if reserves were to increase substantially in Argentina, Bolivia, Brazil, Chile, or Mexico, no more Latin American oil would necessarily be available for export, since discoveries would go initially to reduce current or future imports. A reduction in imports would of course make available more North African or Middle Eastern oil to the United States and other importers. The amounts involved, however, would be less than 1 million barrels a day in the extreme case.

Assuming that Venezuela's exports are maintained at about 3 million barrels a day, that those of Ecuador approach nearly 1 million barrels daily, and that those of Trinidad-Tobago remain at current levels (77,000 barrels a day, excluding reexports of imported oil as products), all of Latin America may by 1980 have a capacity to export some 4 million barrels a day. Net export capacity will remain at about 3 million barrels a day within the region. This capacity is unlikely to change significantly before 1985.

It does not seem likely that this situation will be altered by any concerted action among Latin American countries, either as part

---

*Assumes annual declines of 2 and 4 percent, respectively. The projection in Table 5-6 is equivalent to an annual decline of almost 5 percent.

of their economic union in the Andean Pact or as part of a deliberate oil policy. Economic integration will affect oil supplies only as it increases the demand for oil, either by increasing the rate of industrial growth or by stimulating the petrochemical industry. Although the latter industry is significant in several countries, it is not a major consumer of oil anywhere in the region, and the incremental effect of its growth will probably be slight.*

Representatives of nineteen Latin American countries met in Caracas in August 1972, at the invitation of Venezuela, to discuss common needs and actions concerning energy, particularly petroleum. A second meeting in Quito in April 1973, attended by twenty-one countries, produced an agreement to create a coordinating secretariat and to work toward a Latin American energy organization. Such an organization could presumably have a significant effect on future contractual arrangements by providing technical assistance to less-experienced governments (from Venezuela to Ecuador, for example) and by promoting joint ventures among state oil companies. Particular trade agreements, such as exchanges of oil for other commodities or for tankers or oil-industry capital goods, might also be promoted.[17] These meetings, however, did not resolve the fundamental divergence in viewpoints on prices between exporting and importing nations, and they are not likely to lead to much greater intraregional trade in oil and oil products.

All of the Latin American oil-exporting countries have large import needs and few export alternatives. They are certain to try to increase their export revenues as much as possible, and they will act together only when such action seems economically preferable to individual national policies. All of them welcomed the huge increase in oil prices made possible by the Arab supply restrictions in the winter of 1973-74, and all of them opposed efforts by Saudi Arabia in the spring of 1974 to obtain OPEC agreement to modest price reductions. It appears doubtful that any of the Latin American oil-exporting countries would give more than nominal support to any future OPEC effort to cut back output in order to sustain high prices. They would not want to sacrifice any of their share of the international oil market and would probably rely on the larger Middle Eastern oil-exporting countries to assume the burden of reducing production to support prices.

---

*This industry is well suited to economic integration and is a major component of the planning of the Andean Pact countries (Colombia, Ecuador, Peru, Chile, Bolivia, and Venezuela), but the effect of any integration scheme on exportable supplies will be slight.[18]

LATIN AMERICA AS A CONTINUING
SOURCE OF U.S. SUPPLY

From the standpoint of U.S. interests, one major conclusion emerges from this review of the petroleum situation in Latin America: even by the 1980s, the region will not be able to increase oil exports to the United States much above 3 million barrels a day. To do so, Latin America would have to divert oil supplies away from Canada and Western Europe, which together take about 800,000 barrels a day, and to forestall large sales to Japan from Ecuador or Peru.

Despite the relatively small amounts of oil available, Latin America is still an attractive source of oil for the United States and other importers. Latin America's attractiveness as an oil supplier is of course enhanced by the fact that it is not involved in the Arab-Israeli dispute. Moreover, the region has some significant resources awaiting development, and in which U.S. capital might participate.

In September 1972 the United States began to discuss with the Venezuelan government a bilateral agreement under which the CVP and the private companies would determine the mode of developing and exploiting the *faja*, perhaps by joint enterprises. It is relatively clear why Venezuela wants such an arrangement. It would further limit the power of the private companies and would avoid the domestic political dangers of a "deal" that might appear too favorable to them; it would make it easy for the CVP to borrow abroad for its share of the development, so its share could be larger; and it would give Venezuela preferential access to the U.S. market.

The question of prices under such an arrangement is particularly critical: Venezuela would want at least to match any future OPEC increases, while the United States might want prices guaranteed over the life of the agreement. Other matters would also be difficult: the allowable level of sulfur in imported oil and products, the allowance for variation in supplies to compensate for changes in other sources, or in prices, and the distribution of new refining facilities are examples.

The prospects for realization of a bilateral agreement are uncertain. The terms of future U.S. participation in Venezuelan oil remain to be determined, as does the extent to which future development of Latin America's oil resources will entail government-to-government agreements or be left entirely to negotiations between the foreign oil companies and the governments of the oil-exporting countries. The outcome will be determined as much by the general state of U.S. relations with Venezuela as by the policies of the two governments with respect to oil.

NOTES FOR CHAPTER 5

1. The principal sources of information for this chapter were Joseph Grunwald and Philip Musgrove, "Petroleum," in *Natural Resources in Latin American Development*, Resources for the Future (Baltimore: The Johns Hopkins University Press, 1970); Francis Herron, "Latin American Aspects of Energy Policy," an unpublished background paper prepared for The Brookings Institution (July 1973); Venezuela, Ministerio de Minas e Hidrocarburos, "Hidrocarburos," *Memoria y Cuenta, 1971* and *1972* (Caracas); and Arévalo G. Reyes, "Consideraciones para la explotación de la Faja Petrolifera del Orinoco," *Revista de la Sociedad Venezolanan de Petróleo* (December 1972), pp. 27-39. The paper by Herron includes much unpublished material obtained in interviews with government and oil company officials in Venezuela and with officials of the U.S. Geological Survey.

2. For a history of government-company relations through 1968, with particular reference to Argentina, Mexico, Peru, and Venezuela, see Grunwald and Musgrove, *Natural Resources in Latin American Development*, pp. 257-68.

3. *El Diario* (La Paz), March 16, 1973; *Petroleum Press Service*, March 1973, pp. 98-99.

4. See *Petroleum Press Service*, July 1971, August 1972, January 1973, and February 1973.

5. *Oil and Gas Journal*, August 13, 1973, p. 45.

6. U.S. Geological Survey, *Summary Petroleum and Selected Mineral Statistics for 120 Countries, Including Offshore Areas*, John P. Albers, et al, Geological Professional Paper 817 (Washington, D.C.: Government Printing Office, 1973), entries for individual countries.

7. *The New York Times*, April 29, 1973, p. 11.

8. U.S. Geological Survey, *Summary Petroleum and Selected Mineral Statistics*, entries for individual countries.

9. *Petroleum Press Service*, March 1971 and November 1972.

10. Chapter 8 of the 1972 *Memoria y Cuenta* of the Venezuelan Ministerio de Minas e Hidrocarburos gives the number of wells for gas and water injection and the volume of gas injected over the last decade or more.

11. U.S. Geological Survey, *Summary Petroleum and Selected Mineral Statistics*, entries for individual countries.

12. *Petroleum Press Service*, July 1970, September 1971, and December 1972.

13. Grunwald and Musgrove, *Natural Resources in Latin American Development*, pp. 252-53; U.S. Bureau of Mines, *Minerals Yearbook, 1971*, preprint (Washington, D.C.), p. 18.

14. For a history of Venezuelan policy to 1968, see Grunwald and Musgrove, *Natural Resources in Latin American Development*, pp. 263-68. More recent events are described in *Petroleum Press Service*, February 1970, April 1970, June 1970, January 1971, September 1971, May 1972, July 1972, December 1972, and April 1973.

15. Reyes, "Consideraciones para la explotación de la Faja Petrolifera del Orinoco," describes the various producing and semi-refining processes being tested or considered and their costs.

16. For a recent summary of the *faja* situation and prospects, see *Oil and Gas Journal*, August 13, 1973, pp. 44-45.

17. For a summary of the Quito agreement, see *Platt's Oilgram News Service*, April 11 and April 13, 1973.

18. Martin Carnoy, *Industrialization in a Latin America Common Market* (Washington, D.C.: The Brookings Institution, 1972), Chapters 2, 3, and 4.

Chapter Six

Canada

by Philip H. Trezise

CANADA'S ROLE IN U.S. ENERGY SUPPLIES

After Venezuela, Canada is the most important external source of U.S. energy supplies. In 1973, something like one-fifth, by volume, of U.S. petroleum imports came from Canada; virtually all U.S. imports of natural gas are from Canada. Canada is also a net supplier of electric power to U.S. utilities, and it was at one time and could again be a significant source of uranium for the U.S. nuclear power industry.

At times, Canada has been looked upon as a special case among U.S. energy suppliers. When mandatory import quotas were established for crude oil in 1959, provision was made for an exemption of overland imports from Canada, and from Mexico, on the thesis that those imports, unlike waterborne supplies, would not be susceptible to interruption. Although this exemption was honored somewhat in the breach, and in 1970 was formally rescinded, the view nevertheless endured that Canada could be looked at differently from all other suppliers, including other Western Hemisphere countries.*

*". . . there may be reason to consider Canada more reliable than any other foreign supplier—because of present and potential common energy policies and because of the relatively greater assurance that mainly inland oil movements between the two countries will not be diverted to meet foreign deficits in the event of a world supply shortage."[1]

89

From time to time, indeed, interested parties on both sides of the border have advanced the concept of a North American or "continental" energy system involving across-the-board coordination of U.S. and Canadian energy policies.

The energy links between Canada and the United States are indeed real. Almost all Canadian crude oil and natural gas exports go to the United States. Some U.S. oil refineries depend mainly or entirely on supplies from Canada. U.S.-owned companies are predominant in the Canadian natural gas and oil industries. The inter-provincial pipeline from Alberta to Ontario crosses through Minnesota, Wisconsin, and Michigan, while most of the crude oil for Quebec and Eastern Ontario comes in at Portland, Maine, and goes by pipeline to Quebec. Utilities on either side of the border regularly exchange electric power. And Eastern Canada gets its metallurgical coal and some coal for electric-power generation from U.S. mines. For the most part, these developments have reflected closely the market forces existing in North America.

Nevertheless, as recent events and particularly those of 1973 have shown, U.S.-Canadian energy links do not support the romantic version of "continental" energy relations. As we shall see, Canadian energy policies have been shaped principally on the basis of considerations of domestic interests—and domestic pressures—just as is true elsewhere. Continentalism is not a popular term in Canada. There is no evident reason to suppose that things will be different in the future. Geography and economic forces do tend to tie the U.S. and Canadian energy positions together, but other factors as well shape North American energy relations.

Before considering the kinds of cross-border energy policies that may be in operation in the years ahead, it is prudent to inquire first of all whether there is a basis for expecting Canadian-U.S. energy resource exchanges in large volume in the future.

CANADIAN ENERGY RESOURCES

There are still very great uncertainties about the magnitude of Canada's energy resources, notably its potential reserves of oil and natural gas. For one thing, exploration in the Arctic territories and on the Atlantic continental shelf—both considered to be promising areas for oil and gas finds—was begun rather recently and of course is far from being exhaustive. For another, estimates of exploitable reserves (and exportable surpluses) depend importantly on future prices and price relationships that can only be guessed at now. The discussion that follows takes into account available judgments about physical supplies and about the cost and price considerations that will bear on their exploitation.[2] They provide the basis for speculation about what might happen but, obviously, not for predicting what will happen.

Petroleum

Canada's proved reserves of crude oil are found in the so-called Western Sedimentary Basin, located mainly in the Province of Alberta. Output from West Canada wells was expected by the National Energy Board (NEB) to reach 1.9 million barrels per day in 1973, and to rise slowly to a peak of 2.2 million barrels per day in 1977.

Proved reserves of conventional crude oil (plus natural gas liquids) in the Western Sedimentary Basin are estimated at 9 to 10 billion barrels.[3] A conservative assumption is that another 5 billion barrels will be found by 1985. An earlier, more optimistic forecast, that of Degolyer and MacNaughton,[4] is that another 40 billion barrels are still to be discovered and developed in the basin.

The Canadian government decided in February 1973 that crude oil exports from Western Canada should be controlled so as to conserve supplies for Canadian refineries. In September, Prime Minister Trudeau announced that the inter-provincial pipeline would be extended to Montreal; completion of the project will mean a further Canadian claim on Albertan oil and presumably a reduction in the volume of exports to the United States. In fact, exports from the Western Sedimentary Basin reached a peak of 1.28 million barrels per day in 1973 and have since fallen below 1.0 million barrels per day. NEB projections[5] show exports possibly falling to zero by 1979, a prognosis that may be excessively pessimistic. But taken together, these factors suggest that Western Canada's exports will continue to decline over the medium-term future.

Drilling is under way in northwestern Canada, that is, in the Northwest Territories and the Arctic islands.[6] These are not yet producing areas, and their potential is still to be determined. The Canadian authorities take a moderate view of the prospects of these "frontier areas."* The most recent (1973) estimate is that at a wellhead price of $4, in constant 1972 dollars, their potential is on the order of 7 billion barrels.[8]of economically recoverable oil. At $5,

---

An Energy Policy for Canada cautions that "the information or data base for the frontier area is quite meagre . . . more output from industry sources would be required before [the cost estimates] would justify a high degree of confidence."[7]

their potential increases to 8 billion barrels.[7] On much higher price assumptions—e.g., ten 1972 dollars—the total is about 16 billion barrels, scattered over a number of fields. At this time, and on what are very preliminary data, Arctic oil seems unlikely to reach Canadian or U.S. markets for some years yet.

A third possible source of crude oil is the Atlantic continental shelf off Nova Scotia and Newfoundland. An ambitious program of exploration is under way in this region and some oil has been found. Discoveries to date have not been on a commercial scale, however. The 1973 estimate is that up to 10 billion barrels of oil eventually could be produced at a price of $3 per barrel, and 16 billion or so barrels at $5 per barrel.[9] This east coast crude presumably would replace oil now being imported into Quebec and the Maritime Provinces (at a rate of 750,000 barrels per day in 1972) from Venezuela and the Persian Gulf. But, once more, it is a prospect for a still indefinite future.

In short, the official forecasts for production of conventional crude oil in Canada deal in rather modest numbers. These forecasts are based on cautious, perhaps excessively conservative, estimates of possible discoveries. They must therefore be viewed with reserve. For all that, an optimistic view of the prospects for Canadian net exports of conventional crude oil requires, at a minimum, that we assume that high prices are going to persist for a long time to come.

There remains the potential of the Athabasca tar sands and the heavy oil deposits, located principally in the Province of Alberta. These are believed to contain about 300 billion barrels of recoverable oil, of which 65 billion barrels may be obtained by open-pit mining methods.[10] The remainder would require either deep mining or *in situ* recovery techniques, which have not yet been developed but are the subject of ongoing research.[11] "Synthetic" oil refined from the tar sand bitumen has a low sulphur content (the sulphur is removed in the refining process) and yields high proportions of gasoline and heating oil.

The physical potential of the Albertan tar sands and heavy oil deposits is thus extremely large by any standard, and the recoverable product evidently is of high quality. Supposing that technology for crude-oil recovery at competitive prices can be developed, the tar sands could become, in the words of the U.S. Cabinet Task Force on Oil Import Control, the basis for "continental self-sufficiency."[12] In terms of the next decade or so, however, the tar sands must be looked at skeptically.

Recovery of oil from the tar sands is highly capital intensive. Current estimates are that exploitation of the areas that can be strip mined will require an investment of up to $10,000 per barrel/day of capacity.* A plant with a capacity of 100,000 to 125,000 barrels per day would thus require a minimum capital investment of $1 billion or more. A critical factor in determining whether these plants are to be built is thus the expected price of crude oil over a period long enough to recover these large capital inputs. The one plant in being, that of the Great Canadian Oil Sands Company (Sun Oil), reportedly was not covering capital costs with production of about 50,000 barrels per day, prior to the price increases after October 1973. Estimates given in *An Energy Policy for Canada* are not excessively precise, but they are suggestive:

> For prices somewhere between $3.75 - $4.00 per barrel, a limited number of open pit mining projects become economically attractive. . . . As prices increase, it will be possible to mine both deeper and lower quality oil sands. . . . About 35 billion barrels can probably be recovered for a price of about $6 per barrel. . . . It is very likely that in the coming 15 years, methods will be developed that can extract oil from the deeper oil sands with so-called "*in situ*" methods. . . . It is as yet uncertain . . . but it is generally estimated that rather significant volumes of oil can be recovered for prices between $5 and $8 per barrel.[13]

It is clear that the tar sands will not be a really large new source of oil in this decade. The raising of capital, planning of the necessary complex plant and equipment, assembling construction crews, building the plant, and providing for environmental protection (which raises major difficulties)** are bound to require long lead times. The official Canadian forecast is that output from surface mines might rise to between 300,000 and 500,000 barrels per day by 1980 from the 1973 level of about 50,000 barrels, which would be a sizable feat considering that no new plant construction has yet been begun.[14] A crash development program might hasten matters, but physical—including environmental—problems would still require substantial time for solution.

The New York Times on November 20, 1973, quoted an oil company official as saying that a proposed refinery with a capacity of 100,000 barrels per day would cost $1 billion before operations could begin.

**It has been stated that a plant producing 125,000 barrels of oil per day would have to process 225,000 tons of sand and dispose of 380,000 tons of waste every day; the waste figure is higher because the water used in the refining process is not reclaimed.[15]

Whether the development of *in situ* recovery will greatly hasten the expansion of oil production from the sands is doubtful. Results from the experiments in progress, using stream injection or underground combustion and water flooding to free crude bitumen from the sands, have been called "encouraging."[16] One of the companies engaged in the research has said that it might be ready to ask approval late in the decade for a commercial project (with a capacity of 100,000 barrels per day) at a cost estimated as "similar" to that of an open-pit mining system. At this time, in other words, the *in situ* technique does not promise to be any less capital intensive than surface-mining recovery. Nor does it avoid environmental questions, for the processes now being tested require extremely large volumes of water, for which recycling techniques must be developed. The Alberta government's projection foresees two *in situ* and five surface-recovery plants in operation by 1985, with a total output of 800,000 barrels of oil per day.[17] This level of tar-sands activity, it is supposed, would just compensate for the expected decline after 1979 in output from the conventional crude oil fields.

In brief, the tar sands promise no more than to maintain current levels of Canadian oil production over the next ten years or so. Expectations of major additions to world or North American oil supplies from this source seem bound to be disappointed, barring a technological breakthrough that would make oil recovery cheaper and help resolve the problems of environmental protection as well.

Natural Gas

Canadian natural gas production has developed in parallel with oil output, following upon the discovery of the oil and gas fields in Alberta. Gas output grew from 12.5 billion cubic meters in 1960, to 45.5 billion cubic meters in 1970, and to about 70.8 billion cubic meters in 1972, of which about 40 percent was exported to the United States. Natural gas imports from Canada accounted for 4 percent of total U.S. consumption in 1972. Under present policy, continued or increased exports will have to be justified by new discoveries that will keep Canadian reserves from falling.

In 1973, ultimately recoverable gas reserves were estimated at 22,000 billion cubic meters.[18] Of this total, only 3,400 billion were placed in the Western Sedimentary Basin, another 9,500 billion in the Arctic, and 9,000-odd billion in Atlantic offshore sites. Proved reserves are substantially smaller, of course, and the discovered reserves in the Arctic will not be available until a pipeline or pipelines have been decided upon and put in place. Most of all, however, the commercial prospects for the largest part of the Canadian gas potential are considered to depend on a sharp increase in price. The official Canadian estimate is that if the delivered price in the Great Lakes area was to go up from the present 14-16 cents per 10 cubic meters to $1.25, then 7 to 8 trillion cubic meters of potential reserves would become economical to exploit; at $1.60, the potential would go up to almost 13 trillion cubic meters.[19]

A pipeline from the gas deposits in the Mackenzie Delta through the Mackenzie Valley to markets in the south would be a costly project; the cost of installation is estimated to be on the order of $5 billion (in 1973 dollars).[20] Gas availabilities in the Arctic are now considered too limited to support a pipeline of economic size. The line's commercial feasibility would depend, therefore, on its extension to the Prudhoe Bay field in Alaska, where gas from the North Slope field would be available, once oil production from the Alaskan wells is begun. Here the coincidence of interest seems to be substantial: unless a liquefaction project is undertaken (as has been proposed by a U.S. company, El Paso Natural Gas), the Alaskan natural gas can be brought to market only via a pipeline through Canada; and Canada's gas resources in the Mackenzie Delta can be exploited commercially most readily if the additional North Slope supplies become available in quantities that would make a pipeline down the Mackenzie a feasible investment.[21]

Natural gas has also been discovered offshore, in large quantities, on the islands of the High Arctic. Exploitation of these deposits depends crucially on the development of transportation technology. The problems of laying pipelines in deep, ice-infested Arctic waters have yet to be solved, and alternative means of transport present numerous difficulties, including the high costs of liquefying the gas for shipment.

In sum, the Canadian natural gas potential is quite large, and proved reserves already are substantial. Transportation problems, however, mean that the outlook is poor for expanded Canadian exports of natural gas until late in the decade.

Uranium

The United States at this time embargoes imports of natural uranium while its stockpile is being worked down. Canada is producing and stockpiling uranium as a means of supporting domestic mining operations. Exploration in Canada is at a standstill.

This situation may well be temporary. Canada has an ambitious nuclear energy program, built around its natural uranium, heavy water reactor (the CANDU), which is expected to lead to rapid expansion of domestic uranium consumption. On one projection, the level of domestic use would go from 200 tons (1970), to 1,300 tons in 1980, and to 10,400 tons in 2000 (Table 6-1). Cumulative demand over the rest of this century would come to more than 100,000 tons.

Canadian natural uranium reserves are estimated to be 400,000 tons, available at a price no higher than $15 a pound, with little further exploration.[22] Considerably larger reserves—another 700,000 tons—are believed to be available in the same price range once systematic exploration is resumed. If the recent upward trend in uranium prices continues, Canada could have a large surplus for export.

Table 6-1.   Canadian Economy and Energy Demand

	Actual		Estimated		
	1960	*1970*	*1980*	*1990*	*2000*
Canadian Economy					
Population (millions)	17.9	21.4	25.2	29.8	34.7
Households (millions)	4.4	5.7	7.1	8.5	10.1
Energy Consumption (Btu 10^{15})	3.6	6.4	10.5	16.7	26.3
Energy Annual Increase (percent)	. . .	5.9	5.1	4.8	4.6
Per Capita Energy Consumption (Btu 10^6)	200	300	420	560	760
Canadian Energy Demand					
Coal (million tons)	23	30	47	94	156
Petroleum (million barrels)	315	536	800	1150	1580
Natural Gas (billion cubic meters)	9.4	29.4	59.5	86.4	113.3
Uranium (thousand tons)	. . .	.2	.3	4.2	10.4
Electricity (10^{12} watt-hours)	109	201	390	720	1300
Primary Energy Consumption (Btu 10^{15})					
Coal	0.5	0.7	1.0	1.9	3.1
Petroleum	1.7	3.1	4.7	6.7	9.1
Natural Gas	0.4	1.1	2.1	3.0	4.0
Hydroelectricity	1.0	1.5	2.3	3.0	3.8
Nuclear	. . .	. . .	0.4	2.1	6.3
TOTAL	3.6	6.4	10.5	16.7	26.3
Percentage Distribution of Consumption					
Coal	15.4	11.1	9.8	11.1	11.6
Petroleum	46.0	48.2	44.5	39.8	34.7
Natural Gas	10.5	16.3	19.7	18.2	15.1
Hydroelectricity	28.1	24.2	22.5	17.9	14.6
Nuclear	. . .	.2	3.5	13.0	24.0
	100.0	100.0	100.0	100.0	100.0

Source: E.S. Bell, "The Pressure of Limited Resources on Rising Demand for Energy," National Energy Board, Power Generating Planning (Ottawa, 1972).

Electric Power

In 1972, Canada exported, net, 7.9 billion kilowatt-hours of electricity to the United States, or less than 0.5 percent of U.S. consumption. The Canadian export surplus is expected to remain stable through the 1970s and thereafter to decline to a net of zero by the end of the 1980s.

This forecast is based on a rapid growth of domestic electricity use; demand is expected to grow by 600 percent between 1970 and 2000 and steadily to preempt Canadian power output. Hydroelectric power will grow by less than 200 percent. Few good sites for hydroelectric power remain and environmental opposition to hydro projects is rising. Two major projects are scheduled—on James Bay in Quebec and on the lower Nelson River in Manitoba—and if carried through to completion they may well be the last of their kind; neither is intended or expected to develop a permanent excess of power output for transmission to U.S. markets.

Discussion of a facility to harness tidal flows in the Bay of Fundy is perennial, but not very meaningful. The costs of the project appear to put it outside the competitive range of nuclear power.

Canadian sales of nuclear power generated specifically for the U.S. market are conceivable but unlikely. Even if the efficiency and reliability of the CANDU heavy water reactor exceed the claims made for it, there is room for doubt that either Canada or the United States would be prepared to have Canadian nuclear power plants built to meet U.S. electricity needs.[23]

Since Canadian peak electric-power loads come in the winter and some of those in the United States in the summer, seasonal exchanges between Canada and the United States are both feasible and economic and should continue. But the realistic view is that significant increases in net exports of electric power from Canada to the United States are not in prospect.

Coal

Canada is an importer of U.S. metallurgical coal as well as some steam coal for power plants; imports were 19 million tons in 1972. Canadian coal deposits are principally in British Columbia and Alberta. Some of this coal is suitable for coking and has an export market in Japan. It is badly located in relation to major North American markets, however, and while reserves may be on the order of 120 billion tons, West Canadian coal has little present bearing on U.S.-Canadian energy relations.

CANADIAN ENERGY POLICIES

Factors for Change

The physical availabilities of energy resources put an evident constraint on Canada's energy relations with the United States. A second limiting factor is Canadian energy policy, at both the national and the provincial levels. Canadian views on energy policy are changing, in reflection of changing perceptions of Canadian interests. This already has altered Canadian-U.S. energy relations in important respects and will tend to shape the relationship in the future.

A basic point is that Canada, and particularly the most populous province, Ontario, is highly urbanized and industrial. Like the United States, Canada has been bothered with stubborn problems of inflation and unemployment, neither of which has been mastered. The Canadian labor force is the most rapidly growing of any major industrial country, and both the national and provincial governments cannot fail to be preoccupied with policies to generate more jobs.

The industrial and economic growth of Canada has been accompanied by the growth of national self-awareness, which in one of its forms takes on the color of resentment or hostility toward the United States. It would be a considerable exaggeration to say that Canadian policy or national thinking is dominated by this attitude; it does not override either the numerous political and economic interests that link Canada and the United States or the tradition of Canadian-U.S. friendship. But it would not be off the mark to say that Ottawa's views about proper policy for Canada must be shaped in part by a prudent concern for an appearance of independence from the United States.

These factors are almost certain to cause hesitancy in Canada about deciding on policies toward energy relations between the two countries, as in fact they have. There is a vocal concern that Canada should not be a supplier to the world of natural resources—in the cliche, "a hewer of wood and drawer of water." To be sure, even now Canada's exports are predominantly made up of manufactures, and Canadian economic growth in the future, as is true of the United States also, will be heavily weighted by the output of high-technology industry and, especially, by services. And there is no necessary logic in the proposition that raw material exports are somehow disadvantageous to the exporting nations. But a reluctance to contemplate increased exports of energy raw materials may be no less real for being largely based on a sentimental view of economic life.

If energy resources are to be sold abroad, there will be pressures to maximize the returns to the Canadian public. Tax and royalty policies are certain to be colored by a general concern to prevent excessive profits going to foreign investors in Canadian raw materials or, for that matter, to Canadian private investors.

Any rapid expansion of energy-resource production inevitably would imply large additional U.S. investment in Canada. The current official estimate is that the Canadian energy industry as a whole could require new capital expenditures of from $42 billion to $68 billion (1972 dollars) over the 1970s, depending on decisions about development of the tar sands and oil and gas off the east coast and in the other frontier areas.[24] Since U.S. investors already control more than half the assets of Canada's petroleum industry, additional U.S. investment raises sensitive political issues in an atmosphere in which U.S. economic "domination" has become a matter of political consequence. There is also the point of view that U.S. investments in Canadian natural resources make insufficient contributions to Canadian employment. As the argument goes, these investments reflect the needs of U.S. processing and manufacturing facilities. The resulting captive markets reduce the likelihood of "further processing of Canadian natural resources in Canada."[25]

Still another Canadian objection to energy-resource development on a scale relevant to prospective U.S. requirements has to do with the probable impact of capital inflows on the Canadian exchange rate and through the exchange rate on the export position of Canadian manufacturing industries. Whether or not a new international monetary system will be based on floating exchange rates, Canada is not likely to abandon a flexible parity for the indefinite future. It is reasoned that U.S. investments—either direct or portfolio—will have the effect of pushing up the Canadian dollar's parity, with consequent damage to the international competitive position of Canada's secondary industries. This chain of events would not follow, of course, if U.S. investment flows were matched by imports of U.S. equipment and supplies. But in that case, Canada would have to sacrifice the potential employment gains of producing the equipment at home, a choice containing its own political barbs.

A further issue is the environment. As in the United States, environmental worries have become a part of the Canadian energy debate. To be sure, Canada has had no offshore oil production and no oil spills to excite public attention as the Santa Barbara spill did in the United States. Nor has the development of oil and gas in the Arctic yet reached a level that would threaten the environment. But a domestic controversy has blown up over Quebec's proposed James

Bay hydroelectric project, and British Columbia's alarm at the implications of large oil shipments from Alaska through the Puget Sound to Seattle are surely harbingers of more general misgivings about the environmental threats presented by the energy industry. In particular, energy development in the Arctic territories is certain to be challenged, and could be delayed, by Canadian environmentalists and, as in Alaska, by the claims of the native people concerning the use of their land. Mining of the Albertan tar sands and processing of their bitumen content present very serious environmental difficulties that will have to be dealt with if this resource is to be more fully exploited.

National policy in Canada, as in other countries, must take into account local interests, which are far from being the same from province to province. Alberta, which gets the largest part (some 45 percent) of its provincial revenues from oil and gas royalties, has a producer's interest in high prices and in access to the U.S. energy market. Under the British North America Act, which established the Canadian confederation, the provinces were given the rights to all minerals within their borders, and Alberta made the foresighted decision to retain most of those rights for the province (as did Saskatchewan and British Columbia) and thereby is able to set royalty and production policies. The same North America Act, however, gives Ottawa the power to regulate exports and inter-provincial transportation and so to impinge on Alberta's ability to exploit its market position elsewhere in Canada and in the United States.

Canada's eastern provinces, to draw the most general distinction, have a consumer's interest in cheap energy resources. The national oil policy has heretofore given Quebec and the Maritimes access to imported oil at favorable—i.e., world—prices, while most of industrial Ontario has had to depend on the higher priced West Canadian petroleum for its needs. The radical change in world oil prices, if it is a lasting one, will also change provincial viewpoints in Canada. In that case, Ottawa will have to take a new look at its policies toward energy.

National Policies, Pre-1973

In one fundamental sense, Canadian energy policy has had a consistent theme for a decade and a half. A Royal Commission on Energy was set up in 1957 to look at the outstanding issues, and its findings led to the creation in 1959 of a National Energy Board to regulate the use of energy resources with a prudent concern *for domestic energy needs.* Under this guideline, exports are permissible, in principle, only to the extent that supplies are "excess" to Canadian requirements. The practical guideline for determining the "excess" is the ratio of proved reserves to domestic consumption. This guideline has been applied to prohibit additional exports of natural gas and hydroelectric power and to control and limit exports of crude oil. It is the foundation for the projection that crude oil exports to the United States could fall to zero by 1980.*

Within this broad precept, however, it was initially considered desirable to foster exports, particularly of crude oil, by explicit policy measures. In 1961, the government in Ottawa decided to set aside the Canadian market west of the Ottawa Valley exclusively for domestic producers of crude oil, while Eastern Canada was to remain dependent on imported crude. The idea of a pipeline to Montreal was shelved and the Albertan oil industry was encouraged to find additional outlets in the United States as the basis for an expansion of its production.**

Another major thrust in energy policy was to encourage oil and natural gas exploration in the Arctic as part of a general policy of Arctic development. A set of regulations issued in 1960 was skewed strongly in favor of incentives to exploration. Permits were given, subject only to a requirement for minimum expenditures on the search for oil and gas. If discoveries were made, the companies were able to get long-term leases for a 10 percent royalty (5 percent for the first three-to-five years).

In the nuclear field, Canada chose an independent line of development. National policy is to promote aggressively the Canadian natural uranium, heavy water reactor (CANDU). Power output based on CANDU is expected to gain at the expense of all other forms of domestic energy through the remainder of the century. As noted earlier, Canada has uranium reserves adequate to serve export markets as well as to cover domestic needs, and the outstanding policy is to permit exports, subject to federal licensing standards.[26]

New Directions

Under the basic doctrine of limiting energy exports to amounts consistent with domestic needs, applications for additional natural gas exports have been rejected by the NEB since 1970. In early 1973, controls were put on crude oil exports so as to assure Canadian refineries of continued supplies at "reasonable" prices.

*The range of possible exports is put at from 0 to 1.8 million barrels per day.[27]

**A policy statement in 1963 similarly encouraged the growth of electric power exports to the United States.[28]

Then in September, as world oil prices continued to rise, the decision was taken to end the Ottawa Valley demarcation line and to build pipeline facilities to bring West Canadian crude to Montreal. The expectation is that the inter-provincial line will be extended from its Ontario terminus at Sarnia to Montreal,* with a daily capacity of 500,000 barrels, which if used in full would provide for most of Montreal's present requirements.

With the sharp increase in oil prices, Ottawa turned to an export tax to deal with windfall gains from West Canadian crude oil. Set first at 40 cents a barrel in September, the tax was hiked to $6.40 in January 1974, reflecting the difference between a fixed domestic price of $4.00 and an estimated landed price for oil in Eastern Canada of $10.40.** In March it was decided to allow the domestic price to rise to $6.50, and the export tax was set at $4.00 or slightly more. Under this program, the "first call" on export tax proceeds would be for subsidies to lower oil product prices in Eastern Canada. Thus prices would be roughly equalized nationwide. Alberta's (and Saskatchewan's) royalty revenues would be increased in proportion to the increases in the price for crude oil. And consumption of energy in Canada would be discouraged in greater measure than before, as Ottawa must have judged would be a desirable result.[29]

In the case of natural gas, export prices have been fixed in long-term contracts. With average prices for gas rising in the United States, Canadian claims to higher returns for exported gas are bound to be strengthened. The federal Minister of Energy, Donald MacDonald, suggested in September 1973 that the terms of existing contracts could be revised quite legally to take account of the price of natural gas relative to prices for alternative sources of energy, e.g., petroleum.[30] Since consumers in Canada predictably would object to higher prices, natural gas may provide the next experiment in dual pricing for exports and domestic consumption.

Second thoughts are being addressed as well to development policies for oil and gas and related activities. One aspect of this is the question of an optimum level of energy-resource expansion in Canada. The authors of *An Energy Policy for Canada* looked hard at the possible economic impact on Canada of various assumed levels of energy-resource investment between 1971 and 1980. Their con-

---

*An alternative would be an all-Canada pipeline from Winnipeg around the north shore of the Great Lakes to Montreal. This would be a much more costly project and is thus unlikely to be the one chosen despite political sentiment favoring a wholly Canadian line.

**Saskatchewan in February 1974 raised the wellhead price for its provincial oil to $5 a barrel. The initial effect was to increase domestic refinery demands on Alberta's output.[31]

clusions, carefully qualified for all the uncertainties and short-comings of their model, suggested that a "maximum" ($68 billion) development case—involving both oil and gas pipelines in the Mackenzie Valley, intensive Arctic development, uranium enrichment, and maximum tar sands exploitation—would put severe strains on the country's economic capabilities and would require "a substantial reallocation of human and capital resources even under conditions of slack." On more modest ($50 billion) capital expenditure assumptions, which allow among other things for a Mackenzie Valley natural gas pipeline but for only "gradual" oil sands development, the potential strains are judged to be reasonably tolerable. If this exercise in forecasting is relevant to policy, it will reinforce the case against permitting an all-out response to any pressures for energy-resource expansion. It goes hand in hand, moreover, with the worry that raw-materials development could put upward pressure on the Canadian dollar's exchange rate to the detriment of Canada's manufacturing industries.

Also in question is the matter of lease and royalty policies in the federal territories. The earlier approach, noted above, was to provide significant exploration incentives in the belief that investors would not otherwise risk their money in the more remote regions of Canada. A first modification of this approach was made in 1970, when it was ruled that one-half of each exploration plot would revert to the government rather than being offered for lease to the exploring company. Subsequently, the decision was taken to review the whole system of regulations. "A review . . . is necessary if Canada is to realize the full benefit of the development of its frontier oil and gas resources."[32] The new emphasis, in other words, is toward returns to the nation and away from a development-first philosophy. The official energy study considered various ways to improve the government's returns—more flexible royalty rates, lower depletion allowances, or a two-price system—without plumping for any one.[33] Another alternative has since been decided upon and officially proposed—a national oil company to be formed to explore and develop gas and oil resources (including the tar sands) and perhaps to take over existing private companies as well, thereby, it is hoped, making the government a direct rather than an indirect collector of economic rents on some of the expected oil and gas developments.[34]

CANADIAN—U.S. ENERGY RELATIONS

The foregoing indicates clearly enough that Canadian energy policy has been moving along nationalist lines for the past several years. From Ottawa's point of view—if not Alberta's—energy-resource exports have diminished in importance, while domestic pressures have come to be more compelling. The time has passed when a Canadian political leader could consider exploring the harmonization of energy policies with the United States.* Notions in the United States about Canada's potential contribution to U.S. energy needs have commonly been overblown. In the light of events and a closer look, they need to be scaled down still further. Neither the Albertan tar sands nor Arctic oil is likely to have much bearing on the U.S. energy position for many years to come, partly because of Canadian policies, mostly because these resources are inherently difficult to exploit. Exports of conventional crude oil seem certain to decline once a Montreal pipeline is in place, and U.S. refineries dependent on Canadian crude may in due course have to make adjustments of some magnitude as a result. Canada's surpluses of other energy resources are not prepossessing.

All that said, there is no reason to suppose that the United States and Canada will not have a modest but mutually beneficial energy relationship in the future, as in the past. For some years still, and even after a pipeline has been laid to Montreal, West Canadian crude will help to supply U.S. refineries along the border. If world oil prices continue at relatively high levels, more conventional oil likely will be found in Western Canada to maintain exports a while longer. Ottawa has already raised domestic oil prices. As the world price subsides, Canada's export tax may wither, along with Washington's irritations about it.

Prospects for a pipeline to carry Alaskan and Canadian natural gas down the Mackenzie Valley have improved with a statement by Prime Minister Trudeau that Canada would favor its construction, subject to satisfying environmental and other conditions.[35] The government has said that it is prepared to hold prompt hearings on a right-of-way application and to provide suitable undertakings about the unimpeded movement of Alaskan gas to the United States.[36] A pipeline would seem to accommodate both countries' interests in having access to the gas deposits in the far north. Agreement on it would mark a renewed recognition that the facts of geography do link the two nations together. And if

*When President Nixon and Prime Minister Trudeau met in March 1969, their joint communiqué reported: "In the context of the common interest of the two countries in the expansion of cross-border movements of energy, U.S.-Canadian developments in the matter of oil were discussed at great length. Senior officials will initiate meetings to identify and study areas of common interest in energy matters and work out constructive solutions to current problems. . . ."[37]

considerably more oil is discovered in Alaska, an oil pipeline through Canada might be a preferred alternative to a second Alaskan line.

The newly opened oil refinery at Come-by-Chance in Newfoundland opens another potential area of mutual energy development. On the one hand, the U.S. Atlantic Coast lacks deepwater ports for supertankers. There also are strong environmental objections in the eastern states to new refinery construction along the coast. On the other hand, the Canadian Maritime Provinces have deepwater harbors in being and are interested in attracting investments to help with their severe, chronic unemployment problems. Plans to build two more refineries in the Maritimes must depend on access to U.S. markets, which in turn presumably will turn on the details of Project Independence.

There will no doubt be other instances of obvious common interests bringing Canada and the United States into a close energy relationship. But as it comes to be recognized more fully in both countries that these practical possibilities are nevertheless limited, the quasi-ideological overtones of the cross-border energy discussion presumably will be muted. The United States may stop looking for an illusory continental harmony in energy matters, and Canada may conclude that its energy resources provide quite modest economic or political leverage. Signs that this is the direction of thinking in both countries have already appeared, and the realities argue that it will continue to be so.

NOTES FOR CHAPTER 6

1. U.S. Cabinet Task Force on Oil Import Control, *The Oil Import Question, A Report on the Relationship of Oil Imports to National Security* (Washington, D.C.: Government Printing Office, 1970), p. 70.

2. The most authoritative current data are found in *An Energy Policy for Canada*, in three volumes, issued under the authority of the Minister of Energy, Mines, and Resources (Ottawa: Information Canada, 1973).

3. Ibid., Vol. I, "Analysis," p. 87, Table 1.

4. Degolyer and MacNaughton, *Reports on Estimates of Additional Recoverable Reserves of Oil and Gas for the U.S. and Canada* (Dallas, 1969), p. 24. These estimates are based on Canadian petroleum industry data.

5. National Energy Board, *Potential Limitations on Canadian Petroleum Supplies* (Ottawa, 1972), p. 5.

6. Seventy wells were drilled in 1972 and a like number was projected for 1973. *Wall Street Journal*, October 23, 1973, p. 6.

7. *An Energy Policy for Canada*, Vol. II, "Appendices," p. 90.

8. Ibid, p. 93, Table 3.

9. Ibid., All estimates are in constant 1972 dollars.

10. Ibid., Vol. I, p. 30, Table I.

11. Three companies, Shell, AMOCO, and Imperial, have sizable research efforts in progress on *in situ* recovery techniques. *The Alberta Oil Sands Story* (Government of the Province of Alberta, 1974), pp. 30-31.

12. U.S. Cabinet Task Force, *The Oil Import Question*, p. 45.

13. *An Energy Policy for Canada*, Vol. II, p. 73.

14. *An Energy Policy for Canada*, Vol. II, p. 101. As of 1973, one plant with a capacity of 125,000 barrels per day was at the design stage; the tentative start-up date is January 1, 1978. *Wall Street Journal*, September 9, 1973, p. 31.

15. *Wall Street Journal*, December 5, 1973, p. 46.

16. "Alberta's Oil Sands in the Energy Supply Picture" (mimeo.), address by G. W. Govier, Alberta Energy Resources Conservation Board, 1973 Symposium on "Oil Sands—Fuel of the Future," Calgary, Alberta, September 7, 1973.

17. Ibid., p. 28.

18. Ibid., p. 53, Table 3; the same table gives an alternative and lower 1973 estimate of 16,345.6 billion cubic meters.

19. *An Energy Policy for Canada*, Vol., I, p. 94.

20. Ibid., Vol. II, Appendix B, p. 184, and Annex 6. See also, National Petroleum Council, *U.S. Energy Outlook* (Washington, D.C., 1972), p. 285, which projects total gas-pipeline expenditure, in 1970 dollars, at $5.3 billion between 1976 and 1985.

21. For a discussion of natural gas in the Canadian Arctic, see Judith Maxwell, *Energy from the Arctic: Facts & Issues*, Canadian-American Committee (Washington, D.C.: National Planning Association, 1973).

22. *An Energy Policy for Canada*, Vol. I, p. 119.

23. Ibid., p. 170.

24. *An Energy Policy for Canada*, Vol. I, pp. 209-16.

25. *Foreign Direct Investment in Canada* (Ottawa: Information Canada, 1972), p. 14.

26. *An Energy Policy for Canada*, Vol. II, Chapter VI.

27. Ibid., Vol. I, p. 127, Table 1.

28. Ibid., Vol. II, Chapter VI.

29. Statement by Prime Minister Trudeau, March 28, 1974. (Press Release, Embassy of Canada, Washington, D.C., March 29, 1974.)

30. *Oil and Gas Journal*, September 24, 1973, p. 90.

31. *Globe & Mail* (Toronto), February 20, 1974, p. B-1.

32. *An Energy Policy for Canada*, Vol. I, p. 154.

33. Ibid., pp. 154-55.

34. *Washington Star News*, March 2, 1974, p. G-9; *Washington Post*, December 3, 1973, p. A-14. The Canadian government has a 45 percent equity in Panarctic Oil, Ltd., which has been drilling in the Arctic since 1967.

35. *Wall Street Journal*, December 7, 1973, p. 10.

36. Speech by Minister of Indian Affairs and Northern Develop-

ment, Jean Chretien, at fifty-first meeting of the Council of the Northwest Territories, Yellowknife, January 18, 1974 (Washington, D.C.: Embassy of Canada).

37. U.S. Department of State *Bulletin*, April 14, 1969, p. 324.

Chapter Seven

Other Exporters

In addition to the oil-exporting countries discussed in the preceding chapters, two others—Nigeria and Indonesia—export significant quantities of oil. The present situation and future prospects of the oil and gas industries in those countries are discussed briefly here.

NIGERIA

Production and Reserves

By early 1973, Nigeria had emerged as the world's sixth largest exporter of crude oil. Oil exports increased from 300,000 barrels a day in 1967 to almost 2 million barrels a day in 1972. The chief customers for Nigeria's oil are Western Europe, which buys roughly two-thirds of Nigerian crude exports, and North America. (See Table 7-1.) In 1971, the United States replaced the United Kingdom as the leading individual importer, and by 1972 the United States was importing about 25 percent of Nigeria's crude oil. The high quality of Nigerian crudes (sulphur content averages 0.2 percent), the country's geographical location west of Suez, and its noninvolvement in the Arab-Israeli dispute make Nigeria an attractive source of oil for Europe and North America.

Nigeria is a relative newcomer to the international oil market (the first commercial discovery was made in 1956); oil production rose rapidly, from 67,000 barrels a day in 1962 to nearly 600,000 barrels daily in mid-1967, when the efforts of the Ibo tribe to secede and form a new state of Biafra brought on a civil war. (See Table 7-2.) Prior to the civil war, three major oil companies were producing in Nigeria—Shell-British Petroleum, which has dominated

Table 7-1. Nigerian Crude Oil Exports (in thousand b/d and thousands of US dollars)

	1967			1968			1969		
	Volume	% of Total Volume	Value	Volume	% of Total Volume	Value	Volume	% of Total Volume	Value
Western Europe:	242	80.4	148,302	95	67.4	68,796	346	63.9	240,988
United Kingdom	82	27.4	56,232	38	21.1	27,535	119	22.0	83,731
France	42	14.0	27,378	6	4.5	4,444	58	10.7	39,749
Benelux	34	11.2	23,257	7	12.3	12,214	87	16.0	60,256
West Germany	44	14.7	30,117	..	..	..	8	1.5	5,748
Others	40	13.1	11,318	44	29.5	24,603	74	13.7	51,504
North America:	36	11.9	23,836	20	14.4	15,081	117	21.7	83,860
United States	16	5.2	10,539	11	7.8	8,092	93	17.3	66,150
Canada	20	6.7	13,297	19	6.6	6,989	24	4.4	17,710
Latin America	22	7.4	14,770	26	18.2	19,726	71	13.2	51,486
Africa	..	..	..	..	..	..	6	1.0	3,889
Japan	1	0.3	585	..	..	..	1	0.2	608
Total	300	100.0	187,493	141	100.0	103,603	541	100.0	380,831

Table 7-1. (Continued)

	1970			1971			1972		
	Volume	*% of Total Volume*	*Value*	*Volume*	*% of Total Volume*	*Value*	*Volume*	*% of Total Volume*	*Value*
Western Europe:	745	71.8	509,121	994	69.2	988,441	1,103	62.0	1,105,572
United Kingdom	243	23.4	165,595	266	18.5	266,942	329	18.5	326,566
France	116	11.2	79,464	271	18.9	269,435	285	16.0	297,282
Benelux	194	18.7	134,772	194	13.5	196,290	243	13.7	242,045
West Germany	54	5.2	35,941	65	4.5	65,178	100	5.6	69,698
Others	138	13.3	93,349	198	13.8	190,596	146	8.2	169,981
North America:	169	16.3	120,966	302	20.9	317,212	458	25.7	463,244
United States	138	13.3	100,559	275	19.1	289,198	422	23.7	425,974
Canada	31	3.0	20,407	27	1.8	28,014	36	2.0	37,270
Latin America	115	11.1	78,070	105	7.3	103,099	116	6.5	115,952
Africa	8	0.8	5,844	23	1.6	26,123	26	1.5	25,274
Japan	...	...	...	13	0.9	13,735	76	4.3	76,733
Total	1,037	100.0	714,001	1,437	100.0	1,448,610	1,779	100.0	1,786,775

Source: Nigeria, Federal Office of Statistics, *Nigeria Trade Summary*, various years.
Note: Exchange rate for 1967, 1968, 1969, 1970, 1 £N = $2.80; for 1971, 1972, 1 £N = $3.04

Table 7-2.   Nigerian Crude Oil Production (in thousand barrels)

Year	Total Production	Average Daily Production
1962	24,624	67.5
1963	27,913	76.5
1964	43,997	120.2
1965	99,354	272.2
1966	152,428	417.2
1967	116,553	319.3[a]
1968	51,907	141.8
1969	197,204	540.3
1970	395,836	1,084.5
1971	563,341	1,543.4
1972	657,000	1,800.0

Sources: Nigeria, Federal Ministry of Mines, annual reports of the Petroleum Division, 1962-67, and "Monthly Petroleum Information," 1968-70; *Oil and Gas Journal*, December issues, 1971 and 1972.

[a]In June 1967, before the civil war broke out, production approached 600,000 b/d.

Nigerian oil production from the start; Gulf Oil; and SAFRAP, a French state oil company. While onshore production came to a virtual standstill during the civil war, Gulf, the only offshore producer, expanded offshore production from an average of 57,000 barrels a day in June 1967 to 200,000 barrels a day in 1969. (Only about 25 percent of Nigerian production is offshore.)

The civil war ended with the victory of the central government in early 1970. Exploration and production were quickly resumed and total production of crude oil in 1970 was slightly more than 1 million barrels daily—double that of 1969.

In the postwar period, Shell-British Petroleum, which is responsible for almost all of the onshore production, and Gulf Oil were joined by five other producers: Mobil, Chevron-Texas, AGIP-Phillips, Phillips-NNOC, and SAFRAP-NNOC. The Nigerian National Oil Company (NNOC) was formed in May 1971 to engage in all phases of oil industry activities. Unlike most state companies, NNOC began life as an oil producer. An agreement with SAFRAP gave NNOC an initial 35 percent interest in SAFRAP's four onshore mining licenses. The agreement further provided that NNOC's share would rise as production rose and that a maximum share of 50 percent would go to NNOC when production reached 400,000 barrels a day. Later in 1971, NNOC purchased an effective working interest in AGIP's local affiliate and in AGIP-Phillips's four onshore mining licenses.*

*AGIP is a subsidiary of Italy's state oil company, Ente Nazionale Idrocarburi (ENI).

Since 1956, exploration has been extensive, and over the years sixteen groups have been granted exploration rights. Exploration activity resumed at a brisk pace after the civil war, despite federal legislation in 1970 (Decree no. 51) that provided that each stage of a petroleum operation must be separately authorized and that "special terms" (including government participation) may be imposed on both new and existing prospecting and mining licenses. The law also provided that terms would no longer be fixed when the prospecting license was granted and that a mining lease would be granted only after discovery of commercial quantities of oil. The success ratio of drilling has been high; some 40 percent of all exploratory wells have resulted in new fields, and over 80 percent of appraised and developmental drillings have been successful.

To some extent the Nigerian "oil rush" is still under way, although the rate of increase in production experienced in 1969, 1970, and the first half of 1971 has not continued. The dramatic increases in the postwar years were partially the result of Nigeria's making up for time lost during the civil war. (Immediately prior to the war, several companies were on the verge of producing from newly discovered wells.) While it took a little over two years to jump from 1 million to 2 million barrels a day, most industry experts do not expect production levels to reach 3 million barrels a day until 1978, and possibly not until 1980.

Although petroleum exploration is still in a relatively early stage and the extent of total reserves is not known, the present reserves picture and the geology of Nigeria do not indicate a rapid increase in production beyond the three-million mark. Proved crude oil reserves as of January 1974 were 20 billion barrels. At the current rate of output, proved reserves represent about 20 years of production. The extent of inferred crude oil reserves is estimated by industry sources as another 10 billion barrels. The general view in the industry and among geologists appears to be that Nigeria could, depending on market prices, attain a production level of about 4 billion barrels a day by 1985.

At present, all natural gas produced in Nigeria is a by-product of oil production; only 2 percent is consumed usefully. Proved gas reserves, as of January 1974, amounted to 40,000 billion cubic feet. The potential for export of liquefied natural gas is considered promising.

Government-Company Relations

Nigeria has been a member of OPEC since 1971 and, like other OPEC members, has demanded and received higher payments from the oil companies. Increases in the tax reference prices have been negotiated with companies several times since 1971, and during the 1973 Arab-Israeli war the Nigerian government unilaterally increased prices.

The outlook in Nigeria, as in most major producing countries, is for increasing government control over the oil and gas industries and for continuing efforts to obtain a greater share of oil revenues. As a result of NNOC negotiations in 1972, Nigeria's holding in the Shell-BP refinery near Port Harcourt (the only refinery in the country) was increased from 50 to 60 percent. By the end of 1973, the government had made further movement toward government ownership of the oil industry; in addition to its interests in the Nigerian operations of AGIP-Phillips and SAFRAP, NNOC acquired a 51 percent interest in all new offshore drilling licenses.

During 1972 and 1973 Nigeria, like many other oil producers, was engaged in secret negotiations with the major oil companies over questions of the extent and kind of government control over oil production and marketing—the so-called "participation" issue. These negotiations were suspended in October 1973 when war broke out in the Middle East. Nigeria, however, is likely to arrange matters with the companies so that it can continue to expand revenues. Compared with some producing nations, Nigeria has not been especially threatening to the companies or to consumers. Nevertheless, it is certainly conceivable that Nigeria would participate with other OPEC countries in a production-sharing scheme.

The Push for Development

Nigeria is a very poor country and has a large population. Fortunately, it has abundant resources and is not destined to have a one-industry economy. Even before the oil boom of the 1960s, Nigeria's economy was expanding rapidly. Exports of several agricultural and natural resource products were substantial, and the industrial sector was becoming more diversified.

Since Nigeria became independent in 1960, its governments have generally made more determined efforts toward economic development than has been the case in many other poor countries. Nigeria is also distinguished from many developing countries by the vitality of its business community and the commercial orientation of its government.

Oil revenues have played an important part in Nigeria's development effort. In 1972-73, government income from the oil sector amounted to U.S. $1.37 billion. With higher prices and increasing output, oil seems certain to make an even greater contribution to development in future years. As of mid-1974, the development plan for 1975-80 has not yet been published, but guidelines for the new plan indicate that the government will attempt to take advantage of the opportunity provided by the expanding oil income of the 1970s to forge ahead with development and thereby provide a base for further growth in the 1980s.[1]

Political Prospects

On the surface, at least, Nigeria appears to have made a remarkable recovery from the agonies of the civil war. Despite the bitterness aroused by that long conflict, Biafrans were holding important jobs in the federal civil service, in NNOC, and in high-level appointed posts within a few months after the cessation of hostilities.

The boom resulting from the 1973-74 increased oil revenues casts a warm glow over the country's economic prospects. The hottest political issue in Nigeria today is how these oil revenues should be distributed among the nation's twelve states. This issue cannot be resolved through the political process as long as the military government maintains its ban on political parties. The regime has stated its intention to return to constitutional government in 1976; whether this promise will be honored remains to be seen.

Even if the oil income distribution question could be more or less satisfactorily settled this would not necessarily guarantee political stability. The underlying ethnic and regional separatist forces which led to the Biafran rebellion persist in a number of the states. At the same time, there is little reason to expect political developments that would seriously interfere with the orderly development of Nigeria's oil resources. A military coup is conceivable, but almost any Nigerian government can be expected to have a strong interest in maintaining oil production as a means of obtaining revenues needed for economic development.

INDONESIA

Indonesia—the birthplace of Royal Dutch Shell—is one of the world's oldest oil-producing areas, but its proved reserves are relatively small (about 10 billion barrels at the end of 1972) and its exports have been relatively modest—1.2 million barrels a day in 1973. (See

Table 7-3.) Indonesia is not expected to become a leading exporter unless exploration in offshore areas yields major discoveries. Exports for 1980 are projected at about 2 million barrels per day.

Table 7-3.   Indonesian Crude Oil and Products (thousand b/d)

Exports and Destination	1967	1968	1969	1970	1971	1972
Crude Oil:						
Japan	127.4	188.8	295.8	446.3	482.1	566.0
Australia	83.8	92.0	86.3	46.2	9.6	1.1
USA	60.2	66.8	81.4	68.3	111.1	160.4
Philippines	34.2	35.4	43.1	43.1	37.4	7.6
Trinidad	. . .	2.4	7.7	18.7	14.7	51.8
Europe	. . .	0.6	1.9	. . .	. . .	28.6
Other	. . .	. . .	. . .	1.5	. . .	14.2
TOTAL PRODUCTION	509.0	601.1	740.6	852.4	888.7	1,078.7
Refined Products:[a]						
Japan	19.5	37.2	61.8	76.5	74.2	111.2
Singapore/Bukom	16.7	13.6	12.9	8.3	9.7	3.9
Australia/New Zealand	12.9	11.6	5.4	3.3	1.7	0.1
Netherlands	23.9	8.6	5.7	5.0	0.1	0.1
South America/USA	. . .	0.1	0.6	0.3	1.4	3.7
Others	6.2	4.2	5.9	1.1	1.2	0.9
TOTAL	79.2	75.3	92.3	94.5	88.3	119.9

Sources: Exports — 1967-1968, *Petroleum Press Service*, April 1971, p. 126; 1969-1971, *Petroleum Press Service*, June 1972, p. 211; 1972, *Petroleum Press Service*, July 1973, p. 254.
        Production — 1967-1971, Urusan Economic Dan Statistik, *Bank Indonesia Monetary Bulletin*, September 1972, pp. 180-81; 1972, *Petroleum Press Service*, July 1973, p. 253.
Note: [a]Does not include bunkers.

Indonesia is an important source of oil for Japan and the United States. In 1972, it shipped almost 70 percent of its refined products to Japan—roughly 10 percent of Japan's imports of crude oil and products that year. The United States, which finds the low sulphur content of Indonesian oil very attractive, takes more than half of the remaining crude exports.

Production and Marketing Arrangements

In 1960, a decade before demands for participation shares were pressed successfully by various Middle Eastern producing countries, Indonesia launched its own brand of participation in the oil industry. From the post-independence turmoil of the 1950s, during which nationalist demands were made for drastic revision of the old concession arrangements, the Petroleum Law of 1960 emerged. That legislation provided the basic framework within which the present Indonesian oil industry has evolved.

Under the Petroleum Law, the state company retains all rights to explore and exploit petroleum resources, and foreign companies serve as contractors to the state. The foreign companies operate under various kinds of service contracts, primarily "production sharing arrangements," which normally provide that proceeds from production, after costs, are divided between the state company and the foreign contractor on a 65-35 basis. In addition to delivering crude to the state company for sale to the domestic market, the foreign contractor is required to sell up to 40 percent of total crude output to the state company for sale on the international market. In recent years, Pertamina, the state oil company, has become fully responsible not only for producing and marketing crude and products, but also for supervising the entire range of petroleum activities carried out by foreign oil companies.

The impact of this system of service contracts on the oil industry in Indonesia has varied with the attitude of the government toward foreign investment. The production-sharing arrangements, for example, did not prevent foreign oil companies from being placed under "direct supervision" during the turmoil that prevailed in the last years of President Sukarno's rule. Strong government-industry tensions resulted in declining production in the mid-1960s and in Shell's pulling out in 1965 (leaving Caltex and Stanvac as the main foreign companies operating in Indonesia). The revolution of late 1965 brought to power a government that has worked successfully with the foreign oil companies. Shell resumed operations in 1971, and over 40 companies have entered into service contracts with the government.

Recent Developments within the Industry

Although Indonesia's waters were opened to foreign companies in 1967, it was not until September 1971 that offshore production actually commenced. By that time, virtually all prospective producing areas had been assigned to foreign firms and a boom in exploration had developed. In addition to the numerous foreign oil companies involved in exploration, Pertamina has constructed drilling rigs in potentially lucrative offshore areas. Many of the wells drilled thus far have been dry and no large discoveries have been made, but new production from successful finds could boost output within a few years.

In 1973, Indonesia began exporting about 50,000 barrels of crude oil a day to Japan under a bilateral agreement that may prove to be the most far reaching of all Indonesia's recent agreements relating to petroleum production. Pertamina agreed to deliver, over a period of ten years, 365 million barrels of low-sulphur crude to Japan. In return, the Japanese government is lending Pertamina US $200 million to be used for oil-field development and rehabilitation projects. Added to the proliferation of barter agreements and joint ventures between Pertamina and Japanese firms, this government-to-government arrangement represents a new approach on the part of Djakarta to the use of foreign capital to exploit its oil resources and develop its refining, transport, and marketing system.[2]

Efforts are also being made to develop an Indonesian gas industry. In 1972, initial plans were laid for the eventual export of Indonesian liquefied natural gas (LNG) to Japan and the United States. One project envisages the annual shipment of 7 billion cubic meters of natural gas to Japan. The project would include provisions for sizable foreign investment in construction of a pipeline, a liquefaction plant, and a fleet of LNG carriers.[3] All foreign contacts for the exploitation of natural gas will be on a production-sharing basis that gives the Indonesian government nearly 80 percent of the proceeds.[4]

The growing recognition of the importance of offshore oil and gas production has paved the way for a series of recent agreements between Djakarta and neighboring Asian states designed to promote the peaceful and profitable exploitation of energy resources. In 1971, a tripartite agreement was signed by Indonesia, Thailand, and Malaysia delineating offshore boundaries. This treaty was followed in early 1972 by similar bilateral accords with Malaysia and Australia defining rights to the continental shelf.[5]

Prospects for the Future

Although resource limitations will prevent Indonesia from becoming a leading world producer of oil, the country is nonetheless fortunate in that the quality of its output is excellent, export markets are readily available, and production is steadily increasing. At the same time, an important political issue affecting Indonesia's role as a world energy supplier remains unresolved: the extent of the government's willingness to enter into bilateral agreements involving Japanese capital in exchange for long-term energy supplies.

Indonesia's oil industry is in effect nationalized, but the unique system that has evolved there has elements of free enterprise and is conducive to profit making by all parties concerned. The government's policy of allowing contractors to recover their costs before profits are divided has encouraged foreign firms to bear the financial risks of exploration and development. In 1972-73, the production-sharing agreements between Pertamina and the oil companies provided the government with about 40 percent of its total revenues. Indonesia can be expected to gain an increasingly large share of gross oil profits. (See Table 7-4.)

Table 7-4.   Contribution of Oil Sector to Indonesia's Government Revenues (millions of US dollars)

	1968	*1969/70*	*1970/71*	*1971/72*	*1972/73 (projected)*
Income from oil taxes (profit-share in crude production)	68	127	181	247	497
Receipts from sale of Pertimina products	20	48	83	94	84
Total oil receipts	88	175	264	341	581
Non-oil receipts	396	643	911	1,039	1,382
Total receipts	484	818	1,175	1,380	1,963
Oil receipts as a % of total receipts	18.1	21.4	22.4	24.6	29.0

Source: U.S. Embassy Djakarta, Airgram A-144, June 16, 1972.

Within the broad outlines of the government's policy, several new trends can be distinguished. It is probable that Indonesia's national oil company will try to gain greater proficiency in a wide range of petroleum operations; Pertamina, for example, has recently become involved in refining, transport, and marketing operations. In addition, the government has undertaken steps to persuade foreign contractors to base more of their operations on Indonesian soil.

The Indonesian government can be expected to continue to follow policies designed to achieve the maximum possible production of oil. With a growing population of over 100 million, Indonesia is one of the most densely populated oil-producing nations. The government, with good reason, is pursuing ambitious development plans that could easily absorb all of the oil revenues that are likely to be available in future years.

NOTES FOR CHAPTER 7

1. See Nigeria, Federal Ministry of Economic Development and Reconstruction, *Guideline for the Third National Development Plan 1975-1980* (1973).

2. *Oil and Gas Journal,* September 3, 1972, p. 32.

3. *Petroleum Press Service,* October 1972, p.385.

4. *Petroleum Press Service,* February 1973, p. 74.

5. See U.S. Embassy Djakarta, Airgram A-144, June 16, 1972, p. 12; *Petroleum Press Service* April 1972, p. 150; and Philip Bowring, "Seabed Accord," *Far Eastern Economic Review,* October 28, 1972, p. 41.

Part Three

The Outlook in the Oil-Importing
Countries

Chapter Eight

Western Europe

by Philip H. Trezise

This chapter examines the possible impact of developments in the field of energy on the complex of relations that have evolved between the United States and Western Europe since the end of World War II.[1] Because the availability of energy resources touches on vital interests, the subject has to have important political as well as economic implications for the Atlantic, or transatlantic, relationship. Whether the changed energy situation will have a long-run and adverse effect on relations between Western Europe and the United States is an open question. But that energy issues will be a much larger element in the relationship than in the past seems beyond doubt.

BACKGROUND

Until quite recently, the energy links between the United States and Western Europe were on the whole mutually comfortable ones. Postwar Europe's rapid shift away from coal as a principal energy resource was made possible—was caused—by the easy availability of oil. The major oil companies not only provided the increased supplies of crude that were required but also mobilized most of the massive downstream investment for Europe's refining and marketing facilities. Any misgivings—over the decline of the coal industry, the increasing dependence on imports from the Middle East and North Africa, or the dominant role of the international oil companies in a critical area of national economic life—were overcome by the evident benefits of cheap petroleum. During the 1960s, the price of oil in real terms actually declined, thus widening the advantage of oil over other fuels.

In addition to being cheap, oil was reliably available. The largest part of Europe's oil imports (85 percent in 1972) came from the Middle East and North Africa. In a supply emergency, however, the oil companies could call on spare producing capacity in the Western Hemisphere (in the United States and in Venezuela) to make up any likely shortfall. As late as the 1967 Arab-Israeli war, Western Hemisphere reserves were large enough at least to ease the concern of European energy officials about the availability of North African and Middle Eastern oil supplies.

Over the postwar years, the United States became a chief supplier to Europe of energy resources other than oil. As European coal production fell, an export market appeared on the continent for U.S. coking and steam coal. And the development of the European nuclear power industry was linked to the availability of enriched uranium from the U.S. Atomic Energy Commission, which had a virtual monopoly of supply. Natural gas discoveries, especially in the North Sea in the 1960s, added importantly to European fuel supplies. Nevertheless, oil was established as Europe's predominant energy source and the outlook as the seventies arrived was for that predominance to continue into the eighties and beyond.

European complacency about oil imports was mildly shaken in 1970-71. In the aftermath of the 1967 war and the closing of the Suez Canal, crude oil from the recently developed and nearby Libyan fields had become a very large factor in European imports— by 1970 it accounted for fully one-quarter of the total. When, in 1970, the new revolutionary Libyan government began to reduce output (coming on top of a cut by Syria in the pipeline from Saudi Arabia to the Mediterranean), the consequent loss of a million barrels a day provided the basis for unanticipated increases in prices. The elaborate international company-producer nation negotiations that followed in Tehran and Tripoli in 1971 ratified, and perhaps more than ratified, the changed market situation and resulted in sharply increased prices for the different categories of North African and Persian Gulf oil. By this time, the Western Hemisphere had very little reserve production capacity and the United States was on the verge of becoming the fastest growing importer of all.

Even so, oil remained the least costly source of energy. The reduction in shipments from Libya and eastern Mediterranean points was quickly made up with more oil from the Persian Gulf. Prices probably would have fallen back from the 1971 levels but for the sudden rapid expansion of U.S. import demand and then, in 1972-73, a cyclical upturn in business activity, and consequently energy consumption, in all the industrial countries at once. As it was, the higher cost of oil was a relatively small factor in the European inflationary problem prior to the outbreak of the 1973 war in the Middle East.

While there were misgivings about the shaky prospect for Middle Eastern peace and the implications of that for the security of Europe's oil supplies, the energy policies of the West European countries in September 1973 remained essentially as they had been for the previous decade or longer. That is to say, the prevailing disposition was to rely on oil, provided mainly through the agency of the international firms, as the principal energy raw material and to pursue (or to ignore) the alternatives to oil with all deliberate regard for their relative costs.

The Arab invocation of the oil weapon in October 1973, as frequently forecast as it had been, came as a shock to Western Europe. In its initial form, the Arab program of progressive cuts in shipments threatened to leave the West European peoples immobile, unemployed, and cold, and with no discernible alternatives to oil. The most apocalyptic of the visions of an energy crisis took on a terrifying imminence. In the fact, of course, the use of the oil weapon proved to be considerably less than devastating, for the supply cutbacks stopped far short of reaching their threatened dimensions. Still, a striking demonstration of the power of the oil-producing countries had been given and that, taken together with the consequent escalation of oil prices in late 1973 and early 1974, served to support the proposition that Europe now confronted a radically different energy situation.

A number of questions acquired new urgency. Were there possibilities for reducing Europe's dependence on Middle Eastern and North African oil? Could the economic (or political) power of the West European states as a group be used to obtain assured energy supplies at reasonable costs? Would bilateral arrangements with oil-producing countries be the answer? What role should be allotted to the international companies in the future? Would the Community of Nine now be able to agree on the common energy policy that had eluded the Brussels Commission for years? How should "Europe" or its constituent states cooperate with other consuming countries, most particularly the United States?

MEETING EUROPE'S ENERGY NEEDS

A useful array of data for considering the European energy problem is found in a mid-1972 report of the Oil Committee of the Organisation for Economic Co-operation and Development (OECD), from which Table 8-1 is adapted.[2]

Table 8-1.   Europe's Primary Energy Requirements (millions of metric tons of oil equivalent)

	1960		1970		1980	
	Volume	%	Volume	%	Volume	%
Total requirements	612.0	100	1,041.4	100	1,757.0	100
Indigenous production	407.5	67	409.5	39	749.0	43
Solid fuels (coal, lignite)	353.5		273.0		194.0	
Oil	16.0		23.4		179.0	
Natural gas	11.2		67.7		203.0	
Hydropower	26.0		35.0		48.0	
Nuclear	0.8		10.4		125.0	
Net imports	204.5	33	631.9	61	1,004.0	57
Oil	183.3		596.8		934.0	
Solid fuels	21.1		32.7		42.0	
Natural gas	0.1		2.4		28.0	

Source: Organisation for Economic Co-operation and Development, *Oil: The Present Situation and Future Prospects* (Paris, 1973), p. 42.

Note: Europe here includes the nine European Community countries (United Kingdom, Ireland, Denmark, West Germany, France, Italy, the Netherlands, Belgium, and Luxembourg) and Sweden, Switzerland, Spain, Portugal, Austria, Norway, Finland, and Iceland.

As has been noted in the text, projected energy requirements in 1980 should be reduced to account for the effect on consumption of the higher prices imposed by oil exporters in late 1973 and early 1974.

The period 1960-70 was one of rapid growth in European energy consumption. The average annual rate of increase was 5.5 percent, compared with 4.2 percent in the United States. Growing energy demand was satisfied principally by oil; its share in total energy use almost doubled over the decade, and oil imports grew very fast, at an annual average rate of 12.5 percent. By 1970, Europe had become dependent on imports for more than 60 percent of its total energy resources, and oil accounted for almost 95 percent of the import total.

As will be noted, the Oil Committee's projection for 1980 shows a substantial decline in the growth of energy imports between 1970 and 1980. Even assuming, as the Committee did, an almost unchanged rate of increase in energy use (5.4 percent annually), all energy imports are estimated to grow at an annual rate of slightly more than 4.5 percent, and net oil imports at about the same rate. Of course, the Europe of 1980 would still be heavily dependent on oil from abroad. What is the likelihood that this dependence can be lessened? Or that reliance on Middle Eastern and North African suppliers could be reduced by diversification of import sources?

An obvious point is that the Oil Committee's requirements projections, made in relatively placid 1972, must be scaled down to take account of the higher energy costs that now seem likely to obtain for some years to come. A downward change of one percentage point in the Committee's assumption that energy use would grow at a yearly rate of 5.4 percent would lower 1980 requirements by some 160 million metric tons of oil equivalent (almost 1.2 billion barrels). Equally, it must be supposed that higher oil-import prices will encourage conservation of energy, bring forth more indigenous energy raw materials, and hasten the development of alternatives to oil.

Natural gas has been the most rapidly growing part of Western Europe's energy supplies. Output increased sixfold during the 1960s and has continued to grow. In 1973, the Dutch North Sea gas fields produced 70 billion cubic meters of natural gas and the British fields another 36 billion cubic meters. Output elsewhere in the European Community was more than 42 billion cubic meters, for a total Community production of 148 billion cubic meters, up 45 percent in two years.[3] North Sea production is expected to continue to rise at least until 1980. If energy costs remain high, the Oil Committee's projection that European natural gas output will more than triple between 1970 and 1980 could be on the conservative side.

It is already certain that the North Sea will be a very significant source of crude oil. A number of commercial fields have been found and the drilling under way is sufficiently extensive as to make further discoveries probable. The most recent official British estimate is that by 1980 the U.K. North Sea sector will have proved reserves of 6.0 to 9.5 billion barrels of oil and a daily output in the range of 1.4 to 2.4 million barrels.[4] Reserves in the Norwegian sector are considered to be of the same order of magnitude, and the 1980 production forecast is for 1.2 million barrels a day. A U.S. industry estimate places total North Sea supply at "as much as" 3 million barrels a day by 1980 and at 4 million barrels a day by 1985.[5] Other much more optimistic projections put total North Sea production in the "early 1980s" at 6 million barrels a day.[6] Choosing from among these estimates the figure of 3 million barrels a day, North Sea oil in 1980 would provide about 13 percent of the European requirements foreseen by the OECD Oil Committee. At 6 million barrels a day, Europe's oil imports would fall back to 1970 levels—or taking into account the effect of higher prices on consumption, below that.

That European coal output will increase significantly in response to higher energy prices is not a certainty. Coal and lignite production in Western Europe fell by more than 22 percent during the 1960s, despite heavy domestic subsidies and tight quotas on imports. While large British and German reserves (500 billion metric tons) could be the basis for increased production, manpower shortages will probably be a stubborn obstacle; in the United Kingdom, for example, 39 percent of all mineworkers are reported in the 50-65 age group.[7] Germany's long-term energy program, published just before the October 1973 war, assumed that coal production would continue to fall, from 84 million metric tons in 1972 to 58 million metric tons in 1980, even though costly subsidies would be continued.[8] British policy, as represented by the settlement of the February-March 1974 coal strike, would seem to be to maintain output at the current yearly level of approximately 130 million metric tons if possible; to aim at more would tend to push the average cost of coal production to a level approaching the current costs of oil, which likely is not going to last. It is probably a safe supposition that the decline in European coal production will slow somewhat in the light of the scare about oil imports, even if oil prices fall well below present figures. Still, the OECD Oil Committee's estimate of 1980 production at close to 200 million metric tons of oil equivalent seems a relatively optimistic one.

The Oil Committee foresees nuclear power usage in 1980 at a level equivalent to 125 million metric tons of oil. That figure is not likely to be exceeded despite the energy problems that have arisen since the estimate was made. Long lead times for nuclear plant construction and the troubles plaguing the British nuclear program in particular suggest that the twelvefold increase in nuclear power projected for the 1970-80 period will be achieved only with difficulty. The U.S. Atomic Energy Commission's projections of U.S. and foreign nuclear power development by 1980 have in fact been cut back significantly.[9]

Whether Europe can get large additional energy supplies outside the Middle East and North Africa is also in doubt. There is a modest market for U.S. coal in Western Europe and in time it might be expanded. But European utilities have been wary of the impact of safety and environmental legislation on the availability of U.S. coal. Coal from Eastern Europe, mainly from Poland, is also imported, but this is a very limited source of energy in relation to European needs. Unless world coal and oil price relationships shift decisively and durably in favor of coal, which is by no means certain, it is not likely that coal imports will become an important factor in European energy supplies.

Natural gas imports from the Soviet Union conceivably could be a substantial energy source. The USSR places its proved reserves at 18 trillion cubic meters, more than twice those of the United States, and potential reserves at 100 trillion cubic meters.* Apart from some modest exports to Austria, Soviet natural gas has not yet been sold outside Eastern Europe. Contracts with Italy, West Germany, and France, however, call for exports, nominally to begin in 1974, of up to 18 billion cubic meters a year.** A more intriguing speculation, however, can be taken from a current West German-USSR technical study of a 2.5 meter pipeline from the Siberian fields, which could carry up to 200 billion cubic meters of natural gas, or almost as much as is now predicted to be total European Community consumption in 1980. If Siberian gas can be transported to Europe at prices competitive with other fuels, which would imply large quantities, then a substantial move away from, mainly, oil would be possible, even probable. But this is still a highly speculative prospect. If past delays in pipeline construction between European Russia and Western Europe are indicative, imports of natural gas from distant Siberia are probably well off in the future.***

In sum, the chief realistic possibilities for a significant reduction in Western Europe's dependence on Middle Eastern and North African oil are two: a decline in the rate of increase in energy consumption and, on the supply side, crude oil and natural gas from the North Sea and perhaps from adjacent waters as well. Together, as Chapter 13 discusses in a world context, these forces in time can help to alter quite radically the global supply-demand balance. But Western Europe will continue to be a giant importer of energy resources for the indefinite future. It is to be expected that energy questions will continue to occupy and even to preoccupy European officials and political leaders for a long period ahead. One consideration that is bound to get further attention is that of a concerted energy policy on the part of the nine European Community states.

A EUROPEAN ENERGY POLICY?

The nine European Community nations, which account for some 85 percent of the region's energy consumption, have long had as an objective a "common" policy on energy questions. In pursuit of this objective, the European Commission in Brussels has produced a

---

*See Chapter 11 for a fuller discussion of Soviet natural gas availabilities and prospects.

**Small amounts of Soviet gas in fact began to flow into West Germany in late 1973.

***Soviet deliveries of crude oil to West Germany and France fell below contracted amounts in 1973 and this too will contribute to uncertainties about the reliability of Soviet energy supplies.[10]

number of draft policy statements. None has gone beyond the draft stage. As long as oil was cheap and seemingly assuredly available, interest in policy harmonization was minimal and national policy positions were overriding. First responses to the energy "crisis" of 1973 were no more favorable to a commonality of energy policies. Rather the contrary, for the other Community members were notably cool, in public at least, to the plight of the embargoed Netherlands, and the initial preference was for bilateral rather than Community approaches to the oil-producing nations.

Attitudes may change in favor of concerted action toward the producing countries and the international oil companies and in favor of Community decisionmaking on energy issues. This will require, however, a reconciliation of differing national interests, traditions, and philosophies.

Member Country Positions

At one pole is France, which for many years has espoused and practiced a policy of thoroughgoing governmental intervention in the energy sector. From the late twenties, French planning has aimed at creating a strong domestic oil refining and marketing industry, gaining French participation in the overseas oil concessions, and otherwise limiting France's dependence on the U.S. and British international companies. While the state's monopoly powers were delegated in part to private firms, crude-oil imports were licensed and refiners could be directed to buy oil from approved sources. France has one state-owned integrated oil company, ERAP (Enterprise de Recherches et d'Activités Pétrolières), and a public share of 35 percent in a second firm, CFP (Compagnie Francaise des Pétroles).

During the 1960s, the development of the Algerian oil fields provided France with a substantial volume of "franc" oil—that is, oil produced by French companies, directed to the French market, and paid for with francs. By 1970, Algerian oil accounted for 26 percent of France's total requirements. But the Franco-Algerian arrangement was upset in 1971 when the Algerians nationalized the French holdings. That experience has not deterred France from promoting French oil concessions elsewhere, notably in Iraq, and France presumably would hope to develop a Community policy around the concept of preferred suppliers.

France is poorly endowed with domestic energy resources. The coal industry has been allowed to run down. Known natural gas reserves are modest and production has levelled off. A project in Brittany for producing electricity with tidal power has disappointed its advocates and more ambitious plans for harnessing the tides have been set aside. Most hydroelectric possibilities have been developed. Long-term planning therefore has centered on nuclear generation of electricity. France has the technology for uranium enrichment and is actively pressing for the construction of a European gaseous diffusion plant based on French technology. A French-sponsored group of private and public organizations from Italy, Belgium Sweden*, Spain, and France, Eurodif as it is called, in effect is a competitor of a British-Dutch-German project, URENCO, which will produce enriched uranium using the new centrifuge process. The French government has rejected an invitation to join in the centrifuge project and proposes, with its European partners, to go ahead with a \$2.7 billion plant that will have an ultimate enrichment capacity of 9,000 metric tons.[11]

The United Kingdom is in the un-European position of being in sight, potentially, of substantial energy self-sufficiency by the 1980s. British coal production is the largest in Europe. North Sea natural gas and, in the near future, petroleum, plus nuclear power, could possibly end Britain's dependence on imports. Or, even if doubts about nuclear power development and coal output prove justified, in the early 1980s the United Kingdom might be importing only 20 or 25 percent of its energy supplies.

With one and one-half of the major oil companies British owned, official British views on oil policy naturally have been affected by the claims of the international industry. This may change as the status of the companies changes or as the British political scene changes. Nonetheless, concern for the fortunes of British Petroleum and Royal Dutch Shell is likely to dissuade the United Kingdom from any European policy that is unduly antagonistic to the major oil companies.

More important to British attitudes toward European Community energy proposals, however, is the matter of distribution of oil and gas from Britain's North Sea sector. British law requires that all oil produced in the sector be landed in the United Kingdom and that all gas be offered to the British Gas Corporation at "reasonable" prices (which has meant below market prices elsewhere in Europe). On one not uncommon view of national interest, Britain will choose to keep its North Sea energy resources for its own use, at below world prices so as to benefit British industry. If it does so, conflict with the Rome Treaty and its strictures against discrimination within the European Community would appear to be nearly inevitable.

---

*Sweden has since withdrawn from Eurodif, but Japan has shown an interest in contracting for fuel and possibly in taking a share in the proposed diffusion plant.[12]

Still, precedents for discrimination in favor of domestic interests do exist. The Netherlands is the other Community nation with a substantial domestic energy resource, in the form of natural gas, which provided 42 percent of Dutch primary energy in 1972. Under Gasunie, the mixed public-private company that controls the distribution of most of the natural gas, exports were not only permitted but ran ahead of domestic offtake until long-term export commitments and the growth of home consumption threatened to exceed a prudent depletion rate in the principal producing fields. Dutch authorities then had second thoughts, to the point in 1972 of vetoing a sales contract between an independent gas producer and a West German consortium.* The Dutch have also, however, taken the long-sighted step of raising gas prices at home and to export markets as a way of slowing down the growth of demand.

The Netherlands, in any case, has its own special interests. Royal Dutch Shell is an important one, and it will be bound to influence the Dutch on Community policy. Also, Dutch ports are major transshipment points for crude oil going up the Rhine, and the refineries in the Netherlands are large exporters of oil products. The Dutch have thus not been well disposed toward a narrowly "European" energy regime, and they may be even less so after the signal indifference shown by France and the United Kingdom to the Arab embargo of oil for the Netherlands in 1973.

West Germany's domestic energy resource is coal, an industry that has been in secular decline since the 1950s. Heavy subsidies have merely slowed the reduction of coal output, and for the future the industry will have a continuing problem of attracting workers. Even so, oil has only recently overtaken coal and lignite as an energy source, and the German reserves of solid fuels offer at least a partial alternative to imported supplies of energy raw materials. Government policy continues to favor the use of solid fuels—over oil or natural gas—for power generation.

German oil policy has been to rely heavily on the international companies, which control about 75 percent of the downstream German industry in addition to being suppliers of virtually all of Germany's crude. Partly because of restiveness over the dominant and expanding role of the majors, in 1968 the German independent refiners and marketers, with government encouragement, formed a joint company, Deminex, as a vehicle for obtaining crude through exploration, buying into existing concessions, or purchases of participation oil. Deminex in turn receives, or is entitled to receive, public subsidies for exploration (loans up to 75 percent of

*A compromise was subsequently reached after a German appeal to the Dutch courts.

costs, repayable only if discoveries are made) and for purchases of proved concessions (a 30 percent outright grant). To challenge the dominance of the foreign-owned oil companies, the government is fostering a new downstream combination of private companies (with a 40 percent share for the publicly owned chemical firm, VEBA), which it is hoped will control one-fourth of Germany's refining and marketing capacity.

West Germany's "long term energy program," published in September 1973, called for an extension of the government's activist role in German energy policy, including a readiness to support special arrangements for getting additional supplies of the natural gas which, it is hoped, will be a rapidly expanding source of energy at least through 1980. It stressed also the urgent need for a joint European energy policy. At the same time, the document reaffirmed the desirability of retaining as much as possible of a free market in energy, and it specifically welcomed the continued presence of the international oil companies in the German market—to the point of making them eligible for the first time to receive the state subsidies applicable to tankers built in German shipyards.[13]

Finally, Italy, an energy have-not, traditionally has made cheap energy its first priority. The national gas and oil group, ENI (Ente Nazionali Idrocarburi), in fact was created as a counterweight to the feared "monopoly" position of the majors. With governmental aid, ENI has managed to obtain crude oil concessions that supply about 25 percent of the Italian market, and it has its own monopoly of Italian gas supplies. But ENI aside, Italy has followed a basically nondiscriminatory policy, in accord with the objective of getting low-cost supplies. Now that energy-resource costs have skyrocketed and oil supplies have proved to be vulnerable, Italy no doubt would be ready to accept any European policy that promised to give assurance of secure supplies. Meanwhile, however, Italy, like its Community partners, has pursued bilateral energy deals. These include a number of natural gas projects, among them a pipeline across the Mediterranean from North Africa to Sicily and then through the Messina Strait to the mainland, and extension of a pipeline from Austria to supply Italy with Soviet natural gas. If all goes well with these and other projects, Italy will have succeeded in diversifying its sources of energy materials, although still not effectively away from North Africa and the Middle East.

The Content of Common Policy?

As this chapter is being written in the spring of 1974, the European Community is at a low point in the swings between progress toward greater unity and relapses to unilateralism or nationalism. Even if no differences of member-state interests existed, it would not be a promising time for constructing a Community policy on energy. For all that, the scare that Europe had in the last quarter of 1973 surely will sustain an interest in possible measures *as a Community* to make Western Europe's energy situation a less vulnerable one. Indeed, at a Copenhagen summit meeting in December 1973, the heads of the Community governments agreed on a series of general recommendations on energy policy, and they created a special committee of senior officials to help carry out the decisions that the Council of Ministers presumably would take. But from the summit communiqué to action may be a long path.

Of course, there are a number of relatively non-controversial matters on which a Community that was moving purposefully toward economic and political integration would wish to harmonize national laws, regulations, and policies. It would be logical, for instance, to have common rules for environmental protection, including rules governing nuclear processes and installations. Western Europe is a small place and few environmental risks can be wholly contained within national borders. It does seem that a Community ought to decide that no member state can have a competitive advantage by slighting the need for reducing pollution and safety hazards. Similarly, harmonized technical specifications for energy installations would make sense. Or, in the case of intra-regional pipelines, a single set of regulations would be better than several national sets.

The Community has an official guideline for petroleum stocks that calls for compulsory oil storage equal to 90 days' consumption, to be achieved by 1975. (The European Commission has proposed that a target of 120 days' consumption be studied, but that level of stocks is not going to be a practicable objective in a period of tight supplies and high prices.) Other measures for oil emergencies—such as a common rationing system, the pooling of stocks, or a reserve tanker fleet to cope with an interruption of short-haul oil supplies—could be useful supplements to national stockpiles, but they are still to be undertaken.

No feasible stockpiling program, or other plans to cope with emergencies, can alter very much Europe's dependence on insecure sources for its principal energy resource, oil. In principle, the Community could mobilize and pool its considerable resources to finance domestic (or quasi-domestic) alternatives. One possibility would be to support European coal production on a scale adequate to maintain or even to increase current levels of output. The Community already subsidizes coking coal on a limited scale, and national coal subsidies amount to sizable sums.* Any proposal to increase the present level of aid to the coal industry would certainly incite a controversy between the coal-producing nations and the coal importers, Italy and the Netherlands. But the probable costs of a serious program for halting or reversing the decline of the European coal industry virtually guarantee that this alternative will be considered only in circumstances more extreme than those now obtaining.

Uranium enrichment is a project that seemingly ought to call for Community action. It is virtually certain that demand for nuclear fuel eventually will take up the output of both the large gaseous diffusion plant that France proposes to build and the centrifuge plant being promoted jointly by the United Kingdom, Germany, and the Netherlands. These are costly undertakings, however, and the European Commission has suggested that investments be coordinated and that the Community finance stockpiling of nuclear fuel, if temporary surpluses should develop.[14] The Commissions's ideas have not prevailed and a resolution of intra-Community differences seems distant still.

In the end, of course, oil will continue to be the main source of energy for Europe far into the next decade and very likely beyond that. As has been noted earlier, the North Sea will provide an important and possibly a preponderant share of the increment of European oil consumption between the present and 1980 or 1985; and drilling in other offshore waters—the Celtic Sea and Portugal's continental shelf—may bring in more new fields. This development will contribute to easing the European energy position, since it will help reduce requirements for imports from overseas. Whether it will contribute to Community harmony is more in question. At least while world fuel prices remain high, those Community members that are successful in finding oil or natural gas in their sectors may opt for policies intended to confer competitive benefits on their domestic economies, that is, by setting differentially low prices for the energy resources under national control. That could be a very serious departure from the Rome Treaty's principle of nondiscrimination, which in some circumstances could threaten the whole Community idea. Thus, in another period of severe shortage, a refusal by some members to share available local energy resources could be disastrous for Community solidarity.

Whatever energy resources may flow from the North Sea and adjacent waters, the overwhelming probability is that the West

The Economist (March 2, 1974, p. 15), says that Britain's nationalized coal industry may need more than $1 billion in price increases and direct grants simply to avoid a deficit in 1975.

European nations as a group will remain the largest importer of North African and Middle Eastern crude oil. In most cases, the first response of governments to the Arab oil cutbacks in October 1973 was to seek bilateral deals with oil-producing states. Six months later, after a succession of high-level missions and a flood of publicity, the announced results were not prepossessing. The British government is said to have been assured an additional 200,000 barrels a day of Saudi Arabian crude (Britain imported more than 2.2 million barrels a day in the first half of 1973). France reportedly had contracted with Saudi Arabia for some 180,000 barrels a day of oil over a three-year period (French imports were running at more than 2.7 million barrels a day in the first half of 1973). Prices in the various agreements or proposed agreements have not been made public, but it is likely that the oil producers have been asking for at least the going market price, which might not be a bargain for very long.*

It would seem possible that the Community states might have a stronger negotiating posture acting as a unit rather than as individual nations. The Community is the world's largest trader and the largest importer of oil. Its Commission has suggested that the Community market might be opened, presumably on a preferential basis, to industrial and agricultural products from the oil countries, a proposal that might be attractive to nations like Iran or Algeria.** Or, as the Commission proposed in general terms in April and May 1973, the Community could go to the French system of import licensing and official control over refining and marketing capacity. Then the Community market might be managed so as eventually to maximize bargaining leverage over the oil producers. Under such a system, also, the separate national incentive programs for oil exploration and development might be combined under a Community financing plan.

The 1973 summit meeting at Copenhagen endorsed Community negotiations with the oil-exporting countries on cooperation in the economic development of the exporting countries, on industrial investment in the EC countries by the oil exporters, and on the provision of stable energy supply to EC members at reasonable prices. This is a formula broad enough to cover the French position

---

*L'Express is quoted by the Washington Post (February 14, 1972, p. 12) as saying that the French three-year deal with Saudi Arabia is costing ERAP $2.40 a barrel more than the ARAMCO price of $8.40.

**Whether the Community could offer special tariff arrangements on oil imports to selected producers is problematical. Crude oil enters the Community market free of duty under a GATT rule. Petroleum products are duty free, under the Community's generalized preference scheme, up to quota limits. These limits have been undersubscribed, presumably reflecting the competitiveness of refineries within the Community.

that Europe should seek an oil relationship with producers independent of the United States and the international companies. It is hardly so definitive, however, as to commit its signers to any specific course of action.

COOPERATION WITH THE UNITED STATES

Western Europe does not have as even a nominal alternative a Project Independence from imports. What seems to be an exceedingly optimistic prospectus by the European Commission sees ways to reduce the Community's import dependence from a present 60 percent to 43 percent by 1985.[15] This would require holding coal production at current levels, which is a doubtful proposition, and building nuclear plants to provide 50 percent of the Community's electricity by 1985, a target well above any going estimates. Absent some wholly unexpected development, Western Europe must continue for an indefinite time to get the largest part of its energy from the Middle Eastern and North African oil fields. It is not wildly farfetched for a West European politician to reflect on the situation of an extended oil supply interruption leading to massive unemployment and widespread hardship; or, in a less apocalyptic vein, to consider the administrative and political complexities of having to ration energy supplies. These kinds of considerations must have been a part of the initial European reaction to the Arab actions in the last quarter of 1973. And, inevitably, European leaders will have to make further judgments as to how access to overseas oil can best be assured.

Much of what Western Europe might do, or is likely to do, about its energy vulnerabilities need hardly affect relations with the United States. If the Commission was to succeed in getting closer agreement on Community emergency procedures—stockpiling, rationing, sharing, and so on—no one outside could well object. Subsidies for energy exploration or development are going to be provided everywhere, not only by the Community or the European states individually. Should some European nations turn to the USSR for natural gas, or for enriched uranium, this could not be divorced from political and security questions; but the resulting diversification of supply sources would tend to offset security worries; and the United States itself is permitting its firms to negotiate for LNG imports from the Soviet Union.

The U.S.-owned oil companies of course are an element in the European-U.S. relationship. Together with Royal Dutch Shell and British Petroleum, they still have a dominant position in the

European oil industry, as suppliers, transporters, refiners, and marketers. European attitudes toward them include unease, suspicion, and hostility, along with other points of view. It is possible that the Community members in due course will decide, in part in order to assert control over the major companies, to adopt the French pattern of a managed oil market in which imports come only from officially designated sources and the development of the refining and marketing system is under close official direction. This is not an imminent prospect, however, and in any case the international companies, which have not found their operations in France to be unprofitable, probably would adjust to such a new commercial situation readily enough.

As for the oil company concessions in the Middle East and elsewhere, their status is changing rapidly in response to the pressures of the host countries. There is no reason to suppose that many existing concessionaires will be replaced by new ones, European or otherwise, unless on terms much more favorable to the host countries but no more durable than those already in force. To fear a contest for the concessions seems to be a misreading of trends in the oil-producing states. The participation oil that the oil countries will receive in increasing volume over the next few years likely will give European national companies enhanced access to crude supplies. In a weakening world oil market, participation oil may have to be sold at discount prices and thus would give its buyers competitive advantages. The most sensitive problems in that event, however, will tend to be with and among the OPEC countries as they try to bolster a crumbling price structure, and the consequences for anyone of such a situation are not easily foreseen.

The European Community has a so-called "Mediterranean policy," which has consisted up till now of preferential trade arrangements of quite limited scope with Tunisia, Morocco, Malta, and Israel. Although its commercial importance is modest, the policy has been opposed by the United States both on principle and because it has implications for an eventual zones-of-influence world. For some but not all Community members, on the other hand, it has been a desirable extension of influence into a region historically and economically associated with Europe, and done through the only means available to the European nations, that is, through trade. It is conceivable that the Mediterranean policy could be the prototype for a far more substantial arrangement involving some or all of the region's oil producers and covering oil supply, technology, investment, and trade on preferential bases.

To thus expand on the Mediterranean policy would not be easy. There would be opposition within the Community, on political grounds and also from those producers who would be exposed to more import competition. The reception that might come from across the Mediterranean is not certain either. Prospective Arab associates well might ask for an anti-Israel *quid pro quo* that would be very difficult to provide. But if an effort to build from the existing preferential system was really successful, the result undoubtedly would be to add tensions to relations with the United States. One consequence might be a further strengthening of U.S. ties with Saudi Arabia, and Iran, in something of a struggle for influence in the Middle East. Beyond that, the scenario of future events could be elaborated to the point of including strains on security relations, given that the U.S. Sixth Fleet is the principal element of NATO in the Mediterranean.

At all events, it is evident that none of this is likely to happen quickly. Decisionmaking in Europe is a ponderous process. The advantages and disadvantages of various approaches to energy or oil policy are not at all clear. Unilateral action seems to be of limited worth and multilateral measures are difficult to organize. There will be time, almost certainly, to pursue the beginnings that have been made since the Arab embargo on U.S.-European cooperation in dealing with the energy problem.

A principal objective unquestionably ought to be energy conservation. As is argued in Chapter 13, a check on the consumption of oil in the United States, Western Europe, and Japan would work toward altering radically the bargaining positions of oil consumers and oil producers. For consumers, a more favorable supply-demand relationship would itself be a factor for security, as well as the basis for lower oil prices. As illustrated by events, the Arab production-cutback decisions of 1973 would have been far harder for the Arabs to take if the world oil market had been slack instead of being extraordinarily tight.

Western Europe and Japan alike have reacted negatively to any thought of an importing country bargaining group, an OPIC to match OPEC. They are surely right, if the idea was for a "consumers' cartel" to confront OPEC. The bargaining power of consumers would not be notably enhanced by saying that they were acting as a group, when in fact none would be prepared to refuse to buy the oil his market required. On the other hand, once the market has weakened somewhat, consumer-producer negotiations encompassing price and assurances about supply might be a feasible and desirable enterprise.

There are numerous other potentially useful things for the major consuming areas to do together. Actions to ensure against the risk of supply disruption provide an obvious area for cooperation. If tanker capacity in mothballs makes sense for consumers separately, a common reserve fleet would probably be cheaper. A scheme for emergency sharing of supplies is bound to be difficult to agree upon, given the differing degrees of import dependence among the consuming regions, but the effort is worth making. Any credible indications that consumers do have plans for coping with supply cutbacks would be bound to be deterrents to arbitrary actions by producers.

Although joint research or coordination of research programs may promise more than is likely actually to be delivered, the field is wide and some useful cooperative actions may be possible. Joint investments in both conventional and unconventional, and high-cost, sources of hydrocarbons, such as oil shale or tar sands, would be anything but simple to organize, but then the possibilities have not even been examined in any serious way. International cooperation in the nuclear field not only may contribute to energy supply; given the outlook for the spread of nuclear materials, it may be a condition for the survival of a habitable planet.

It is not the intention here to offer a prediction as to whether the energy problem will further divide or more closely unite Western Europe and the United States. Obviously, political and economic forces are pulling in both directions. It is a fact, nevertheless, that few imaginable happenings short of military aggression could more gravely threaten normal economic and political life in the Atlantic countries than a prolonged disruption of energy supplies. If they find themselves unable to reach even a minimum of common understanding and policy on energy issues, then the prospects for Atlantic relations must be viewed as being poor.

NOTES FOR CHAPTER 8

1. Much of the data and analysis in this chapter has been drawn from a background study, *Western European Energy Policy*, prepared in 1973 for The Brookings Institution by the European Community Institute for University Studies in Brussels.

2. *Oil: The Present Situation and Future Prospects* (Paris, 1973), p. 42.

3. *Petroleum Press Service*, August 1973, p. 302.

4. *Petroleum Press Service*, June 1973, pp. 204-205.

5. *The Oil and Gas Journal*, November 26, 1973, p. 32.

6. Peter Odell, "Oil and Western European Security," *Brassey's Annual* (New York: Praeger Publishers, 1972). *Petroleum Press Service*, June 1973, p. 205, reported a similar estimate by the senior chairman of DeGolyer & MacNaughton, an oil consulting firm. And Cazenove & Co. in *The North Sea: The Search for Oil and Gas and the Implications for Investment* (London, 1972),

p. 95, offers two "hypotheses," the higher of which is the same 6 million barrels a day, with the comment that "this assumes that the continuing exploration in the North Sea will prove a succession of new fields, as is usually the case with major oil producing provinces."

7. *The Economist,* March 2, 1974, p. 16.

8. *Petroleum Press Service,* October 1973, pp. 367-69.

9. *Wall Street Journal,* March 12, 1974, p. 7.

10. *The Economist,* March 16, 1974, p. 92, and *The Oil and Gas Journal,* July 15, 1974, p. 25.

11. *Petroleum Intelligence Weekly,* December 2, 1973, p. 2.

12. *The Economist,* March 30, 1974, p. 55.

13. *Petroleum Press Service,* October 1973, pp. 367-69

14. *Petroleum Intelligence Weekly,* December 3, 1973, p. 2.

15. *Washington Post,* April 12, 1974, p. D-9.

Chapter Nine

Japan

Japan

Only the United States, the Soviet Union, and possibly China consume more energy than Japan, and Japan's use of energy, like its gross national product (GNP), has been growing very rapidly. No major industrialized nation is more dependent on imported energy.

JAPAN'S ENERGY ECONOMY[1]

During the 1960s, Japan's energy consumption more than tripled, and in 1972 it reached the equivalent of 1.7 billion barrels of petroleum. About 70 percent of this enormous quantity of energy was generated by oil; coal, which provided one-half of Japan's energy in the mid-1950s, accounted for only one-fifth of the total in the early 1970s. (See Table 9-1.) Japan's reasons for substituting oil for coal were similar to those of other industrialized countries making the same move: oil was relatively cheap, convenient, and readily available.

Table 9-1. Changes in Primary Energy Supply in Japan (percentages)

	1955	1960	1965	1970
Coal	49.2	41.5	27.3	20.7
Petroleum	20.2	37.7	58.4	70.8
Electric power (hydro & nuclear)	21.2	15.3	11.3	6.7
Others	9.4	5.5	3.0	1.8
Total	100.0	100.0	100.0	100.0

Source: Saburo Okita and Kenichi Matsui, "Japan's Energy Trade," paper presented at Conference on World Trade Policy, Maidenhead, England, April 7-10, 1973.

The shift from coal to oil caused a sharp increase in Japan's dependence on imported energy. In 1955, Japan obtained three-quarters of its energy supplies at home; in 1972 three-quarters of its energy was imported, principally in the form of crude oil and refined products.[2]

In 1972, Japan obtained about 82 percent of its imported oil from the Middle East, the same share provided by that region in 1955. The regional percentage figures, however, obscure the dramatic increases in the share of Iran and the concurrent decline in the shares of Saudi Arabia and Kuwait.* (See Table 9-2.)

Table 9-2.   Japan's Energy Imports by Principal Sources of Supply, Selected Years (thousand barrels of oil equivalent)

Supplier	1955	1960	1972
Crude oil:			
Saudi Arabia	28,073.9	34,817.5	304,121.7
Kuwait	6,890.2	87,593.5	230,675.1
Iran	2,638.8	7,403.3	607,363.8
Other Middle East	2,638.8	27,780.7	147,479.6
Indonesia	4,837.8	22,429.8	199,009.5
Sarawak	3,738.3	5,790.7	35,697.1
Other	439.8	10,628.5	44,639.7
Total	49,257.6	196,444.0	1,568,986.5
Coal:[a]			
United States	12,250.0	25,480.0	80,531.5
Australia	49.0	6,860.0	99,396.5
Canada	0	2,695.0	36,970.5
Communist countries	1,813.0	4,998.0	19,379.5
Other	294.0	1,470.0	1,347.5
Total	14,406.0	41,503.0	237,625.5
Liquefied Natural Gas:[b]			
United States	. . .	. . .	1,320,000.0
Brunei	. . .	. . .	120,000.0
Total	. . .	. . .	1,440,000.0

Source: United Nations, Department of Economic and Social Affairs, *World Energy Supplies*, Statistical Papers, Series J (New York, various years).

Note: Table does not show imports of refined products. In 1971, slightly less than one-sixth of refined products consumed in Japan were imported.

[a]Includes lignite and coke.

[b]In thousand cubic meters.

*This shift does not necessarily reflect a Japanese desire to reduce dependence on Arab oil. The attractiveness of Iran as an export market and the high sulphur content of some Saudi oil could also have caused the Japanese to favor Iran.

Most of Japan's oil imports are supplied by the major international oil companies. In the Japanese fiscal year ending March 31, 1971, the majors provided 60.9 percent of crude oil imports. Japanese firms supplied only 8.5 percent, down from 9.8 percent in the previous year. The balance was principally accounted for by independent U.S. companies. Slightly over 1 percent came from the Soviet Union.[3]

Japan is also heavily dependent on outside sources for its other energy supplies. Most of its domestic coal is unsuitable for use by the steel industry. Coking coal is therefore imported, mostly from the United States, Australia, Canada, and the Soviet Union.

Japan imports all of its uranium ore; Canada provides about half of total requirements, and South Africa and France (selling ore from Niger) are the other major suppliers. (See Appendix Table A-11.) Japan is at present totally dependent on the U.S. Atomic Energy Commission for uranium enrichment services.

Imports of liquefied natural gas (LNG) from Alaska began in 1969. Brunei was added as a source in 1972, and other arrangements to import this fuel are being made. (See Table 9-3.)

Table 9-3.   Plans for Importing LNG into Japan

| Status/Source | Annual Quantities | | Time Imports Start |
	1,000 tons	Billion Cubic Meters	
Imports started:			
Alaska	960	1.4	1969
Brunei	5,150	7.2	1972
Contract concluded:			
Abu Dhabi	3,000	4.2	1976
Under negotiation:			
North Sumatra	7,000	9.8	1977
Iran	2,000	2.8	1980
Sarawak	6,000	8.4	1980
Australia	5,000	7.0	?
Soviet Union			
(Yakutsk)	7,000	10.0	?

Source: "Japan's Energy Problems," an unpublished study prepared for The Brookings Institution by the Japan Economic Research Center and the Institute of Energy Economics (Japan), August 1973.

FUTURE ENERGY REQUIREMENTS AND SOURCES OF SUPPLY

Growth of Total Requirements

Japanese government projections made before the October 1973 war in the Middle East foresaw declining rates of growth in

GNP, increases in the shares of GNP going into private consumption and public investment, and a decrease in the share of private capital formation. As a consequence, energy requirements were also expected to grow less rapidly and to lag somewhat behind GNP growth rates. Nevertheless, the anticipated increase in Japan's energy requirements was quite impressive. By 1985, total requirements were projected to reach roughly 5.7 billion barrels in oil equivalents. Per capita energy consumption in 1985 would then equal that of the United States in 1973.*

These estimates of future energy requirements must almost certainly be reduced as a consequence of the Arab oil embargo and supply restrictions of 1973-74 and the associated huge increase in oil prices. These developments may retard Japan's rate of economic growth and therefore its energy consumption in the near term. Over the longer run, changed consumer attitudes and government policies may reduce the amount of energy consumed per unit of national output. High prices may be expected to encourage energy conservation by individuals and industrial firms. In addition, the government can probably be counted on to encourage a shift toward less energy-intensive industrial development in order to reduce the nation's high expenditures for imported energy.**

The effect of these changes on Japan's total energy requirements in 1980 and 1985 cannot easily be estimated. For illustrative purposes, however, reducing the earlier estimate of the annual rate of growth in total requirements from 7.5 percent to 6.5 percent would seem both conservative and realistic. (See Table 13-4.)

Sources of Energy

Oil will continue to be Japan's major source of energy for many years, but its share of total energy consumption will peak in the mid-1970s and decline thereafter as the shares of nuclear power and other energy sources (principally liquefied natural gas) grow. The long decline in the importance of coal will probably continue, although perhaps at a slower rate.

---

*If, as appears probable, both the rate of growth of Japan's GNP and the GNP elasticity of demand for energy continue to decline after 1985, the slowing down in the increase in energy requirements will also continue. Necessarily speculative calculations done by the Institute of Energy Economics (Japan) and the Japan Economic Research Center project Japan's total energy requirements at the equivalent of 8.7 billion barrels of oil in the year 2000, or about 1.5 times 1973 U.S. per capita consumption.[4]

**The tendency of high energy-consuming industries to build new capacity outside Japan for environmental reasons, which was apparent before the October 1973 war, will also contribute to a slowing down in the growth of energy requirements.[5]

The projected shift in the composition of Japan's primary energy supply will probably not be fundamentally altered by reactions to the October 1973 war and related events. All of the modest reduction in total energy consumption posited above for illustrative purposes should probably be subtracted from oil consumption, but this would still leave oil by far Japan's most important source of energy in both 1980 and 1985. (See Table 9-4.)

Table 9-4. Primary Sources of Japan's Energy Consumption (percentage distribution)

	1972 *Actual*	*1980* *Estimated*	*1985* *Estimated*
Primary electric power (hydro and nuclear)	3.6	10.2	18.0
Coal	22.2	12.0	10.8
Petroleum	72.6	71.3	62.6
Gas and Other	1.6	6.5	8.6
Total	100.0	100.0	100.0

Sources: 1972 actual figures: United Nations, Department of Economic and Social Affairs, *World Energy Supplies 1969-1972*, Statistical Papers, Series J, No. 17 (New York, forthcoming). 1980 and 1985 estimates are based on Table 13-4.

In contrast to what may happen in the United States, large absolute increases in pre-October 1973 projections of consumption of coal and gas are not likely. Japan's coal and gas resources are very limited, and increased imports of these fuels would require long lead times and probably also large overseas investments. Any marked acceleration in nuclear power plans also appears unlikely because of both long lead times and environmental problems.

The shift in Japan's sources of energy will be seen most dramatically in the generation of electricity. As Table 9-5 suggests, nuclear power plants could be supplying over 40 percent of Japan's electricity in 1985, and plants burning gas another 13 percent. These percentages would be even greater, if—as was suggested above—higher prices cause a slowdown in the growth of petroleum consumption for power generation as well as other purposes. The estimate for nuclear power in Table 9-5 assumes that about 60,000 megawatts of nuclear-generating capacity will be in place by 1985. This projection, however, could prove to be somewhat optimistic in view of current problems in locating sites for nuclear plants.

Energy Imports

Japan's dependence on imported energy will almost certainly increase. Japan has almost no oil or uranium and very little

Table 9-5.   Forecast of Electricity Generation by Japanese Power
Companies (10^9 kilowatt-hours)

	1970	*1975*	*1980*	*1985*
Demand	315 (100.0)	477 (100.0)	689 (100.0)	903 (100.0)
Hydro	74 (23.5)	87 (18.2)	95 (13.8)	126 (14.0)
Thermal	236 (74.9)	328 (68.8)	384 (55.7)	383 (42.4)
Coal	46 (14.6)	21 (4.4)	21 (3.1)	21 (2.3)
Petroleum	186 (59.1)	282 (59.1)	291 (42.2)	242 (26.8)
LNG	4 (1.3)	25 (5.2)	73 (10.6)	121 (13.4)
Nuclear	5 (1.6)	62 (13.0)	210 (30.5)	394 (43.6)

Source. "Japan's Energy Problems," an unpublished study prepared for The
Brookings Institution by the Japan Economic Research Center and the Institute
of Energy Economics (Japan), August 1973.
Note: Figures in parentheses show percentage of total.

gas, and its domestic coal output is expected to continue to decline.
Imports of these fuels must therefore rise sharply over the next
decade. Only large discoveries of oil and gas in nearby offshore areas
could check the rise in import dependence to roughly 95 percent by
the early 1980s.

In 1972, Japan imported 4.8 million barrels of oil per day.
For 1973, imports have been estimated at 5.4 million barrels per
day.[6] Before the October 1973 war, oil imports were projected to
rise to nearly 9 million barrels per day in 1980, and to over 11
million barrels per day in 1985. If it is arbitrarily assumed that,
because of higher prices and government policies, the growth in
energy consumption will lag one percentage point behind previous
projections, and if all of this reduction is applied to oil imports,
Japan's oil-import requirements would be reduced to less than 8
million barrels per day in 1980 and to less than 9 million barrels per
day in 1985.

Even before the October 1973 war, Japan planned a rapid
growth in LNG imports as shown below:

1972 (est.)	6.1 million tons
1980	18.0
1985	30.0

Larger imports are possible as a consequence of the war and
associated events. A large part of the imported LNG will be used in
thermal power plants. Some, however, will be fed into big city
gas-supply systems for household use. Table 9-3 summarized the
planned sources of LNG.

Japanese requirements for nuclear fuel will also steadily
increase. Future requirements for uranium ore were projected before

October 1973 as follows:

1980	10,000 short tons
1985	13,000
1990	16,000

The associated requirements for enriched uranium were estimated to be:

1980	5,000 separative work units
1985	8,000
1990	11,000

All of the uranium ore will presumably be purchased abroad, and most of the enrichment services will probably be performed outside Japan. (Japanese policy on uranium imports is discussed below.)

JAPAN'S ENERGY POLICY

Impact of the Arab Embargo and Supply Restrictions of 1973-74

Even before the October 1973 war in the Middle East, Japan was acutely aware of its high degree of dependence upon imported energy, especially oil. The Japanese feared that war or political instability in the Middle East would disrupt their major source of oil, or that some less well-defined contingency would cut their long, vulnerable oil supply line running from the Persian Gulf, across the Indian Ocean, through the narrow Strait of Malacca, and up the coast of East Asia. The Japanese also feared that, even if Japan was not directly involved in some future crisis affecting the production and marketing of oil, the major oil companies would give Japan a lower priority than the United States and the United Kingdom in allocating the reduced supplies of crude oil.

Japanese energy policy before October 1973 sought to deal with these concerns through several means:

— Reducing dependence on oil principally through the increased use of nuclear power.
— Diversifying sources of oil through exploration activities and supply arrangements in many parts of the world.
— Bringing a larger percentage of Japan's oil and uranium supplies under the control of Japanese firms.
— Relying on diplomacy to fend off potential threats to Japan's sources of oil and oil supply line. In the Middle East, Japan pursued a policy of low visibility and neutrality in the Arab-Israeli dispute.

When the Arab oil-exporting countries announced their selective embargo and production restrictions on October 17, 1973,

Japan's worst fears appeared about to be realized. Japan, it was true, was not on the Arabs' list of embargoed countries, but neither was it on the "friendly" list. The Japanese saw themselves competing unsuccessfully with other nations in the neutral category for a share of the reduced total supply of oil. When the initial production cutback of 10 percent was increased to 25 percent in early November, Japan's concerns were heightened.

In retrospect, the Japanese government's handling of the crisis was remarkably cool and deliberate. Perhaps because Japan's oil stocks were sufficient for nearly two months of normal consumption, the government initially called for only modest reductions in energy use by industry and the general public. Legislation giving the government emergency powers to impose compulsory controls on energy consumption was not enacted until more than two months after the onset of the crisis.[7]

The Japanese government concentrated its efforts on getting the Arabs to declare Japan a friendly nation. For a time, it appeared that the Arab price for this action might include steps, such as breaking diplomatic relations with Israel, that would endanger Japan's ties with the United States.[8] Japan therefore attempted to steer a middle course.

On November 22, the Japanese government issued a formal statement that called on Israel to withdraw from all Arab territories occupied in the 1967 war and that emphasized Japan's support for "the rights of the Palestinian people for self-determination." The statement also somewhat obliquely supported the legitimacy of the state of Israel by expressing "respect for the integrity and security of the territory of all countries in the area." The statement ended, however, with the ominous declaration that "the Government of Japan will continue to observe the situation in the Middle East with grave concern and, depending on future developments, may have to reconsider its policy towards Israel."[9]

The U.S. government reacted to the Japanese statement with only a mild expression of "regret."[10] The statement—and intensified Japanese diplomatic activity in Arab countries—soon began to yield the desired results from the Arabs. On November 28, Japan was among the group of nations exempted from a further 5 percent cut in oil shipments from Arab countries, and on December 25, the Arabs declared Japan a friendly nation.

Japanese satisfaction over this diplomatic success was submerged, however, by a new concern. Two days earlier, the oil-exporting countries of the Persian Gulf had unilaterally raised the posted price of crude oil to a level that required the oil companies roughly to double the market price, f.o.b. the gulf. Japan had been relieved of the threat of a serious physical shortage of oil, but it now faced the unexpected problem of meeting a sharply increased oil-import bill at a time when it was already running a deficit in its basic balance of payments.[11]

Although undoubtedly a heavy blow to the Japanese, the price increase occurred at a time when slowed economic growth was holding other imports in check, but the total value of exports was still rapidly rising. The main short-term effect of the increased oil-import bill was to force Japan to choose between reducing its long-term capital outflow (including aid and investment in the less developed countries) and financing the outflow by drawing on foreign-exchange reserves. Beyond the immediate future, the impact of higher oil prices on Japan will depend on broad international considerations (see Chapters 13 and 14).

The question to be asked here is what effect will the events of the fall and winter of 1973-74 have on Japan's energy policy. From the perspective of the summer of 1974, it appears that the October 1973 war, the Arab embargo and supply restrictions, and the sudden upward leap in oil prices will strongly reinforce the already existing Japanese desire to diversify sources of energy in order to avoid excessive dependence on any single source of supply. Even though the major oil companies appear to have treated Japan fairly during the period of reduced Arab production, the Japanese may also pursue with increased resolve their goal of placing a larger share of their total oil imports in the hands of Japanese-controlled firms. This is so because, somewhat paradoxically, the recent good behavior of the foreign companies underlined how serious the crisis would have been if the companies had discriminated against Japan.

Two new elements may be introduced into Japanese energy policy. Arab use of the oil weapon quickly taught the Japanese that neutrality and a low posture are not enough. Japan can therefore be expected to pursue a more active diplomacy in the Middle East and other oil-exporting areas and to use trade and investment agreements to support its foreign policy goals. This will not be an easy course, however, since Japan has had little experience in dealing with the complexities of Middle Eastern politics.

Conservation will also probably become a lasting part of Japanese energy policy, although how vigorously the government will pursue energy-saving measures is not yet clear. Government encouragement of a shift to less energy-intensive industries can be expected to continue.

Japan's Area of Choice in Energy Policy

Whatever the consequences of the events of late 1973 and early 1974 on Japanese thinking and behavior with respect to energy, the nation's area of choice remains much what it was before the October 1973 war.[12] In attempting to assess what Japan may do in the field of energy in future years, it is instructive to examine in turn Japan's alternatives to oil, possible new sources of oil, and prospects for increasing its control over its energy supplies.

Prospects for materially changing the energy-supply pattern projected for 1980 and 1985 in Table 9-4 are not good. Most of Japan's hydroelectric potential has been exploited; the serious obstacles to speeding up the nuclear power program have already been noted; geothermal energy is still an unknown quantity; and solar power and controlled nuclear fusion are for the more distant future.

Larger imports of LNG are possible, but the increase in imports projected above is already substantial and only enormous imports could make much difference in Japan's energy supply pattern. This leaves coal as the remaining possibility.

Apart from coal imported for metallurgical purposes, Japan's demand for coal has declined steadily since reaching a peak of 43 million metric tons in 1961. The domestic coal industry is plagued with production problems and labor shortages and is heavily subsidized by the government. Its sole remaining market survives only as a result of a government request to the electric power industry that it accept coal deliveries in return for compensatory measures to make up for losses incurred.[13]

Even though the future of domestic coal is not promising, imported boiler coal could in theory reduce Japan's dependence on Middle Eastern oil. Whether a policy of relying more on imported coal for electric-power generation would make economic sense, however, depends on future prices of coal and oil. Before the large increases in oil prices in late 1973 and early 1974, imported coal was considerably more costly than oil in terms of caloric content. Even if imported coal is (or becomes) competitive with oil as a result of high oil prices, several factors work against a rapid, large-scale substitution of coal for oil: coal's adverse impact on the environment, the relatively greater cost of constructing coal-burning power plants, and the heavy investment that may be required to develop large new sources of coal.

Despite these problems, it is entirely possible that, in the interest of diversifying energy sources, Japan will import increasing quantities of boiler coal in the late 1970s and 1980s. Possible sources of such imports include Australia, China, the USSR, the United States, and Canada.

Since alternatives to oil provide only limited possibilities, reducing Japan's dependence on the Middle East for oil is largely a matter of finding new sources of oil in other parts of the world. Japanese companies, usually in partnership with one or more foreign companies have shown increased interest in exploring for oil in the East China Sea and the waters closer to Japan itself.[14] Drilling is under way off several parts of Japan's long coast, but exploration in the East China Sea has been retarded by conflicting claims to the continental shelf (see Chapter 12).

The Japanese government has facilitated the exploration of offshore areas by granting drilling concessions, conducting gravimetric and seismic surveys, and extending financial support through the Japan Petroleum Development Corporation. Expectations are, however, modest. Experts of the Ministry of International Trade and Industry (MITI) estimate ultimately recoverable offshore reserves at 850 million tons (6.2 billion barrels).[15] Production may not reach 1 million barrels per day before 1980.[16] Japanese oil companies also have entered into joint ventures for oil in more distant areas, including Indonesia, Nigeria, and Peru. Results thus far have been disappointing.

Japan has shown a strong interest in importing energy materials from China and the Soviet Union. The Japanese appear to believe that the advantage of reducing dependence on the Middle East outweighs any risks involved in dependence on Communist countries.

Japan began small-scale imports of oil from China in 1973, importing a total of 1 million tons (7.3 million barrels) in that year. More substantial exports of oil from China to Japan are possible in the late 1970s and 1980s. (See Chapter 12.)

Quite large exports of Soviet oil, gas, and coking coal to Japan could begin in a few years, but far from trivial political and commercial obstacles must first be overcome. Three major projects— Tyumen, Yakutsk, and Sakhalin—are under discussion between Japan and the Soviet Union. (Participation by U.S. companies in all three is possible.)

The Tyumen oil project has been under discussion since at least 1966. Under this project, as originally formulated, Japan would extend a 20-year credit in the amount of U.S. $1 billion (subsequently raised to $1.7 billion) to finance a pipeline from the Tyumen oil fields in West Siberia to Nakhodka on the Pacific coast opposite Japan. The loan would be repaid in oil. The Soviets at one time contemplated gradually increasing annual shipments to 40 million tons (293 million barrels) by the early 1980s, but in 1973, the estimate was lowered to 25 million tons (183 million barrels or about 500,000 barrels per day).[17] In March 1974, the Soviets introduced a new complication into the negotiations by proposing that Japan extend a credit of U.S. $3.3 billion to help finance a new railroad that would move Tyumen oil from a point northwest of Lake Baikal to Sovetskaya Gavan on the Pacific opposite Sakhalin.[18]

Earlier Japanese linking of the Tyumen project and other major commercial deals with settlement of their claim to several islands northwest of Hokkaido, which the Soviets have occupied since the end of World War II, has apparently been dropped. The major political obstacle to the Tyumen project is China's opposition. In an interview with visiting Japanese journalists on March 11, 1973, a senior Chinese official objected strongly to the planned pipeline on the ground that it would supply Soviet forces that might invade China.[19] It is also suspected that the Chinese are disturbed by the fact that building the pipeline involves constructing a new road parallel to the Sino-Soviet border. If so, they would presumably object even more strongly to the construction of a new east-west rail line. In any event, it can be assumed that the Chinese disapprove of any development that improves Soviet-Japanese relations or strengthens the economy of Siberia, and the Tyumen project would do both.

The Chinese could make a major issue of their objections to this project, but they will probably decide that to do so would damage their relations with Japan without any certainty of achieving the desired result. They are more likely to be stimulated to try to increase their own exports to Japan. China's decision to begin exporting oil to Japan could have been in part a reaction to the possibility of a Japanese-Soviet oil deal.

The Yakutsk gas and coking coal project is really two separate projects involving the development of resources in the same part of the Soviet Far East. The Soviets are requesting a Japanese credit of $300-$350 million to develop the Yakutsk coal fields. Japan would be repaid in coking coal at the annual rate of 5 to 6 million tons over a 10-year period beginning in 1979 or 1980. In March 1974, a Japanese business group and the Soviet coal export agency agreed on the rough outlines of this project, but details remained to be negotiated.[20]

The Yakutsk gas project is similar to the Tyumen oil project but much larger and for that reason less likely to go through soon. The required investment for this project could be as high as $4 billion, with Japan and private U.S. interests contributing equal amounts. The credit would be used to build a 1500-mile pipeline from the Yakutsk gas field to Nakhodka, a liquefaction plant at that point, and LNG tankers to carry gas to Japan and the West Coast of the United States. Repayment would be in gas over a 20-year period at the annual rate of 14 billion cubic meters (10 million tons of LNG).

If the Tyumen oil project materializes, the Yakutsk gas project probably would be too big for Japan alone. The Yakutsk gas reserves, moreover, are not confirmed. In April 1974, Japanese negotiators agreed to extend a credit of U.S. $100 million to finance exploration of the Yakutsk gas fields, subject to the United States' providing a like amount.[21]

The Sakhalin oil and gas projects have gone through several changes. The Soviets at one time proposed a gas pipeline from Sakhalin to Hokkaido, but Sakhalin's gas reserves turned out to be disappointing and their development is now apparently linked in Soviet thinking with the development of the larger Yakutsk deposits.

The Soviets originally proposed onshore drilling for oil; they now talk about offshore drilling, which would require the participation of a U.S. company, for technical reasons. A Gulf Oil Corporation geologist went to Sakhalin in the summer of 1972 and reportedly found the locale promising. The Soviets have proposed a Japanese credit of $230 million to finance a thorough exploration of offshore resources. The Soviets and Japanese are, however, far apart, and prospects for agreement are uncertain. The Soviets want exploration to be conducted on a straight contractual basis, which would mean that Japan would get only first priority on purchasing any oil that may be found. The Japanese want a fixed percentage of the oil.

Japan's efforts to diversify its sources of oil also frequently serve another policy objective: increasing national control over oil supply. Japan's publicly proclaimed goal is to obtain 30 percent of its oil supply from Japanese firms or firms in which there is a major Japanese interest.* Despite vigorous efforts, however, Japan remains dependent on foreign oil companies for 90 percent of its oil.

The chosen instrument to achieve increased national control (and diversified sources of supply) is the Japan Petroleum Development Corporation (JPDC), a wholly owned government corporation subject to the policy guidance of the Minister of International Trade and Industry. The law establishing JPDC in 1967 authorized it to invest or loan money in overseas exploration for oil and gas, to guarantee private investment in oil and gas exploration and production, to give technical advice, and to conduct geological investigations. The JPDC both responds to proposals by private Japanese companies and searches out opportunities on its own initiative. The JPDC is currently concentrating on Southeast Asia and

*This goal was apparently first enunciated in 1967 in a report to the Minister of International Trade and Industry by the Advisory Committee for Energy and was confirmed in a MITI white paper on energy issued in September 1973.[22]

the Persian Gulf, but it is also involved in projects in Alaska, Canada, Central America, Africa, and Australia.

An increased role for JPDC appears likely as part of an effort to strengthen the Japanese oil industry. Criticism of the present fragmented condition of the industry is widespread in informed Japanese quarters, and sentiment exists for creating a large, vertically integrated oil company. One possibility under informal discussion is to convert JPDC into such a company. Another is to entrust the formation of a Japanese "major" to one of the giant conglomerates, such as Mitsui or Mitsubishi, both of which already have affiliates in the oil business. Whatever course is adopted, the Japanese can be expected to move cautiously. They are acutely aware of their dependence on the international oil companies and will try to avoid antagonizing them.

Japan's policy with respect to uranium bears some similarity to its policy on oil. Little need is felt to diversify sources of uranium ore, but, as in the case of oil, Japan has set the goal of bringing 30 percent of its uranium ore supply under the control of firms in which there is a major Japanese interest.

Under the guidance of a "project team" representing both government and private industry, and with some governmental financial assistance, Japanese firms are prospecting for uranium in Canada, Australia, and Africa. Although prospecting results thus far have been meager, Japan should have no difficulty in meeting its future requirements for uranium ore by continuing to make long-term contracts with its current suppliers.

More concern is felt over future supplies of enriched uranium and over Japan's total dependence on the U.S. Atomic Energy Commission for enrichment services. The realization that by 1983 the AEC's present facilities will be inadequate to meet the rising demands of present customers throughout the non-Communist world has increased Japanese interest in alternative sources of supply.

A Japanese study group is examining the relative advantages of U.S. and French gaseous-diffusion enrichment technology. Japan has also received a special invitation to join in financing a plant that would use the new centrifuge technology being developed by the United Kingdom, the Netherlands, and West Germany. A strong predisposition appears to exist, however, in favor of U.S. technology. Japan would probably prefer to enter into a partnership with private U.S. interests to build an enrichment facility using U.S. gaseous diffusion or centrifuge technology.* This is so even though certain

*A joint communiqué issued by President Nixon and Prime Minister Tanaka on August 1, 1973, stated that the two leaders "agreed that the two governments should exert their best efforts for the satisfactory realization of a joint venture" in the uranium enrichment field.[23] In mid-1974, the feasibility of a joint enrichment project was under study by a private U.S.-Japanese group.

security restrictions would have to be accepted and the facility would probably have to be located in the United States.*

Eventually, Japan hopes to create some domestic enrichment capacity based on its own technology. Research is under way on both the centrifuge and the gaseous-diffusion methods, but much greater resources are being devoted to the former.

Purchasing enriched uranium from the Soviet Union is another possibility that has aroused interest in Japan. A semi-private mission of Japanese nuclear experts visited the Soviet Union in June 1973 and was told that the Soviets would sell enriched uranium to Japan if a governmental agreement on atomic energy could be worked out.[24] As of mid-1974, however, no concrete actions appear to have been taken in that direction.

Oil and International Cooperation

Japan faces something of a dilemma in trying to decide how to structure its international relations in dealing with oil problems. On the one hand, it fears becoming the "odd man out" in the club of largely white industrial nations on this or any other major issue; it specifically fears a no-holds-barred competition for oil supplies with the United States, in which the United States might take advantage of Japan's dependence on the big U.S. oil companies. On the other hand, Japan sees danger in too-close association with the other major oil-importing countries and feels that it gains special political and economic advantages in dealing independently with the oil-exporting countries.

A Japanese policy appears to be emerging that carefully balances these conflicting considerations; the policy has three major components: limited cooperation with other major oil-importing countries; increased economic ties with the oil-exporting countries (and continued disassociation from U.S. support for Israel); and advocacy of an all-embracing organization of oil importers and exporters.

Masao Sakisaka, president of the Institute of Energy Economics, spelled out the limitations on cooperation among importers in a paper written in 1973.[25] "First of all," he wrote, "it must be made clear that the importing nations in holding talks at the OECD, do not take a stand similar to a sort of consuming nation league aiming to suppress the demands of the OPEC." Rather, he continued, the aim of such discussions should be to avoid an "oil war" among importers, establish a system of "mutual accommoda-

*Building the facility in Japan would create security problems in view of the weakness of Japan's laws for the protection of classified information. The high cost of electricity in Japan is an additional compelling reason for locating the plant elsewhere.

tion" in time of temporary shortage, provide economic aid to oil-producing countries, and study means of developing new sources of energy, new energy technology, and capital.

Japan's increased direct dealings with the non-Communist oil-exporting countries are of course part of the previously discussed policy of seeking greater national control over oil supplies. As the position of the international oil companies declines and the oil-exporting countries gain a greater say over the disposition of their oil, Japan will encounter, and act on, more opportunities to purchase oil directly and to enter into joint ventures with the national companies of the oil-exporting countries. Thus far, however, the Japanese have been slow to enter into barter agreements. They have sought to gain political and economic advantages, including more assured access to oil, by offering to finance large industrial and other projects in oil-exporting countries. But they have usually not linked these offers to the supply of specific quantities of oil.*

The Japanese concept of cooperation among both importers and exporters of oil was apparently first fully articulated in December 1972 in an article by Director General Toyama of MITI's Mine and Coal Bureau.[26] Mr. Toyama called for an international conference of oil importers and exporters that would create (and apparently maintain continuing surveillance over) an international oil organization. This organization would include within itself a world oil development fund and a world energy resources technical research institute. One of the functions of the international oil organization in Toyama's proposal would be to study "a formula for the international adjustment of oil demand and supply plans," which suggests an effort to arrive at an international commodity agreement for oil. Other functions include promotion of exploration and development, cooperation in emergencies, development of other fuels, and joint environmental activities.

The strong Japanese desire to avoid being locked into a grouping of oil-importing countries in potential confrontation with the oil-exporting countries was made clear at the Washington Energy Conference in February 1974. In his formal statement to the conference, Japanese Foreign Minister Ohira declared in part:

My delegation is participating in this Conference of major oil-consuming countries, anticipating that it will be the first step in building a

---

*This policy may be no more than commercial prudence, given the current high price of oil. It may also reflect a desire to avoid the displeasure of the U.S. government, which has taken a strong stand against special barter deals for oil.

harmonious relationship between the oil-producing and consuming countries. . . . Japan feels it is of primary importance to realize, as early as possible, a constructive dialogue with the oil-producing countries. . . .[27]

Reducing Vulnerability to Supply Interruptions

Japan is not now able to protect its long oil supply line from the Persian Gulf. The cost of creating a navy capable of performing this task would be prohibitive and would arouse widespread fears of resurgent Japanese militarism. Japan therefore relies on its diplomacy, rather than its limited military power, to prevent any serious threat to its oil supply line from arising.*

The Japanese have recognized for some time that an oil stockpile is a necessary form of insurance against supply interruptions. Current stocks of crude oil and refined products are sufficient for over two months' normal consumption. Japan also favors creating an emergency allocation system covering all members of OECD. Efforts before the October 1973 war to bring this about were, however, frustrated by disagreement between the United States and Japan (backed by many European countries) over whether the emergency plan should provide for the sharing of domestically produced, as well as imported, oil. The United States modified its position at the Washington Energy Conference in February 1974,** and both the United States and Japan joined the other conference participants (except France) in a communiqué that called for "a system of allocating oil supplies in times of emergency and severe shortages."[28]

CONCLUSIONS

There is little that Japan can do to check the rise in its dependence on imported energy to about 95 percent of total energy consumption by the early 1980s. Development of offshore oil and gas resources in Northeast Asia (assuming that they are located in areas over which

---

*A possible nonmilitary threat is posed by the claim of Indonesia and Malaysia that the Malacca Strait is not an international waterway. Enforcement of this claim is not likely, however, to take the form of closing the strait, but only that of applying safety regulations and possibly tolls to ships passing through it. This problem and the physical limitations of the Malacca Strait explain Japanese interest in constructing a canal across the Kra Isthmus in Thailand.[29]

**In his opening remarks at the conference, Secretary of State Kissinger stated: The United States declares its willingness to share available energy in times of emergency or prolonged shortages. We would be prepared to allocate an agreed portion of our total petroleum supply provided other consuming countries with indigenous production do likewise.[30]

Japan will be able successfully to assert control) will at best make only a difference of a few percentage points.

Faced with this situation—and with the experience of the 1973-74 Arab oil embargo and supply restrictions very much in mind—Japan can be expected to intensify its efforts to diversify its sources of imported energy. These efforts may, among other possible achievements, result in fairly large imports of oil and gas from the Soviet Union, and possibly also from China, in the late 1970s and early 1980s. But these imports and possible increases in imports from Africa and Latin America will not eliminate Japan's overwhelming dependence on the Middle East for oil. Japan will remain heavily dependent on the United States for uranium enrichment services, but will probably diversify somewhat its sources of enriched uranium by entering into contracts with West European suppliers and possibly also with the Soviet Union.

Japanese policymakers will increasingly be torn between the danger of cooperating with the West and being caught again in an anti-U.S. oil embargo, and the danger of being isolated in a future scramble for scarce energy supplies. As a consequence, Japan's energy policy may for a time contain seemingly inconsistent elements. On the one hand, Japan will seize the opportunities to make special oil supply arrangements that are created by the displacement of the international oil companies by the national companies of the oil-exporting countries. On the other hand, Japan will support some forms of cooperation on oil policy among oil-importing nations and will actively promote cooperation with the United States in the field of nuclear energy.

A partial solution to Japan's energy policy dilemma might be provided by the creation of an all-embracing international organization within which Japan could, with reduced risk, pursue a flexible policy combining just the right mixture of independence and cooperative action. The tentative Japanese proposal for an international oil organization mentioned above may therefore become a prominent part of Japanese energy policy whenever the international environment is regarded as sufficiently favorable.

NOTES FOR CHAPTER 9

1. This section relies heavily on a paper on "Japan's Energy Trade" by Saburo Okita and Kenichi Matsui, which was presented at the Conference on World Trade Policy, Maidenhead, England, April 7-10, 1973.

2. United Nations, Department of Economic and Social Affairs, data to be published in *World Energy Supplies, 1969-1972*, Statistical Papers, Series J, No. 17 (forthcoming).

3. *Petroleum Intelligence Weekly,* February 26, 1973, p. 1.

4. This section on future energy requirements and sources of supply is based in part on an unpublished study of "Japan's Energy Problems" prepared for The Brookings Institution by the Japan Economic Research Center and the Institute of Energy Economics (Japan) in August 1973. Unless otherwise indicated, data presented in this section are drawn from that study.

5. See *The Economist,* July 28, 1973, p. 76, for a discussion of the plans of the Japanese steel industry to build new plants abroad.

6. *The Impact of the October Middle East War,* Hearings before the House of Representatives, Subcommittee on the Near East and South Asia of the Committee on Foreign Affairs, 93rd Cong., 1st sess., October and November 1973, p. 159.

7. *Japan Times,* December 22, 1973, p. 1.

8. *New York Times,* December 11, 1973, p. 34.

9. *Japan Times,* November 23, 1973, p. 1.

10. *Japan Times,* November 25, 1973, p. 1.

11. The discussion here of the impact of higher oil prices on Japan is based in part on "Balance of Payments: Recent Trends and Problems" (mimeograph) by Kazuo Nukazawa, assistant director, International Economic Affairs Department, Keidanren (Japan Federation of Economic Organizations), Tokyo, January 24, 1974.

12. This discussion of Japan's alternatives in energy policy is based in part on off-the-record interviews with informed persons in Tokyo in February 1973. Other general sources used include: *The Interim Report by the Petroleum Subcommittee of the Advisory Committee for Energy* (Tokyo: Ministry of International Trade and Industry, 1971); *On Basic Problems of Energy, Centering on Petroleum* (Tokyo: Federation of Economic Organizations [Keidanren], 1973); Masao Sakisaka, *"Energy Crisis" and What Japan Should Do About It,* Institute of Energy Economics (Tokyo, 1973); and the previously cited paper, "Japan's Energy Problems," Japan Economic Research Center and the Institute of Energy Economics (Tokyo, 1973).

13. Japan Economic Research Center, *"Japan's Energy Problems,"* p. 26; and "The Energy Policy of Japan," *OECD Economic Outlook,* no. 48 (October 1970), p. 18.

14. See especially, "Japan Looks Offshore," *Petroleum Press Service,* December 1970, p. 445.

15. *Ibid.*

16. Japan Economic Research Center, *"Japan's Energy Problems,"* p. 30.

17. *Washington Post,* October 18, 1973, p. A-34.

18. *Asahi Evening News* (Tokyo), March 28, 1974, p. 1.

19. *Yomiuri* (Tokyo), March 12, 1973.

20. *New York Times,* March 18, 1974, p. 45.

21. *Washington Post,* April 27, 1974, p. A-20.

22. Ministry of International Trade and Industry, *Interim Report,* p. 24.

23. *Washington Post,* August 2, 1973, p. A-1.

24. *Washington Post,* June 22, 1973, p. 31.

25. Sakisaka, *"Energy Crisis" and What Japan Should Do About It.*

26. "International Organ for Energy Resources Needed," *Nihon Keizai* (Tokyo), December 22, 1972.

27. U.S. Department of State, Washington Energy Conference, Document 9, February 11, 1974 (unofficial translation), p. 8.

28. Ibid., Document 17 (Rev. 2), February 13, 1974, p. 4.

29. *Nihon Keizai* (Tokyo), July 13, 1973.

30. U.S. Department of State, Washington Energy Conference, Document 6, February 11, 1974, p. 7.

Chapter Ten

The Less Developed Countries

by Robert M. Dunn, Jr.,

Energy is of critical importance to the economic growth of the less developed countries (LDCs), and the vast majority of these countries depend at least in part on imported energy, principally oil. The huge increase in oil prices in 1973-74 created especially severe problems for those LDCs that, even before the dramatic price rise, were having difficulty earning the foreign exchange needed to pay for their essential energy imports.

Energy consumption by the LDCs is increasing rapidly. In 1960, this group of nations accounted for only about one-tenth of the non-Communist world's total energy use; by 1973 their share of the total approached one-eighth. In providing for their rising energy requirements, the LDCs face the same problem as do the advanced nations in determining the proper balance between fossil fuels and nuclear power, but because the LDCs can less afford to make mistakes, they must make that decision with particular care.

PATTERNS OF ENERGY CONSUMPTION AND IMPORT REQUIREMENTS

Historically, energy consumption in the LDCs has closely followed changes in gross national product (GNP).[1] Studies of the relationship between economic growth and energy demand in the LDCs suggest that a 10 percent increase in GNP will produce an increase in energy demand of 13 to 16 percent.[2] In the developed countries, a 10 percent increase in GNP would typically be accompanied by only about a 9 percent rise in energy requirements. This decided difference between the developed and less developed countries

163

reflects the fact that the early stages of economic development usually involve a sharp shift away from agriculture, which uses relatively little energy per dollar of output, to manufacturing, which is far more energy intensive. A fully developed economy has typically completed this shift and is often moving toward a larger role for the services sector of the economy, which uses relatively less energy.*

Consumption patterns among various sources of energy changed significantly in the 1960s. Between 1960 and 1968, the use of oil in the LDCs increased from 29 percent of total energy consumption to 43 percent, while coal consumption declined from 56 percent of total energy to 34 percent.[3] Some interesting regional differences exist in that the share of oil in Latin America actually declined slightly (72 percent to 69 percent) due to a sharp increase in the use of natural gas (8 percent to 18 percent). In the less developed non-Communist parts of both Asia and Africa, however, the share of oil increased significantly. As for other energy supplies, the share of coal declined everywhere, the share of natural gas increased everywhere, and the combined shares of nuclear and hydroelectric power declined slightly in Latin America and increased slightly elsewhere. (Projections of future nuclear power production in the LDCs can be found in Table 16-1, Chapter 16.)

Although a few LDCs are major oil producers and exporters and a few others are roughly self-sufficient in oil, the vast majority of the LDCs must import all or most of the oil they consume.** Together they imported 2.4 million barrels per day in 1970,*** considerably more than double the amount that they imported in 1960.[4] About 80 LDCs are net importers of oil, including most of those with large populations, such as India, Brazil, and Pakistan. Before the recent large increase in oil prices, the oil-importing LDCs accounted for about 80 percent of the total GNP of the LDCs,[5] so their situation is far more typical than that of the oil-exporting LDCs or the few that produce about what they need.

Imports accounted for about 67 percent of the oil consumption of the oil-importing LDCs in 1970.[6] Both consumption

---

*The LDCs, however, are typically moving away from the use of unmeasured, non-commercial energy sources, such as firewood and animals, so the statistics overstate somewhat the actual increase in energy use.

**The LDCs as a whole (not including China) are also not well endowed with coal; they possess only 10 percent of the world's proven reserves. Dependable information on natural gas reserves is not available. Considerable unused hydro capacity is known to exist in many LDCs (most notably Brazil), but some of it would be relatively expensive to develop and is far from markets.

***Not including bunkers.

and imports grew at the impressive rate of over 8 percent per year in the 1960s. As recently as 1960, refined products constituted 45 percent of LDC oil imports, but the construction of a number of refineries in the LDCs during the last decade has reduced refined products to only 5 percent of the total.

THE IMPACT OF HIGH OIL PRICES

The balance-of-payments situation of many oil-importing LDCs was being seriously threatened by increasing oil costs even before the outbreak of the 1973 Arab-Israeli war. Table 10-1 indicates the extent of the price increases in the 1970-73 period. In early 1974, the price rose to about $8 a barrel, plus transportation costs, and threatened little short of unmitigated disaster for many of the LDCs.

Table 10- 1.   Oil Prices for Importing LDCs

	1970	*1971*	*1972*	*1973 (est.)*
Cost (free on board)	$1.34	$1.74	$1.93	$2.42
Shipping Costs	$0.57	$0.57	$0.59	$0.59
Price (cost, insurance, freight)	$1.91	$2.31	$2.52	$3.01

Source: Author's estimate.

Note: Shipping costs are based on charter rates for the relatively small tankers (35,000 tons) that can use the ports of most LDCs; these costs could be somewhat lower if the so-called supertankers could be unloaded in LDC ports.

If it is assumed that both the total oil requirements and the oil imports of the LDCs grew at an annual rate of 8.5 percent between 1970 and 1974 (which would reflect a continuation of the consumption growth rate of the 1960s and an increase in the rate of growth of domestic production), the early 1974 price would imply a total oil-import bill in 1974 of almost $10 billion, six times what it was in 1970. Moreover, even if the total value (in current dollars) of the exports of the oil-importing LDCs continued to grow at the recent annual rate of 10.8 percent, total exports in 1974 would amount to only about $34 billion. Oil imports would therefore use up roughly 30 percent of the export proceeds of this group of countries.

It is important to realize, however, that the oil-importing LDCs differ greatly in their ability to adjust to higher oil prices. Some (e.g., Brazil) have rapidly growing economies and booming export sectors. Others (e.g., Malaysia) are at least temporarily in a strong financial position because the prices of their principal exports have risen rapidly. Still others (e.g., Taiwan) may be able to pass on a large part of increased oil costs to foreign purchasers of their manufactured products.

The hardest hit LDCs are those with slowly growing economies and stagnant export sectors. In this category are the populous countries of South Asia, the drought-afflicted countries of Sub-Saharan Africa, and some of the poorest countries of Latin America. The countries of South Asia were widely believed to be in serious trouble when oil was selling for under $2 a barrel, in that slow economic growth, stagnant exports, drought, and rapidly growing populations combined to suggest grim futures. The current price of oil makes their crisis immediate rather than long term.

Brazil and India probably represent two extreme cases. Brazil has had such rapidly growing exports (10 percent annually from 1960 to 1972 and almost 24 percent between 1965 and 1972) that even relatively high prices for oil should not be a major problem. In 1972, oil imports amounted to less than 10 percent of total export receipts, and this percentage might even fall by the end of the decade if the recent pace of export growth is maintained or even approached. India, on the other hand, faces what appear to be impossible problems. Export receipts grew at a rate of only 1.2 percent between 1960 and 1972. Whereas oil costs amounted to about 20 percent of exports in 1973, an early 1974 study by the International Bank for Reconstruction and Development (IBRD) predicted a 1974 oil-import bill of $1.5 billion for India, which would consume over 50 percent of expected export receipts.[7] With very limited foreign-exchange reserves, large debt-servicing requirements, and vital needs for other imports (such as capital equipment), India is in awesome trouble in 1974 to say nothing of 1980 and 1985. The only bright side to this otherwise grim picture is that India is considered to have good prospects for finding more oil and is known to have the ability to produce far more coal. Unfortunately, both will require large amounts of capital, and it is not clear where that is to come from.

Pakistan's situation is like India's in that oil imports are a major drain on export receipts, and stagnant exports indicate that this drain will worsen. Korea's prospects are more like Brazil's, in that rapid export growth should cover even the effects of recent oil price increases. Korea is, however, dependent on exports of manufactured goods to the industrialized countries, and a recession or an epidemic of protectionism in the United States or Western Europe could present major problems.

The effect of high oil prices on the LDCs may change with the passage of time. It is therefore useful to consider separately the possible impact of the increased cost of imported oil in the next few years and over the somewhat longer run.

The Short Run—1975-77

Difficulties in finding substitutes for oil quickly may prevent the LDCs from reducing the growth rate for both total oil requirements and oil-import requirements below 8.5 percent annually in the near term. If so, consumption and imports would rise in the following manner in the period 1975-77. Two sets of estimates for import costs are shown: one based on the early 1974 average delivered price of $8.60 per barrel, c.i.f. (cost, insurance, and freight), and one at $6.60 per barrel to reflect the possibility that the price of oil may fall.

	1975	*1976*	*1977*
Consumption, thousand b/d	5,400	5,900	6,400
Imports, thousand b/d	3 700	3,900	4,200
Import costs, in billions, at $8.60 bbl.	$11.6	$12.2	$13.2
Import costs, in billions, at $6.60 bbl.	$8.9	$9.4	$10.1

In 1970, the average c.i.f. price of crude oil to the LDCs was about $1.90, so three-fourths of the estimated import costs shown above are the result of price inflation rather than growing consumption. The impact of these cost increases is put into perspective when compared with the total net foreign aid disbursements of $8.6 billion to the LDCs in 1972 by the sixteen members of the Development Assistance Committee of OECD. Even in the unlikely event that the oil-importing LDCs manage to maintain the 10.8 percent rate of growth of exports (at current prices), 1977 export receipts would still be only $41.9 billion, and oil-import costs would (at early 1974 prices) use up nearly one-third of those receipts.

To add to the gloom, the 1973-74 price increases have had some side effects that will worsen the plight of many countries. Oil is a major input for fertilizers, and a combination of previously existing fertilizer shortages and the high cost of oil has produced rapid price increases for fertilizer. In some areas, it is apparently unavailable, while in others it can be had only at very high prices. Those LDCs that are in the midst of a "green revolution" have switched to strains of grain that are heavily dependent on fertilizer, and the current scarcity-price situation can be expected to create serious food shortages within a year or two. India and the Philippines are expected to be particularly hard hit.

There is also a virtual certainty of at least a modest economic downturn and perhaps a serious recession in the United States, Western Europe, and Japan, in part because of the high price of oil. This would reduce, in turn, the demand for imports of a wide range of primary products and probably reduce prices sharply; this, of course, would worsen the terms of trade for many LDCs and could make the burden of increased oil prices even heavier for them. Countries such as Malaysia (which exports rubber and tin), Chile (copper), and Ghana (cocoa) have enjoyed increases in export prices that have at least in part offset increased oil-import costs, but a recession in the industrialized countries could bring the prices for these and other primary products down again quickly. The key question is whether reduced demand would also bring down oil prices or whether the major oil exporters would be able to keep prices high by restricting production. No confident prediction is possible, but in a mild recession the price of oil might not fall to the same extent or as rapidly as would the prices of other primary commodities.

The Long Run—1980-85

Although the situation facing the oil-importing LDCs in the short run is grim, the prospects for the longer term are not quite so bad. The importance of time derives from the rather long lags between major oil price changes and the full response of both buyers and sellers to the new prices. If the price of imported oil remains at its early 1974 level for a period of time, many LDCs will be able to reduce oil imports through a combination of increased domestic oil production, the development of substitute energy sources, and conservation.

In recent years, the oil-importing LDCs have produced about one-third of their oil needs. Current prices for oil will, however, make additional exploratory efforts worthwhile, which in turn could be expected to raise this figure considerably after a lag of a few years. The success of such efforts will vary from one country to another, but the continental shelf of Asia is considered to be a particularly promising area for drilling, so that India, Bangladesh, and Pakistan might be importing much less oil by the 1980s, if the price incentive for exploration remains strong and if government policies encourage investors to provide the necessary risk capital for intensive drilling.

In addition, many LDCs can reduce oil consumption by developing alternative energy sources—coal, gas, and hydropower. India, for example, has large coal deposits that could be extensively developed if sufficient external capital is available both to open new mines and improve the rail system for moving coal to markets. A number of other oil-importing LDCs, e.g., Argentina, South Korea, and several countries in Africa, also have the potential to produce more coal, but again time and capital are required. Brazil has large amounts of unused hydro capacity, and Pakistan has natural gas and lignite. And over the longer run, increasing numbers of LDCs can be expected to turn to nuclear power for part of their energy needs.* All of these energy sources can be used to produce electricity, which would reduce the use of oil for that purpose. Transportation systems, however, would remain dependent on oil unless railroads electrify or revert to steam engines or economic electric buses and trucks are developed soon. Nevertheless, the elimination of oil as a fuel for electricity production could help considerably, and in some countries increased domestic production could then cover much of the remaining need for oil.

The most important reason for modest optimism for the 1980-85 period is that it is unlikely that even the best efforts of the oil-exporting countries will succeed in maintaining early 1974 prices into the eighties.** Worldwide exploratory and development efforts, movements to replace oil with other energy sources wherever possible, and conservation measures are likely to reduce significantly if not cripple the market for oil at early 1974 prices. Although a return to prices in the $2 to $3 range (f.o.b. Persian Gulf) is unlikely, prices of $4 to $6 a barrel seem distinctly possible.*** The question is how long it will take these market forces to develop, and then, just how far prices will fall. Since the world has never experienced such sharp increases in the price of oil, history and econometrics cannot provide the answers to these questions. All that is possible is to suggest some intuitively plausible answers and to examine the implications of each. What follows is just such a process and cannot be taken as a hard prediction.

The prospects for the oil-importing LDCs in 1980-85 can be viewed as a trade-off between price and a combination of reduced oil consumption and increased domestic production. At prices even close to those prevailing in world markets in early 1974, the LDCs can be expected to have made great efforts to restrain oil consumption (through a combination of conservation and the development of alternative energy sources) and to have encouraged

*The considerations that LDCs must weigh in choosing between fossil fuels and nuclear power are discussed later in this chapter.

**See Chapter 13 for an analysis of possible trends in the international oil market.

***All prices in this discussion are in 1973 dollars and hence are assumed to rise in current dollars with the world price level.

domestic exploration for oil. At prices of about $6 a barrel, there would be less, but still considerable, incentive to pursue both of these courses. A price of $4 a barrel might suggest continuation of past consumption patterns, but considerable exploratory efforts would still have to be made to offset the growth in consumption and to hold imports to their present share of total consumption.

Table 10-2 shows estimates of oil consumption, imports, and import costs for the LDCs in 1980 and 1985 at per barrel prices of $4, $6, and $8 (plus $0.60 in transport cost at each price level). At $4, it is assumed that consumption will rise at an annual rate of 8.5 percent from 1970 to 1980 and 6.5 percent from 1980 to 1985. At $6, consumption growth drops to 5.5 percent from 1970 to 1980 and to 3.5 percent from 1980 to 1985. At $8, consumption grows only 4.5 percent a year from 1970 to 1980 and 3.0 percent from 1980 to 1985. The percentage of oil consumption produced domestically is also assumed to vary with price, ranging from 35 percent at $4 per barrel to 45 percent at $6 per barrel and 50 percent at $8 per barrel.*

All of the estimates produce 1980 oil-import costs of $8 to $9 billion; 1985 costs fall between $10 and $12 billion. Since 1974 costs have been estimated above at almost $10 billion at current prices, and costs for 1975-77 are projected at $9 to $12 billion a year, the estimates for 1980 and 1985 suggest that by the mid-eighties the LDCs' situation should be considerably improved—if their exports continue to grow at an acceptable rate.

The LDCs must maintain rapid export growth through 1980-85 if they are to avoid serious problems in financing oil and other important imports. With total 1972 exports of $27.5 billion, an impressive 5 percent real rate of growth for exports would produce 1980 exports of $39.8 billion (in 1974 dollars) and 1985 exports of $51.9 billion.** When these figures are compared with the oil-import costs projected in Table 10-2, it becomes obvious that even with the flattening out of oil costs in the 1980-85 period, oil will remain a major drain on LDC export receipts. If the real rate of growth of exports was to fall significantly below 5 percent, the LDCs would be in serious trouble. The success of the LDCs then depends crucially on what happens in the industrialized world; a long recession that reduced the demand for LDC exports could be disastrous. An attempt by the oil-importing industrialized countries

*See the notes to Table 10-2 for further explanation.
**The 5 percent real rate of growth of exports is derived from the 10.8 percent rate of growth in 1973 (in current dollars) and an assumed 6 percent rate of world inflation.

Table 10-2. Estimated Oil Consumption, Imports, and Import Costs for the Oil-Importing LDCs for 1980 and 1985 at Various Prices of Oil

	1980	*1985*
Consumption (million b/d)		
at $4.60 per barrel	8.0	10.9
at $6.60 per barrel	6.2	7.3
at $8.60 per barrel	5.6	6.5
Imports (million b/d)		
at $4.60 per barrel	5.2	7.1
at $6.60 per barrel	3.4	4.0
at $8.60 per barrel	2.8	3.2
Import Costs (billion dollars)		
at $4.60 per barrel	8.7	11.9
at $6.60 per barrel	8.2	9.6
at $8.60 per barrel	8.8	10.0

Source: Author's estimates based on text discussion.
Notes: All dollar figures in this table are in 1973 dollars.

Actual figures for 1970 were:

consumption	3.6 million b/d
imports	2.4 million b/d
import costs	$1.7 billion

Import costs include $0.60 transportation costs.

The consumption figures for the $4.60 case are consistent with the pre-October 1973 projections presented in Table 13-1, and those for the $6.60 case are consistent with the revised projections of Table 13-4. In both cases, projections for all LDCs have been reduced to remove the projected consumption of net oil-exporting LDCs. The $8.60 case was derived by assuming a growth rate of 4.5 percent from 1970 to 1980 and 3.0 percent from 1980 to 1985. In all three cases, the slower growth of oil consumption in the period 1980-85, as compared with 1970-80, reflects the anticipated more rapid shift from oil to nuclear power and other fuels (principally coal) in the 1980s.

to protect their payments balances from the effects of oil price increases through a retreat to protectionism would also cut LDC exports and produce the same results.

MEANS OF MEETING INCREASED OIL-IMPORT COSTS

The LDCs face the prospect of very sharp increases in oil-import costs during the next few years, followed by a leveling off of costs as the combined result of a reduced rate of growth in oil consumption, increased domestic production, and downward pressures on world oil prices. The key question is how are the LDCs going to get through the next four or five years that must probably elapse before these reactions take effect.

One obvious approach would be for the LDCs to seek increases in loans from the World Bank and the International Monetary Fund (IMF) as a means of financing imports of oil and other essential commodities. The sums required, however, are quite large, and the most needy LDCs are very poor credit risks. They will in fact probably find it impossible to meet current debt-servicing schedules in the face of greatly increased oil-import costs. Moreover, since additional loans would merely finance normal oil consumption, they would not increase the borrowers' productive capacity and ability to meet additional interest charges. Any new loans would therefore have to be extended on "soft" or highly concessional terms (low interest rates and many years to repay) or the lending agency would have to recognize the high likelihood of defaults.

Concessional loan funds are, however, in short supply. The World Bank must pay commercial rates for the money that it lends, and its soft-loan affiliate, the International Development Association, has encountered great difficulty in obtaining the grants from member governments that are needed to replenish its funds. The industrialized nations might conceivably pay part of the interest charges on World Bank loans to LDCs,* but the LDCs would appear to have little leverage in trying to get aid donors to take this novel step. The LDCs can, however, hope in the near future for some increase in loans on non-concessional terms from the World Bank group, coupled with a willingness to extend old loans when they become due.

The IMF has induced several of the oil-exporting countries to lend it money to establish a special "oil facility" that will in turn make loans to oil-importing countries running balance-of-payments deficits because of high oil prices.** These loans will, however, have to be repaid within seven years. Interest rates, according to the managing director of the IMF, will be higher than on normal fund drawings, but "somewhat below" the cost of alternative sources of funds.[9]

---

*The industrialized countries might raise funds for such a subsidy by selling part of their gold stocks for a profit in commercial markets. The political feasibility of such a move by governments holding substantial amounts of gold is, however, at least open to question. They are much more likely to use their gold as security for loans to meet their own balance-of-payments problems.

**As of mid-1974, a total of $3.4 billion had been made available to the new facility.[8]

Seeking an increase in traditional bilateral foreign aid is of course another approach open to the oil-importing LDCs. There appears to be little enthusiasm, however, among the industrialized nations for substantial increases in foreign aid to pay the increased oil-import bills of other nations. The industrialized nations are preoccupied with their own difficulties in coping with higher oil prices and face other major economic problems, including chronic inflation. There may also be some reluctance to move too quickly to help finance LDC oil imports for fear that such a step would be perceived as tacit acceptance of the new oil prices and might also reduce both market and political pressures for price reductions. Nevertheless, it is conceivable that effective leadership could elicit a positive response from the developed countries by convincing them of the dangers to the international economy and to world political stability if a large number of the LDCs suffer greatly increased hardships from reduced output.

The oil-exporting countries, and particularly those exporting countries with revenues greatly in excess of their immediate needs, might also appear to be logical sources of aid for the oil-importing LDCs. Although the initial response of the oil exporters to pleas from the oil-importing LDCs was negative, the pressure for some help in the form either of concessional loans to finance oil imports or a lower price for oil to less developed countries has continued and with some success. In March 1974, Iran announced a decision to make some funds available to the World Bank for concessional loans and also announced soft loans for two-thirds of the cost of Iranian oil to a few countries, including India.[10] And in early April, shortly before the opening of the special session of the UN General Assembly on raw materials, OPEC announced its intention to set up a fund to make loans to those LDCs particularly hard hit by the oil price increases.[11] By mid-1974, however, the size and mode of operation of the special fund had not yet been determined.

The special UN session called in general terms for establishment of a fund to aid the oil-importing LDCs. The special session also requested the Secretary General of the United Nations to launch an "emergency operation" to enable the countries most seriously affected by the high prices of oil and other commodities to maintain their essential imports for the next 12 months.[12] The Secretary General responded by appealing to 44 countries, including the industrialized countries (both Communist and non-Communist) and the major oil-exporting countries, for emergency assistance.[13]

Until reasonably adequate assistance on oil imports to the

poorest developing countries is forthcoming, the pressures for positive responses from the oil exporters, the international financial institutions, and the industrialized countries can be expected to continue. Given their differing economic situations and political orientations, the oil-importing LDCs will not find it easy to concert their efforts, but if their difficulties worsen in the next year or so, a unified LDC pressure group seeking relief from high oil prices could develop from sheer necessity.

NUCLEAR ENERGY VS. FOSSIL FUELS

Besides efforts to expand domestic oil production and to use more coal in those countries that either have coal deposits or could economically import it, the principal alternative to increasing oil imports for many LDCs is seen to be the expansion of nuclear power. Given present technology, nuclear power can substitute for oil only in the generation of electricity.

Electricity requirements have been growing very rapidly in many LDCs.[14] The IBRD estimates that the developing countries will need new electricity generating capacity of from 150,000 megawatts to 200,000 megawatts between 1973 and 1980, which suggests capital requirements of about $35 billion for generating facilities and $50 billion for transmission systems.[15] Most of this electricity is needed for industrial production rather than individual consumption, so most of the additional capacity must be created if the growth of the LDCs is not to be restricted. Expansion of electricity capacity by the amount suggested here would require an annual rate of investment of over $10 billion. Total GNP in the LDCs was only $373 billion in 1970. This means that the LDCs would need to invest over 2.5 percent of 1970 GNP each year to produce adequate electricity. Given the other pressing demands on limited available capital, meeting this goal will be difficult and a shortage of electricity will be a problem for at least some LDCs in the next decade.

The present debate is over whether the limited capital available ought to be used for nuclear or fossil-fueled plants or a combination of both.* Issues centering around nuclear power in less developed areas are often more complex and controversial than in industrialized nations. Differences in levels of industrial and social development, wide divergencies in the availability of conventional

*Chapter 16 deals with nuclear power problems in a broader context.

energy sources, and uncertain future power needs not only raise questions about the size, type, and scheduling of nuclear energy projects, but also the more important question of whether some less developed nations should consider introducing nuclear power at all into their electric power systems.

Argentina, Brazil, India, Korea, Mexico, Pakistan, and Taiwan already have nuclear power plants in operation or under construction, and a recent study by the International Atomic Energy Agency (IAEA) is quite optimistic about the future market for nuclear reactors in these and other LDCs.[16] The IAEA study suggests a large market for reactors in the larger LDCs during the 1980s and consequently implies huge foreign aid or loan requirements for the construction of nuclear power plants.

The issue of reactor size is an important aspect of the debate over the suitability of nuclear power for LDCs. Nuclear power plants become considerably more efficient with increased size, and this has led to the argument that only the LDCs with fairly large electrical grids can use nuclear power efficiently and safely. A widely cited general rule has been that no single generating unit should represent over 15 percent of the capacity of a grid, and this means that—if the minimum size of an efficient reactor is about 800 megawatts—only the few LDCs with grid capacities of over 5,000 megawatts can reasonably use nuclear power. (By 1975, for example, Pakistan's total electric generating capacity is expected to be at about 4,200 megawatts.)

Many proponents of nuclear power for the LDCs argue, however, that reactors as small as 200 to 300 megawatts are economical in many situations, particularly if the world price of oil is $5 per barrel or higher. Moreover, the efficiency of small units could be greatly improved if the manufacturers would put more effort into producing standardized small reactors rather than emphasizing only large custom-designed units. The IAEA study mentioned above strongly supports this view that relatively small reactors can be economic in countries where attractive alternative sources of fossil fuels do not exist.[17]

Even if efficient units in the 200 to 300 megawatt range are possible, nuclear power probably still makes sense only for the LDCs with sizable electrical grids.* Thailand was the smallest

---

*Under the rule of thumb cited earlier, a 300-megawatt unit could be used only in a system with a total capacity greater than 2,000 megawatts; even by 1980, many LDCs will not have achieved this capacity. Also, a country able to absorb only one or two nuclear units might not be able to afford the highly paid technical and managerial talent needed to operate such units.

country the IAEA study predicted to be a fairly certain candidate for nuclear power in the 1980s.[18] Unless large and expensive regional grids are constructed, the smaller countries of Africa, Asia, and Latin America will have to remain dependent on fossil fuels for the foreseeable future. Nonetheless, the 15 to 20 LDCs that probably can use nuclear power account for the vast majority of the population and GNP of the LDCs. Thus, nuclear power is potentially a major factor in the developing world.

Fossil-fuel prices and the proximity of alternative fossil fuels to major energy markets can be deciding factors in determining whether or not a nuclear plant is economic. In the past, nuclear-power plants were considered competitive with oil-burning units in only a handful of LDCs. However, if the price increases of early 1974 remain in effect, oil will no longer be competitive in most countries unless coal and nuclear-fuel prices also increase tremendously. As for coal, if deposits are located close to major cities, a small or medium-size nuclear reactor will typically not make sense economically, but if the coal must be shipped some distance to reach generating plants or if expensive long-distance transmission systems must be put in place to move the electricity to markets, the advantage shifts significantly to nuclear power. One study of costs in India suggests that if a generating plant can be built over a coal mine and the electricity used nearby, coal will be about 10 percent cheaper than nuclear power, but if the plant is 800 kilometers from the mine, nuclear power becomes about 12 percent cheaper.[19] This calculation obviously depends in part, however, on the efficiency of the transport system.

Capital costs are also important in determining the relative advantages of fossil and nuclear power. Nuclear-powered plants require from 1.5 to 2.5 times as much capital per unit of electricity generated as fossil-fueled plants. Since small nuclear plants are even more capital intensive than large ones, the difference in capital costs is especially favorable to fossil fuel when construction of small electric power facilities are under consideration.

Since capital is scarce in the LDCs, high interest rates ought to be used for planning purposes, which would tend to add further to the attractiveness of small fossil fuel plants over small nuclear plants. This advantage is at least partially offset, however, by the eagerness of the countries and firms that produce reactors to sell their systems and their consequent willingness to provide liberal credit arrangements for LDCs buying nuclear power plants. Each of the various competing producers of reactors appears to be convinced that if a few successful plants can be put into operation quickly, the prospects for selling more plants will be greatly improved. The Canadian government, for example, has provided soft loans to potential LDC customers for its CANDU heavy water reactor to get it in operation abroad. Canada recently financed a 137-megawatt plant for Pakistan on decidedly concessionary terms, and it is unlikely that Pakistan could have raised money as easily for a fossil-fuel plant.

The U.S. Export-Import Bank (Ex-Im) and comparable banks in other supplier nations tend to promote nuclear exports to less developed countries. If it wishes, Ex-Im can finance or assist in financing virtually the entire cost of a nuclear project—covering up to 90 percent in direct loans and guaranteeing the remainder through local banks. Furthermore, because the bank sees its job as promoting U.S. exports, its criteria for loans are generally limited to judgments regarding the technological viability and credit worthiness of a proposed project.

The World Bank, on the other hand, is willing to consider nuclear-power financing proposals for less developed nations, but it insists on evaluating such requests in the context of their overall economic and developmental impact. The bank therefore looks into prospects for growth, the size of the supporting electrical grid, the merits of alternative power sources, and the availability of trained personnel. Although the World Bank's interest rates are comparable to Ex-Im's, it has had the reputation of taking a very conservative view toward nuclear power projects and has received relatively few requests for loans for such projects. However if fossil-fuel costs remain high and reliability of supply uncertain, the World Bank may find nuclear power acceptable (or even preferable) in some countries and may become involved on a modest scale in the financing of nuclear power plants.

Those LDCs that do decide to launch nuclear energy programs must choose between reactors using enriched uranium and those using natural uranium. Dependence on foreign supplies of enriched uranium is seen by some as a potential problem for the LDCs and hence as a strong argument for natural uranium, heavy water reactors of the CANDU variety. A number of LDCs, including India, have enough uranium ore for their needs, but none has, or is likely to have soon, the ability to produce enriched uranium.

Enriched uranium, light water reactors are generally thought to be more efficient than natural uranium systems, but Canada claims that its CANDU system is fully competitive due to its extraordinary reliability. In addition, if an LDC has serious balance-of-payments problems, it may assign a high shadow price* to foreign-exchange expenditures for planning purposes, which would increase the cost of imported enriched uranium and make reactors using lower cost domestic natural uranium more attractive. It is probably impossible to reach a firm conclusion on the relative costs of enriched versus natural uranium systems that will hold for all LDCs. It does seem clear, however, that a strong desire to be as independent as possible of foreign sources and controls may lead some LDCs to favor natural uranium reactors.

India (initially with Canadian help) has developed a virtually autonomous nuclear power program built around natural uranium reactors, indigenous uranium resources, and plants for fuel fabrication and reprocessing. Only a few developing nations, however, will be able to manufacture their own power reactors or supply their own fuel or reprocessing services in the near future. But the nuclear energy needs of these nations can readily be satisfied through equipment purchases and technological services from a growing number of reactor firms in industrialized nations seeking a share of the market in developing nations. As other nations compete with the United States for exports of enriched uranium and enrichment technology (notably centrifuges), light water reactors may become more attractive to the LDCs. Whether natural or enriched uranium reactors dominate the LDC market, however, at least some of the LDCs will eventually begin to achieve independence in all stages of the commercial nuclear cycle, as India has already done.

CONCLUSION: ENERGY POLICY ALTERNATIVES FOR THE LDCS

The LDCs face the fundamental problem of obtaining the energy needed for their development from the most economical and reliable sources, whether domestic or foreign. So long as oil prices remain high, the most pressing energy problem for many LDCs will be finding substitutes for imported oil. Many LDCs are also at or near the stage of development at which they must determine the proper balance between fossil fuels and nuclear power in their energy economies.

In the short run, there is very little that the LDCs can do to reduce their dependence on imported oil. Increasing domestic

*Shadow prices are estimates of economic costs that are used for planning when market imperfections produce prices that fail to represent real costs.

energy production takes time, even if the large amounts of capital required can somehow be raised. Conservation measures offer far more limited possibilities for the LDCs than for the developed world since relatively little gasoline is used for private transportation; oil is used primarily in vital internal transport systems, in fertilizer production, and as part of other important production processes. Oil consumption can be reduced significantly in the short term only if important productive activities come to a halt. But both fertilizer production and internal transport, for example, have important forward linkages, and if they stop, a great deal more in the economy stops.

Thus, the LDCs in some sense must have roughly the amount of oil they have been consuming, and if their needs can be expected to grow at past rates, for the short run, the question then becomes how can they pay for it. The solution to this problem is largely outside the control of the LDCs themselves. Only a few LDCs can expect to cover their greatly increased oil-import bill through their export earnings and borrowing on commercial terms. The poorer LDCs must rely on action by the industrialized nations, international financial institutions, and those oil-exporting countries with revenues excess to their current needs to furnish additional concessional aid. It is too early to judge whether the large amount of aid required will be forthcoming. The only way that the LDCs can influence the outcome is to continue to publicize their serious financial problems and to increase their diplomatic pressure on those nations that are in a position to help.

Over the longer run, the LDCs will have to make greater efforts to develop domestic oil resources and to substitute other forms of energy for oil. Offshore oil prospects appear particularly good for a number of Asian and African countries. Increased exploration for oil can be encouraged by enacting appropriate foreign-investment laws. The fact that some LDCs will not want to grant concessions to foreign companies need not create serious difficulties. Exploration and exploitation can be conducted under service contracts or through joint ventures with either the international oil companies or the national oil companies of the developed countries.

Although other fossil fuels and hydroelectric power can—with sufficient time and capital—partially substitute for oil in some countries, the distribution of world coal reserves and potential hydroelectric sites is such that nuclear power represents the principal existing alternative to oil for many LDCs. The decision on whether to install nuclear power plants depends on total planned capacity of a country's electric power grid, the competitiveness of capital and operating costs with those for conventional power plants, and whether the LDC wants to devote scarce domestic technical talent (or depend on expensive, highly trained foreign technical personnel) to nuclear power.

At the oil prices prevailing in early 1974, nuclear power plants were generally regarded as cheaper than oil-fired plants for the larger LDCs. However, given the long lead time required for a nuclear power plant to become operational, it is impossible to know which type of plant will produce electric power more cheaply some eight or ten years hence.

Policy decisions on nuclear power in LDCs ultimately should depend on whether the country in question feels that the oil market will be unstable and prices high for some years to come and whether nuclear power is regarded as sufficiently reliable, safe, and efficient to justify commitment of scarce capital and technical resources. In order to facilitate the development of reactors suitable to the needs of developing countries, the LDCs could, within the IAEA, press for research and development efforts to develop efficient small reactors and to develop standardized reactor equipment that would then lower capital costs and simplify technological requirements.

NOTES FOR CHAPTER 10

1. Sam Schurr, ed., *Energy, Economic Growth, and the Environment*, Resources for the Future (Baltimore: The Johns Hopkins University Press, 1972), p. 180.

2. Ibid., pp. 182-83.

3. Ibid., p. 187.

4. United Nations, Department of Economic and Social Affairs, *World Energy Supplies, 1961-1970*, Statistical Papers, Series J, No. 15 (New York, 1972), Table 2.

5. Alirio Parra, "Some Considerations on the Demand and Supply of Petroleum in the Seventies in Developing Countries," paper presented at the Interregional Seminar on Refining in Developing Countries, New Delhi, January 22-February 3, 1973, p. 39. The United Nations Series "J" data for LDC oil imports for 1970 agree almost exactly with the Parra data cited in this chapter. It appears that the United Nations is the source of his numbers.

6. United Nations, *World Energy Supplies, 1961-1970*, Table 2.

7. Working paper, International Bank for Reconstruction and Development, cited in *New York Times*, March 11, 1974, p. 10.

8. *New York Times*, September 30, 1974, p. 1.

9. *New York Times*, May 7, 1974, p. 1.

10. *New York Times*, March 20, 1974, p. 41.

11. *New York Times*, April 8, 1974, p. 1.

12. *New York Times*, May 2, 1974, p. 1.

13. United Nations, Department of Economic and Social Affairs, Resources and Transport Division, "World Energy Requirements and Resources in the Year 2000," in *Peaceful Uses of Atomic Energy*, in 15 vols., Proceedings of the Fourth International Atomic Energy Conference, Geneva, September 1971, United Nations and the International Atomic Energy Agency (New York, 1972), Vol. 1, p. 308.

14. International Bank for Reconstruction and Development, "Nuclear Energy and Electric Power Programs in the Developing Countries," in *Peaceful Uses of Atomic Energy*, Vol. 6, p. 426.

15. International Atomic Energy Agency, *Market Survey for Nuclear Power in Developing Countries: General Report* (Vienna, 1973), pp. iii and 6-9. This study included a rather involved technical approach to the trade-off between nuclear and fossil-fuel plants. Its main defect is that a number of LDCs apparently refused to allow their markets to be surveyed. India and Brazil, for example, are not included. This survey predicts that only 8 percent of the generating capacity of the LDCs will be nuclear powered in 1980, but it suggests that this percentage will rise rapidly in the following decade.

16. Ibid., pp. 6-9.

17. Ibid., pp. 5-7.

18. V. A. Sarabhai, K. T. Thomas, V. N. Meckoni, and K. S. Parikh, "Impact of Nuclear Technology in Developing Countries," in *Peaceful Uses of Atomic Energy*, Vol. 6, pp. 384-85.

19. International Bank for Reconstruction and Development, "Nuclear Energy and Electric Power Programs," in *Peaceful Uses of Atomic Energy*, Vol. 6, p. 429; and International Atomic Energy Agency, *Market Survey*, p. 39.

Part Four

The Roles of the Soviet Union and China

Chapter Eleven

The Soviet Union

The USSR—traditionally largely self-sufficient in energy and engaged in world energy trade only peripherally—is undergoing an "energy crisis" of its own. Although richly endowed in hydro-carbons, uranium, and waterpower, the country is having difficulty developing its resources rapidly enough to meet growing domestic and export requirements. This problem will become more critical toward the end of the 1970s as domestic and export demands continue to increase and as established coal mines and oil and gas fields approach depletion. A number of large oil and gas fields have yet to be exploited, but they are located mostly in the eastern part of the Soviet Union, far from major industrial centers and ports and where the terrain is generally difficult and the climate harsh.

ENERGY EXPORTS AND DOMESTIC REQUIREMENTS

The USSR for a number of years has been a net exporter of energy supplies.[1] A substantial part of its fuel exports goes to other Communist countries, particularly in Eastern Europe. In 1971, Soviet energy exports to Communist countries amounted to $1.1 billion in comparison with exports of $700 million to non-Communist countries.[2] Trade with Eastern Europe has been viewed as an important element of Soviet hegemony: it confers political power on the Soviet Union—but at a price. The six East European countries now closely associated with the USSR* are exporters of manufactures which, in general, are of a quality only the Soviet Union is willing to accept, and they depend on net imports of

*Bulgaria, Czechoslovakia, East Germany, Hungary, Poland, and Romania.

foodstuffs and raw materials, including fuels, from the USSR. In view of its supply limitations, the USSR will be hard put to choose between energy exports to Eastern Europe and hard-currency areas.

The sellers' market for fuel in the non-Communist world is particularly tempting for the USSR. Hard-currency earnings are needed to pay for imports of advanced equipment and technology, not to mention the imports of food in recent years. Heavy transfers of Western technology are currently seen by Soviet technocrats as the way to modernize the economy with a minimum change of system and ideology. The energy sector itself is in need of modernization and is being asked to earn through exports the wherewithal for its own development and that of other sectors of the economy.

In 1971 and 1972, about one-fourth of total Soviet exports to non-Communist industrialized countries was accounted for by fuels. However, given rising domestic and East European energy requirements and the enormous difficulties associated with developing oil and gas fields in new areas, the capability of the USSR to maintain—let alone increase—fuel exports to hard-currency countries throughout the 1970s is questionable. The growth in energy output declined from an average annual rate of 8.5 percent in 1955-60, to 7.3 percent a year over the next five years, to slightly over 5 percent during 1965-70, to between 4 and 5 percent in 1971-72. Energy consumption, on the other hand, grew by 6 percent annually in 1960-65 and by 5.8 percent annually in 1965-70.[3]

Real GNP increased by 5 percent annually in the first half of the 1960s, by 5.6 percent between 1965 and the excellent crop year 1970, and by little more than an average of 4 percent in the three-year period 1971-73, in which there were two poor crop years and one good year.[4] In general, GNP grew less than energy consumption. Energy-GNP comparisons, admitting all of the uncertainties of the concept, can provide overall signposts for future developments. The Soviet GNP can be expected to increase in the 1970s by between 4.5 and 5 percent annually. Such growth requires a greater energy output than is to be expected during the decade. If energy is not to become a constraint upon Soviet economic development, the USSR will have to improve the efficiency of its energy sector both in output and fuel utilization or reduce net exports of energy products, and possibly both.

MAJOR COMPONENTS OF ENERGY SECTOR

The three major components of the Soviet energy sector—coal, oil, and gas—are described below. Nuclear power is not included in the

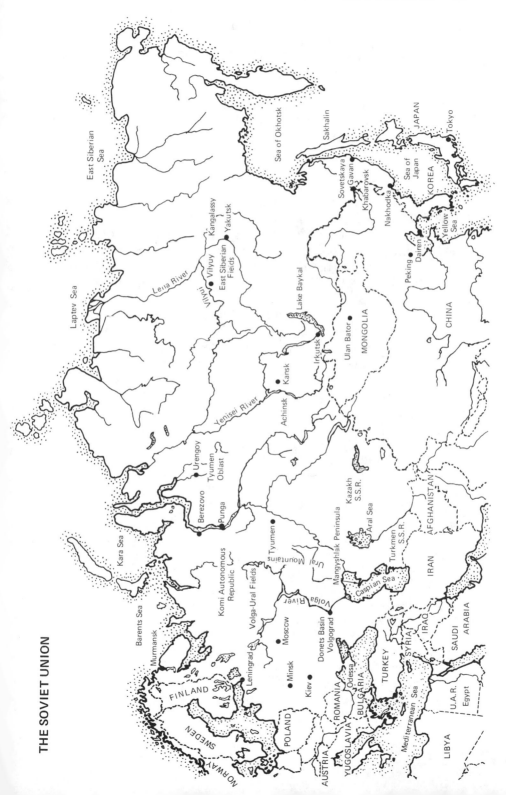

THE SOVIET UNION

discussion because the Soviet nuclear power program is still small and, despite rapid expansion, will remain of marginal importance over the next ten years or so.[5] Similarly, hydroelectric power accounts for a very small portion of Soviet energy consumption and also is omitted. Together, these two sources of energy accounted for less than 2 percent of Soviet energy production in 1970. (Table 11-1 gives a breakdown of Soviet production, consumption, and trade by major energy sources.)

Coal

In 1971, the USSR, traditionally a coal-based economy, derived about 40 percent of its total energy consumption from coal; oil and gas contributed 33 and 26 percent, respectively.[6] In about 1950, coal accounted for approximately 60 percent of Soviet energy consumption. The change over the twenty-year period reflects a major effort to alter the domestic fuel balance. Coal production in the 1960s and early 1970s grew only slightly (about 2 percent annually in terms of tonnage, but only 1.5 percent in terms of heat content), while oil and gas production and consumption soared.

The USSR is believed to possess more than half of the world's total coal resources. Soviet coal deposits have been estimated at roughly 7-8 trillion metric tons; of this, some 200 billion tons appear to be proved.[7] Soviet output in 1971, in hard-coal equivalents, consisted of 188 million metric tons of anthracite and bituminous coal and 273 million metric tons of lignite and other low-grade coal, a total of 461 million metric tons. In terms of world production, the USSR was second only to the United States, which produced 505 million metric tons.[8]

Soviet coal has a number of disadvantages as a major energy source. One problem is quality. While some of the coal is of a high grade, over the years the thermal value of the raw coal generally has declined significantly. Another disadvantage is location. The major mines (including those recently begun to be worked) are vast distances from other raw material resources and from engineering and manufacturing industries.

The fundamental problem, however, is cost. One Soviet energy expert cites the following production cost comparisons (in terms of metric tons of hard-coal equivalents): coal from a major producing area west of the Urals, the Donets Basin, which has deposits at great depths, has a production cost of about 15.3 rubles per ton. Coal from Kansk-Achinsk (which can be surface mined) in Central Siberia costs about 1.8 rubles per ton. (The comparable costs of West Siberian oil and natural gas were 6.2 and 2.2 rubles, respectively.) Transportation costs, however, make the cheap Siberian coal very dear—shipment over a distance of some 3,000 kilometers raises the cost eight or nine times. In Moscow, Donets coal has a delivered price of 18.4 rubles, and Siberian open-pit coal costs 16.1 rubles; the delivered price of Tyumen (West Siberia) gas is 10 rubles (also per metric ton of hard-coal equivalent).[9]

Over the years, the USSR appears to have followed somewhat ambiguous policies with respect to coal. The government for some time has set low production goals for coal relative to production capability. (The goals for the current five-year plan are likely to be exceeded.) Until the mid-1960s, the government subsidized coal by maintaining artificially low prices in order to encourage consumption—a policy at odds with low targets for production increases—and to economize on the use of oil and gas needed for export. Subsidies were discontinued in 1967, and the government turned to the West, Japan, and Eastern Europe for mining equipment and technology to modernize the industry. The government, for example, is encouraging Japanese interest in developing the Yakutsk coal deposits. In the future, the government could decide to expand coal output and increase domestic consumption in order to allow greater exports of oil and gas.

In 1972, the USSR exported about 6 percent of its hard coal output and 12 percent of its coking coal production; coal exports in that year totaled some 28 million metric tons, including 9 million tons of coking coal. (See Table 11-2.) Two-thirds of its exports go to other Communist countries; the rest, consisting of coking coal, is shipped to Japan and, in smaller quantities, to Italy, France, and Austria, among others. The long haul by rail makes the shipments uneconomic, but the USSR has trade obligations to its client states and is also willing to subsidize exports that earn hard currencies.

Eastern Europe is in effect self-sufficient in coal; in 1972, using hard-coal equivalents, the area produced 190 million metric tons of hard coal and 169 million tons of lignite, and consumed roughly the same amount.* Eastern Europe imports Soviet coal, while Poland exports hard coal to the West. Poland has become the fourth largest coal producer in the world, after the United States, the Soviet Union, and China. Of its total output of 150 million metric tons in 1972, Poland exported 18 million metric tons to hard-

---

*The resource endowment of the East European countries varies greatly. Poland has ample hard coal; East Germany and Czechoslovakia possess large quantities of lignite; but the coal resources of Bulgaria, Hungary, and Romania are negligible.

Table 11-1.   Soviet Production, Trade, and Apparent Consumption of Energy (in billion metric tons of hard-coal equivalent)

	1950	Increase 1950-60 in % per annum	1960	1965	Increase 1960-65 in % per annum	1970	Increase 1965-70 in % per annum
Coal and Lignite:							
Production	224.7 (77.8%)	6.1	407.6 (59.3%)	450.6 (46.1%)	2.0	472.6 (37.6%)	0.9
Net exports	-8.9	0.8	9.6	18.6	14.1	20.5	2.0
Apparent consumption	233.6	5.5	398.0	432.0	1.7	452.1	0.9
Oil:							
Production	54.8 (18.8%)	14.6	212.2 (30.9%)	347.6 (35.6%)	10.4	504.2 (40.1%)	7.7
Net exports	-3.8	29.5	50.3	120.3	19.1	124.2	0.6
Apparent consumption	58.6	10.7	161.9	227.3	7.0	380.0	10.8
Natural Gas:							
Production	8.2 (2.8%)	22.0	61.4 (8.9%)	169.1 (17.3%)	22.0	263.7 (21.0%)	9.3
Net exports	...	...	...	...	...	-0.1	...
Apparent consumption	8.2	22.3	61.4	169.1	22.5	263.8	9.3
Hydroelectric Power	1.6 (0.6%)	14.9	6.4 (0.9%)	10.2 (1.0%)	9.9	15.6 (1.2%)	8.8
Nuclear Power	...	...	...	...	...	0.4 (0.03%)	...
Total energy production	289.3	9.1	687.6	977.5	7.3	1,256.5	5.2
Net exports	-12.7	16.8	59.9	138.9	18.3	114.6	-3.8
Apparent consumption	302.0	7.6	627.6	838.6	6.0	1,111.9	5.8
Population (in millions)	178.5	...	212.4	229.6	...	241.6	...
Energy consumption per capita (metric tons)	1.7	5.8	3.0	3.6	3.7	4.6	4.9

	1975	Increase 1970-75 in % per annum	1980	Increase 1975-80 in % per annum	1985	Increase 1980-85 in % per annum
Coal and Lignite:						
Production	534.4 (33.4%)	2.5	604.9 (30.9%)	2.5	685.3 (27.1%)	2.5
Net exports	23.0	2.3	25.2	1.8	29.7	3.3
Apparent consumption	511.4	2.5	579.7	2.5	655.6	2.5
Oil:						
Production	671.6 (42.0%)	5.9	797.7 (40.8%)	3.5	1,067.3 (42.1%)	6.0
Net exports	171.6	6.7	137.7	-4.3	196.8	7.4
Apparent consumption	500.0	5.6	660.0	5.7	870.5	5.7
Natural Gas:						
Production	373.0 (23.3%)	7.1	522.0 (26.7%)	7.0	732.4 (28.9%)	7.0
Net exports	11.4	14.1	40.0	14.3	87.5	8.3
Apparent consumption	351.6	5.9	482.0	6.5	644.9	6.0
Hydroelectric Power:						
Production	18.8 (1.2%)	3.7	25.3 (1.3%)	6.0	33.2 (1.3%)	5.0
Nuclear Power	2.8 (0.2%)	48.0	6.0 (0.8%)	16.5	16.0	21.0
Total energy production	1,600.6	5.0	1,955.9	4.1	2,533.3	5.3
Net exports	216.0	13.5	202.9	-1.2	314.0	9.1
Apparent consumption	1,384.6	4.4	1,753.0	4.8	2,219.3	4.8
Population (in millions)	253.7	...	267.0	...	281.5	...
Energy consumption per capita (in metric tons)	5.5	3.6	6.6	3.7	7.9	3.6

Sources: United Nations, Statistical Office, *World Energy Supplies*, Series J (New York, various years). N. K. Baybakov, ed., *Gosudarstrennyy Pyatileniy Plan Razitiva Narodnogo Khozyasta SSR Na 1971-1975 Gody* [State five-year plan for the development of the USSR national economy 1971, 1975] (Moscow, 1972).

U.S. Department of Commerce, Social and Economic Statistics Administration, *Estimates and Projections of the Population of the U.S.S.R., by Age and Sex: 1950 to 2000*, Series P-91, No. 23 (Washington, D.C., 1973).

Narodnoye Khozyaistvo SSSR 1922-72 [National economy of the USSR 1922-1972] (Moscow, 1972).

Note: Figures in parentheses in output columns indicate the percentage of total energy production accounted for by each source of energy.

Table 11-2.   Energy Exports of the USSR

Destination	1962	1963	1964	1965	1966	1967	1968	1969	1970	1971	1972
								Coal and Coke (1000 metric tons)			
Total	19,039	20,161	21,914	20,610	24,388	26,151	25,817	28,075	29,386	28,660	28,110
East Europe:	12,676	12,880	13,965	13,018	15,269	15,251	13,408	14,446	16,027	18,005	17,754
Hungary	820	1,562	1,929	1,579	1,468	984	801	807	937	974	967
Poland	991	1,194	1,203	1,184	1,169	1,143	1,308	1,108	1,125	1,192	1,191
Romania	...	...	...	...	...	950	1,096	1,126	1,241	1,474	1,611
West Europe:	3,090	4,364	4,247	3,934	5,314	5,543	5,178	5,972	6,211	5,310	5,149
Austria	882	972	969	894	921	816	837	851	911	844	807
Benelux	...	...	...	...	299	284	310	350	307	194	359
Denmark	...	...	...	...	913	644	666	737	660	539	447
France	949	1,804	1,713	1,570	1,465	1,497	1,266	1,311	1,526	1,420	1,191
Italy	1,021	1,194	1,213	1,023	1,360	1,845	1,715	2,074	2,039	1,769	1,636
Sweden	237	394	352	447	280	388	462	557	681	474	640
Other	...	...	...	...	76	69	78	92	87	70	69
Other Importing Countries:	3,093	2,886	3,446	3,524	3,797	5,138	6,281	6,759	6,363	4,684	4,674
Finland	938	873	1,266	1,182	1,141	1,070	975	1,082	1,047	1,090	1,099
Yugoslavia	1,008	1,057	1,092	1,109	1,067	999	1,149	1,078	1,139	1,144	1,086
Japan	1,147	956	1,088	1,233	1,589	2,379	2,730	3,200	2,855	2,450	2,489
U.A.R.	...	...	...	...	...	471	477	498	537	...	...
All others	180	31	256	134	8	219	950	901	785	661	533
						Petroleum — Crude and Total Products (1000 b/d)					
Total	819.58	901.21	952.57	1037.90	1354.07	1571.94	1701.48	1863.00	1967.04	2071.92	2113.72
East Europe:	256.97	297.53	329.03	397.72	507.80	554.66	651.87	760.28	808.91	899.32	983.38
Bulgaria	31.44	32.89	26.68	26.22	80.68	89.93	110.53	130.65	140.70	159.98	159.77
Czech.	80.74	92.32	103.29	127.90	138.69	159.27	168.44	201.00	211.05	237.18	259.29
E. Germany	48.99	61.52	79.13	98.95	127.09	125.89	151.82	177.85	187.78	208.60	230.76

Table 11–2. Energy Exports of the USSR (continued)

Destination	1962	1963	1964	1965	1966	1967	1968	1969	1970	1971	1972
Hungary	34.78	37.79	43.19	50.10	59.61	68.45	76.03	86.13	95.66	101.61	111.14
Poland	60.99	72.99	76.69	94.53	101.72	14.10	145.02	164.65	173.72	195.55	222.42
Romania	0.03	0.02	0.05	0.02	0.01	0.02	0.03	...	...	...	...
West Europe:	*260.09*	*296.74*	*305.18*	*298.25*	*457.66*	*602.49*	*601.10*	*637.94*	*672.19*	*671.40*	*658.26*
Austria	7.16	9.69	10.76	9.45	14.01	14.46	16.89	13.43	21.17	22.68	19.43
Benelux	...	...	...	...	3.11	9.93	25.62	24.29	54.66	73.77	99.49
France	15.48	22.50	24.06	32.40	53.88	59.18	55.23	54.27	50.24	90.45	62.31
W. Germany	38.48	44.51	59.47	51.92	90.84	112.30	119.60	117.41	126.66	125.77	131.82
Italy	142.24	155.02	154.67	147.64	178.43	240.18	238.49	215.07	205.02	180.90	168.84
Sweden	52.69	60.90	55.10	56.35	75.70	76.57	96.91	94.02	96.85	91.84	87.70
Others	4.04	4.12	1.12	0.49	41.69	89.87	48.35	119.45	117.59	86.99	88.67
Other Importing Countries:	*280.00*	*278.97*	*290.46*	*299.77*	*351.70*	*355.80*	*384.68*	*393.39*	*418.08*	*425.06*	*402.51*
Finland	58.56	69.62	84.40	90.14	121.00	129.87	151.28	162.47	156.26	172.20	173.41
Yugoslavia	9.36	14.60	17.07	20.33	23.33	30.16	50.33	50.25	54.27	57.89	68.30
Japan	59.88	62.04	72.14	78.01	83.41	65.84	55.56	44.22	54.27	66.33	20.10
U.A.R.	26.76	19.59	15.06	15.53	20.84	23.32	20.77	20.47	32.94	...	...
Cuba	88.13	84.82	91.64	95.00	102.32	106.47	106.77	115.77	120.34	128.64	140.70
China	37.31	28.30	10.15	0.76	0.80	0.14	0.15	0.21	...	...	...
All others	22.52	27.97	27.90	42.16	36.91	58.99	63.82	71.39	67.86	76.14	69.57
Natural Gas (million cubic meters)											
Total	8.51	8.52	8.37	11.09	23.45	104.18	48.98	75.47	93.49	129.05	143.64
Austria	...	...	...	...	...	...	4.02	22.14	27.09	40.46	46.27
Czech.	...	...	...	...	...	75.13	16.64	25.17	37.99	46.44	54.88
Poland	8.51	8.52	8.37	11.09	23.45	29.05	28.32	28.15	28.40	42.14	42.49

Source: *Foreign Trade of the USSR*, for the specified years.

currency countries. Polish plans for 1980 envisage an output of 200 million metric tons and exports to the West of 25 million tons.

Oil: Domestic Supply and Requirements

The "ultimately recoverable" petroleum resources of the USSR are generally considered to be vast. The U.S. Geological Survey, for example, in a 1973 study estimated that the USSR had between 100 and 1,000 billion barrels of ultimate resources (including offshore).[10] The United States, Venezuela, Algeria, Saudi Arabia, Iran, Kuwait, and Libya were the only other countries in this category.

The Soviet Union does not use the Western oil industry concept of "proved reserves" and does not publish statistics of any kind about oil exploration. The *Oil and Gas Journal* estimated Soviet crude oil reserves as of January 1973 at 75 billion barrels,[11] but did not explain the derivation of the estimate. A 1971 study by Robert E. Ebel, a Western energy expert, indicated that the ratio of reserves to current annual output in the USSR is roughly the same as in the United States, i.e., 10:1.[12] Both the *Oil and Gas Journal* figures and Ebel's 1971 reserve-production ratio are unavoidably very crude estimates. Nevertheless, the geology of the country, both offshore and onshore, and the past and planned development of the oil industry indicate that the USSR possesses petroleum resources that are more than sufficient to supply domestic requirements for the remainder of the twentieth century and beyond.

Crude oil production in the Soviet Union increased almost six fold between 1950 and 1963, at an average annual rate of nearly 14 percent. During this period, planned production targets were usually either fulfilled or surpassed. In 1963, the rate of growth started to decline; by 1970 it had fallen to 6.8 percent, and by 1972 to about 6 percent. A principal reason for the decline in the production growth rate apparently is the rapid, uneconomic exploitation (e.g., poor recovery techniques and out-dated equipment) of low-cost oil reserves in the production areas located near the main population and industrial centers between the Volga River and the Ural Mountains. Between 1965 and 1970, Volga-Urals production increased less than 2 percent annually, and by 1970, it was close to stationary. The Volga-Urals share in total oil output, slightly over 70 percent in the early 1960s, is scheduled to decline to 44 percent according to the 1975 plan. This figure may actually be higher, however, because production in West Siberia—the area soon to be developed intensively—has fallen behind plan targets.

Because of the apparent infeasibility of expanding production significantly in established oil regions, Soviet oil production is on the move. In 1960, less than 5 percent of total output came from east of the Urals; the figure was 18.2 percent in 1970, and the planned target for 1972 was 36.6 percent. The immediate future points to West Siberia, in particular the large Tyumen Oblast and its rich Samotlar deposit and other promising fields; the Mangyshlak Peninsula (on the eastern shore of the Caspian Sea) in Kazakhstan; the Turkmen Republic in Central Asia; and the Komi Republic in the European north. In the more remote future, the promising areas are the northern stretches of the Tyumen Oblast, as well as East Siberia between the rivers Yenisey and Lena, the area north of the Caspian Sea, and various offshore regions—possibly around Sakhalin in the foreseeable future and farther north later on. But all these new locations, vast distances from the centers of domestic and foreign consumption, involve a move into a barren, inhospitable wilderness and the tapping of oil deposits with many unfavorable features, including low-gravity ratings and high sulphur and paraffin content.

Estimates of average West Siberian crude-oil production costs at wellhead vary widely among available sources. One difficulty in making such estimates is that the ruble is purely a domestic currency, and Soviet price relationships differ greatly from those in the Western world. Robert Ebel estimated 1970 West Siberian production costs at 3.8 rubles per metric ton, or 0.52 ruble per barrel.[13] A. Probst, a Soviet energy expert, computed the 1970 cost of Tyumen oil at 0.59 ruble per barrel.[14]

Based on Probst's estimates, and at the official rate of exchange prevailing in 1970 ($1.11 for 1 ruble), the cost of Tyumen oil in West Siberia would be $0.65. At that time, the ruble was generally undervalued for producers' goods and the dollar was overvalued. Allowing for a 20 percent dollar devaluation, Tyumen oil would have had a 1970 wellhead cost of roughly $0.79.* Thus, in terms of production costs, Soviet oil in West Siberia falls somewhere between U.S. costs (estimated at roughly $1.60 per barrel for 1970[15]) and Middle Eastern oil-production costs, which were

---

*While converting costs in rubles to dollar values by means of the official exchange rate is not a very satisfactory way to arrive at the dollar value of Soviet oil-production costs, it appears to be the only method available, short of computing individual production costs for each economy. The exchange-rate calculation offered here is, at best, an extremely crude estimate of oil-production costs in West Siberia. The $0.79 per barrel costs estimate compares with an estimate of Soviet oil-production costs of $0.80 made by Adelman and Bradley in the mid-60s.[16]

estimated at between $0.10 and $0.20 a barrel in 1968.[17] Because of the vast distances between Siberian oil resources and markets and also because of the difficult climate and terrain, transport costs are also important. Probst estimates that transmission of Siberian crude oil over a distance of 1,860 miles by 40-inch pipeline adds about 20 percent to wellhead costs.[18]

The growth of Soviet petroleum consumption has slowed along with the slowdown in the growth of output—the rate of consumption growth has declined from roughly 11 percent annually between 1950 and 1963, to about 7.5 percent between 1963 and 1970, to roughly 6 percent in 1971 and 1972. (See Table 11-1.) This reflects to a degree the slowdown of the economy as a whole and also the government's desire to make oil available for export by referring domestic demand to coal and to the rapidly increasing supply of natural gas.

Oil: Patterns of Trade

After the initial post-World War II years of net imports from occupied countries came an enormous push to develop the oil industry in the Volga-Urals region. Exports began to soar; by the late 1960s they accounted for 28 percent of domestic production. Since then, exports have leveled off as a result of declining production growth rates and rising domestic requirements.

Soviet oil has in the past commanded high prices in the USSR's client states. After the huge increases in international oil prices in 1973-74, the prices specified in the long-term agreement covering Soviet oil exports became bargains for the importers. In all probability, however, these agreements will eventually be renegotiated to bring their price provisions more in line with international prices.

Soviet exports of crude oil and oil products to Eastern Europe have increased significantly since the early 1960s—from 256,000 barrels a day in 1962 to over 980,000 barrels a day in 1972, about two-thirds of Eastern Europe's total consumption that year. In return for supplying its client states in Eastern Europe with oil, the Soviet Union looks to them for some of the capital investment needed to develop new oil fields. Thus far, Czechoslovakia and East Germany have obligated themselves to provide machinery on credit for the West Siberian oil fields. Discussing oil and gas deliveries to Hungary and other East European countries, N. Baybakov, head of the Soviet State Planning Commission, promised that "the major part of Soviet mineral oil and natural gas exports, similar to the practice of earlier years, will be directed to the socialist countries in the next five years as well. It will be the same in the more distant future, too." But, added Baybakov pointedly, the envisaged expansion of Soviet oil and gas output "is limited only by investment funds. In addition to the development of deposits, construction of pipelines also takes a lot of money. . . . Cooperation by the socialist countries in the field of development should be increased."[19]

In addition to Eastern Europe, the USSR has had to supply its Far Eastern associates, which means hauling petroleum long distances by rail or (for North Vietnam) by rail and ship. In past years, China bought Soviet petroleum products and some crude oil, but those shipments ceased with the souring of Sino-Soviet political relations in the early sixties.

Cuba annually requires between 37 and 41 million barrels of oil, about one-tenth of Soviet shipments of oil to client states in the recent past. The oil, chiefly crude, is shipped in Soviet tankers that return in ballast. This is an expensive undertaking and there has been speculation about a Soviet-Venezuelan rapprochement, which would enable Soviet tankers on the Cuban route to take Venezuelan oil to Scandinavia. Another possibility would be to supply Cuba with Venezuelan oil, with or without the USSR as a middleman.

Because of the pressing need for Soviet oil for domestic consumption, some efforts have been made to supply the client states with oil from outside the Soviet realm. The amounts have up to now been small. In 1971, Egypt a recipient of Soviet oil products at the same time—supplied 15.3 million barrels of oil to East Germany, Cuba, and Yugoslavia; Egypt has also sold oil to Bulgaria and Hungary in recent years. Algeria has shipped 1.5 million barrels on Soviet account to Bulgaria and Yugoslavia, and Algeria, Libya, and Syria have supplied small quantities directly to Bulgaria. Iranian oil has been shipped to an East German refinery for processing on a Western firm's account, and Romania—although a net exporter of petroleum—has imported small amounts of oil from Iran and Libya in recent years. In 1972, the USSR obtained slightly more than 66 million barrels of crude oil from the Middle East and North Africa (20 to 30 million barrels more than in 1971) to meet requirements in Eastern Europe and at home and to provide a temporary market for oil from recently nationalized fields.* Virtually all of the oil

---

*Eastern Europe as a whole possesses a considerable refining capacity (reported in 1970 as 232 million barrels per day in Czechoslovakia, East Germany, Poland, Hungary, and Bulgaria), which can be used for refining Middle Eastern and North African oil, as well as for Soviet crude.

imported from the Middle East and North Africa has been purchased through bilateral trade (involving lines of credit) or direct barter arrangements, under which the producing country receives Soviet or East European industrial and military goods. Since 1973, however, OPEC countries have begun to insist not only on much higher prices, but also on payment in hard currency.

Apart from small exports of oil and oil products to various countries in the Third World (primarily Egypt), Soviet shipments to non-Communist countries have been concentrated in Western Europe. Soviet exports to Western Europe increased from 260,000 barrels a day in 1962 to 658,000 barrels a day in 1972 (see Table 11-2). Given Western Europe's fast rising consumption, however, the proportion of total oil consumption in Western Europe supplied by the USSR declined from close to 9 percent in 1960 to 4 percent in 1971.[20] Italy became a large importer of Soviet oil in the late 1950s, absorbing at that time one-third of the Soviet exports to Western Europe; in 1972 Italy's share was about one-fourth. Other principal West European importers are West Germany, Sweden, and France. In 1972, Finland, which purchases nearly all of its oil from the USSR, imported more Soviet oil than any single West European nation and was the Soviet Union's leading customer outside of Eastern Europe.

Oil exports to non-Communist countries have remained fairly stable as a share of total Soviet oil exports, but their composition has varied over time. Crude oil accounted for only 18 percent of the small total in 1955, rose to 60 percent by 1965, and then receded to 52 percent in 1971. Kerosene and diesel oil (the latter at times in short supply in the USSR) have been prevalent among the products exported. In view of the USSR's refinery problems—capacity limitations, high costs, and low-quality products—exports of crude oil are preferable from a Soviet economic point of view.

In 1972, Soviet oil exports to non-Communist countries netted almost $600 million in hard currency. This was about 20 percent of total hard-currency receipts in foreign trade. Oil exports to the developed non-Communist countries scarcely changed between 1970 and 1972, but oil prices increased both in terms of the U.S. dollar and in real terms, and Soviet revenues have increased correspondingly. Oil trade promises to remain a profitable venture for the Soviet Union.

Natural Gas

Soviet exploration for natural gas (in deposits where it is separate from oil) has made great strides since the late 1950s, and estimates of reserves have soared. Radio Moscow reported in 1973 that "prospected gas deposits in the USSR amount to about 18 trillion cubic meters, approximately twice as much as those of the United States. It is calculated that potential reserves amount to some 100 trillion cubic meters."[21]

Gas is plentiful in the USSR and of good quality, but, like coal, it is not conveniently located. The heavily populated and industrialized regions in the central, northwestern, and western parts of the country have either no gas fields or insufficient deposits. The Ukraine produces gas but its reserves, like those of most principal gas-producing fields in the North Caucasus, the Volga regions, and central Asia, are declining. As with oil, the new fields are distant and located in difficult terrain. Their development and the installation of pipelines will require tremendous capital investment and new and complex technologies.

The future lies farther east—chiefly in the northern portion of the Tyumen Oblast in West Siberia. In addition, there is new development in Kazakhstan (the Mangyshlak Peninsula) and in Turkestan. There also exist, close to the Arctic Circle, three enormous gas fields (Urengoy, Zapolyarnyy, and Medvezhe) and several smaller fields. Urengoy alone is believed to hold 4 trillion cubic meters of natural gas.[22] Gas is currently being extracted at Berezovo and Punga, also located in the huge Tyumen Oblast.

Three pipeline projects to exploit the Tyumen fields for domestic consumption have been launched. Of these, one—a major line to the south—is now in use. A second major line (to the west) is not yet completed, and the third pipeline—a relatively minor one—is not in operation because of technological difficulties.

In East Siberia, gas has been extracted on an experimental basis at Ust-Vilyuy near Yakutsk, and there is also gas on Sakhalin. The gas deposits of both areas are regarded as candidates for development for liquefied natural gas (LNG) exports to Japan and the United States. In the more remote future, Soviet gas exploration will turn to the plateau between the rivers Yenisey and Lena and to the continental shelf of the Soviet Arctic. Vast amounts of gas appear to exist beneath the rather shallow waters (roughly 75 feet deep) of the Barents Sea, in the Kara and Laptev Seas, and in the East Siberian Sea.

In 1972, the USSR produced nearly 20 percent (approximately 221 billion cubic meters) of the total world output of natural gas—a 35-fold increase since 1950. Nevertheless, the rate of increase in Soviet output has been declining. Whereas the annual average rate of growth between 1965 and 1970 was 9.3 percent, it is expected to average about 7.0 percent during the 1970s.[23] The rate of increase in domestic consumption of gas between 1965 and 1970 was the same as the rate of production increase, 9.3 percent annually. Similarly, the projected decline in production growth rates between 1970 and 1975 can be expected to result in reduced rates of domestic consumption. (See Table 11-1.)

The major reasons for the declining production growth rate (the five-year-plan targets for the industry have always been unfulfilled by a wide margin) include overall weaknesses in the planning system and the enormous technological difficulties associated with tapping the vast resources in extremely cold regions. Undertaking gigantic projects under strong time pressures, the use of outdated equipment, and inadequate gathering and distribution systems contribute to the difficulties. Finally, a plan to use large-diameter pipe (100 inches), which in principle should lower transmission costs, has not been implemented thus far because the technology for manufacturing and laying the pipe has not been mastered. Nevertheless, given the strong demand for energy in adjoining countries, natural gas production should increase, particularly if Western technology is applied to its exploration, extraction, and transmission.

Probst provides several cost-of-production estimates for Siberian gas. The 1971 cost of Tyumen gas is estimated at 1.5 rubles per 1,000 cubic meters, or approximately U.S. $1.65 at that time. (This converts to about U.S. $0.05 per 1,000 cubic feet.) For Central Asian gas, the cost is 3.7 rubles per 1,000 cubic meters or roughly U.S. $4.07.[24] (This amounts to roughly U.S. $0.12 per 1,000 cubic feet.) Transportation costs depend on pipe diameter and increase linearly with distance. Probst estimates that Tyumen gas moved by 48-inch pipe to Moscow costs 6.8 rubles per 1,000 cubic meters. The corresponding price in Minsk, from which gas moves by pipeline to Eastern Europe and to Austria, is 11.4 rubles.

Despite its vast natural gas deposits, the USSR was a slight net importer of natural gas in 1970. Imports began in 1967, when Afghanistan started delivering natural gas in repayment for a Soviet loan. Over a period of 25 years, approximately 600 billion cubic meters of Afghan gas are to be piped into Soviet Central Asia, which as a major gas producer itself will be moving its own surplus to consuming areas farther north. The price for Afghan gas was fixed at U.S. $0.16 per 1,000 cubic feet (28 cubic meters).

Iran began to send gas to the USSR at the end of 1970 as part of a 1966 agreement providing for the delivery of a Soviet steel mill and other equipment, both industrial and military. In return,

Iran promised to deliver 140 billion cubic meters of gas to the USSR between 1970 and 1984. The gas originates in fields east of the Persian Gulf port of Abadan and is piped north through Iran to the east coast of the Caspian Sea.

The original contract price of the Iranian gas was also very low, U.S. $0.19 per 1,000 cubic feet, but in 1973, rising gas prices prompted negotiation of a 30 percent price increase. The Afghan contract price was subsequently raised. By 1980, shipments from these two suppliers will probably equal roughly 5 percent of total Soviet production. The USSR can use the inexpensive gas for domestic consumption, though it does not depend on its delivery, and export its own higher priced gas to hard-currency markets.

Five of the six East European associates of the USSR have been importing Soviet gas by pipeline since the 1960s (see Table 11-2) and will require increasing amounts of Soviet imports in the future. Romania produces all of the gas it requires (in 1972, 22 billion cubic meters) and a small surplus for export. Poland and Hungary are small producers, 6 and 4 billion cubic meters, respectively, in 1972. Bulgaria, Czechoslovakia, and East Germany produce practically no gas.

As of 1973, Austria and West Germany were the only Western countries importing Soviet gas. A 1968 agreement provided for Austrian deliveries of pipe (manufactured in cooperation with West German firms) and other equipment in exchange for Soviet gas. Between 1971 and 1990, an annual quantity of approximately 420 million cubic meters is to be exported to Austria at a price originally set at roughly U.S. $0.30 per 1,000 cubic feet. Payment will probably continue to be largely in the form of industrial equipment from Austria. West Germany, using a branch of the main pipeline crossing the Czechoslovak border between Cheb and Marktredwitz, is to receive amounts of gas increasing over a twenty-year period from 0.5 billion cubic meters to 3 billion cubic meters annually. Gas deliveries to West Germany began in October 1973, after some delays in completing the pipeline. It appears that West Germany originally obtained the Soviet gas at a price of about U.S. $0.35 per 1,000 cubic feet; the price may subsequently have been raised, but specific information on that point is lacking.

FUTURE SOVIET OIL AND GAS TRADE WITH NON-COMMUNIST COUNTRIES

The Outlook for Oil

The Soviet Union's 1975 target for crude-oil output of 3.7 billion barrels almost certainly will not be met. The shortfall by 1975 may be as much as 255 million barrels. This would mean that after oil for anticipated domestic use and export to client states is set aside, only 292 million barrels would be available for export to non-Communist countries—some 36.5 million barrels less than was available in 1970-72.

In an attempt to maintain its oil exports to hard-currency areas, the USSR in mid-1973 openly encouraged its East European associates to increase imports from other sources. The official journal of the USSR's Ministry of Foreign Trade declared the new policy:

> The steadily growing demand of Council for Mutual Economic Assistance (CMEA) member countries for oil and the desire of these countries to improve their total fuel oil and power supply mainly through oil consumption can be met not only by oil deliveries from Russia via the Friendship pipeline . . . but also by supplies from . . . the Middle East and North Africa.[25]

Accordingly, a number of East European countries have begun to negotiate bilateral deals (primarily on barter terms) with Middle East producers for state-controlled oil. It can be expected that for the next few years the USSR will squeeze by and maintain exports to non-Communist countries in physical terms, but at increased prices.

Beyond 1975 the picture darkens for several years. The effort to move into an insufficiently explored East with an inadequately developed industry is likely to result in a further deceleration in the growth of production, perhaps to between 3 and 4 percent annually. This would mean a Soviet output of between 4.0 and 4.2 billion barrels by 1980.[26] If consumption was to grow by between 5 and 6 percent a year, apparent domestic consumption would be about 3.3 billion barrels by 1980, leaving roughly 800 million barrels for export to both client states and the non-Communist world, only a small fraction of recent exports.

Soviet leaders are likely to be anxious to retain oil and oil products as earners of international currencies (and, in specific cases, as political bargaining chips). It is more than likely, therefore, that measures will be taken to make more oil available for export, if not in the short run, then by the early 1980s. The task could be tackled in various ways. An attempt would certainly be made to conserve fuel, though not much should be expected from this effort. Consumption could be directed to coal and gas, the former involving a move toward a more costly fuel, the latter toward a cheaper source; both moves would require additional investments and imports of equipment. Efforts also would be made to accelerate oil production. Here enters the Soviet interest in obtaining large-scale foreign aid for the extraction, transportation, and refining of oil (and for the nearby gas). The hoped-for *deus ex machina* is the foreign capitalist. The feat required, particularly drilling and laying pipes in difficult terrain, is such that only U.S. and Canadian technology can help.

Foreign help might consist of increased shipments of equipment, with some technical assistance provided at the beginning, or joint ventures involving one or more of the major non-Communist industrialized countries. One large joint project that has been under consideration for several years concerns a 48-inch pipeline to carry up to 500,000 barrels a day of West Siberian (Tyumen) oil some 4,000 miles east to the Pacific port of Nakhodka for Soviet use in the east and for export to Japan. Pipelines are now under construction from West Siberia farther east toward Irkutsk, but the proposed project would require a parallel line from the Tyumen fields to Irkutsk and a further extension of over another 2,700 miles from Irkutsk to Nakhodka. The cost would exceed $3 billion, and the construction job could not be completed before the early 1980s, even in the unlikely event that agreement to go ahead were to be reached soon. An alternative would be joint Soviet-U.S.-Japanese exploration of the offshore oil that exists around Sakhalin and, at a later date, Soviet-U.S. offshore exploration in the Soviet Arctic Northeast. (See Chapter 9 for further discussion of possible Soviet-Japanese deals.)

Another alternative would be increased Soviet imports, either for Eastern Europe and other client states, for the Soviet Union's own use, or for reexport to the West. Leaving aside the possible supply of Cuba from Latin American sources, Soviet and East European purchases could be made in the Middle East and North Africa. The Soviet Union has established interests in the oil-rich Middle East and is going to do its best to expand its influence there. (For a full discussion of this subject, see Chapter 3.)

Over the years, the Soviets have obtained Middle East oil and gas in exchange for long-term credits totaling several hundred million dollars, technical assistance in oil production, and military equipment and other goods. Parties to such deals have included Algeria, Egypt, Iran, Iraq, Libya, and Syria. The Soviet deal with Iraq in 1972 illustrates how the Soviets would like this system to work. The USSR imported 29.2 million barrels of crude oil from Iraq, paying for it with Soviet equipment, then sold most of the oil (or Soviet oil that the Iraqi oil replaced) for hard currency, which was used to purchase advanced equipment from the West. It is unlikely, however, that arrangements of this kind will continue on more than a modest scale in the future. The producing states will ask, as Iraq already has, that the USSR pay in hard currency like any other purchaser of their oil and gas, for the Soviet market is less essential to them than other markets and they prefer to choose their own imports, usually from Western countries.

The Soviets are not likely to get into the international oil business on a greater scale than at present as middlemen, buying Middle Eastern oil and transporting it to widespread markets in the West, Japan, Eastern Europe, or the Third World. Soviet port facilities, pipelines, and tankers are not capable of moving substantially larger shipments of oil and gas, certainly not on the scale of future world oil requirements. Moreover, the producing and consuming countries can deal directly; they do not need a Soviet middleman. The producers would need a strong political or economic incentive to replace the private Western oil companies with a Soviet state monopoly. They are much more likely to prefer tanker fleets and downstream operations under their own control and ownership. In sum, between now and the early 1980s, the USSR is not likely to become a major importer from the Middle East and North Africa or to increase significantly its present modest exports to industrialized non-Communist countries, except possibly to Japan.

Prospects for Natural Gas

Agreements for the future sale of Soviet gas have been entered into with Italy, West Germany, France, and Finland. Italy contracted in December 1969 for gas deliveries that were supposed to begin with 1.2 billion cubic meters in 1973 and then to rise to between 6 and 10 billion cubic meters in later years, but deliveries will not begin before sometime in 1974. Finland is to begin receiving Soviet gas in 1974 and the amount is planned to reach 1.4 billion cubic meters by 1979. Sweden and Switzerland will probably also be among the USSR's future gas customers.

The gas deals in force or still in the making appear sound. The West Europeans are intent upon diversifying their energy sources as much as possible, and large-scale imports of Soviet gas would enable them to reduce their dependence on oil. The West Europeans are interested in exporting equipment and other commodities to the USSR in exchange for gas, and the USSR is even more interested in obtaining West European goods; in fact, some of those goods are vital for the completion of the pipelines that are to move gas to Europe. Soviet gas, once the West Siberian fields are developed and connected by pipeline, will come from far away but from abundant deposits and through large-diameter pipes, and its price should be competitive. The gas trade should be mutually advantageous for the Soviets and West Europeans.

Although natural gas production in the remainder of the decade will not duplicate the rapid development of the industry in the late 1950s and early 1960s, gas will still be a Soviet growth industry. If gas output between 1972 and 1980 increases at a rate slightly below the revised goal for 1973 (7.7 percent) and below actual performance during 1967-72 (7.6 percent a year), then 1980 output will be about 390 billion cubic meters. By 1980, imports will have risen to 14 billion cubic meters or, perhaps after some renegotiation with Afghanistan and Iran, by more, while exports may reach 44 billion cubic meters—divided more or less evenly between Eastern and Western Europe, which could amount to about 10 percent of Western Europe's total gas consumption in 1980. This would permit apparent domestic consumption to increase over the 1970s as a whole from 198 to 360 billion cubic meters, or by an average annual 6.2 percent. At 1973 prices and currency values, the 1980 export total would yield roughly half a billion dollars in sales to Eastern and Western Europe.

Liquefied Natural Gas Projects

The USSR's difficulties in developing the gas fields in its eastern regions, in combination with the rising demand for gas in the industrialized countries, have also sparked Soviet interest in applying Western and Japanese technology and equipment in the development of Siberia in return for gas. However, these possibilities raise practical questions for all concerned and, in particular, for the Western countries and Japan. First, are the Soviet deposits under discussion ample enough for the purpose at hand? Second, will costs and prices be reasonable? Third, when would the fuel become available and over what period? Fourth, are there undue political risks? And fifth, how do the costs, risks, and prospective returns compare with investments in the development of energy resources in other parts of the world? The answers may differ from case to case.

The deals under consideration envisage U.S. help for the export of LNG from West Siberia to the U.S. East Coast and U.S.-Japanese help for LNG shipments from East Siberia to Japan and the U.S. West Coast. To satisfy U.S. interests, further exploratory drillings would have to be made in West Siberia, but it is generally believed that gas in the required quantity and quality exists in the ground. The first of these projects (North Star) would involve the annual transmission of some 20 billion cubic meters of Urengoy gas (2 billion cubic feet daily) through a 48-inch pipeline to Murmansk. In Murmansk, the gas would be liquefied and shipped in special LNG tankers to the U.S. East Coast. Implementation of these plans would take the rest of the 1970s or longer; repayment of the debt in gas would take the USSR twenty-five years.

Estimates of the U.S. capital required for the project have risen over time. In 1973 the figure was close to $7 billion,[27] and it can be expected to rise substantially. Of that amount, $2.3 billion is allotted for the development of the gas field and the pipeline to Murmansk. The liquefaction plant and port facilities would cost another $1.5 billion. LNG tankers with a capacity of 120,000 cubic meters cost more than $100 million each. Twenty tankers would be required, for a total outlay of $2.0 billion or more. Finally, a minimum of $30 million would have to be spent on port facilities and deliquefaction installations in the United States. The cost calculations—taking into account the investments just listed (but not the expected supplemental expenditures)—yield a price of U.S. $1.25, c.i.f., for 1,000 cubic feet on the U.S. East Coast, more than double the price of domestic natural gas delivered by pipeline to the same area in mid-1974.

The proposed pipeline to Murmansk would be used only for exports; the U.S. negotiators declined a Soviet request to design a pipeline for domestic, as well as, export purposes. But in the process of developing the gas field and constructing the pipeline, the transfer of U.S. techology would enable the USSR to improve its domestic operations as well as export capabilities. After repayment of the debt to the U.S. firms (and to the U.S. Export-Import Bank), the Soviet government would be able to continue exporting gas on its own account.

If the project were cancelled for some reason before the pipeline was completed, both partners would stand to lose, but inevitably the U.S. side far more than the Soviet side. Once the line was ready, of course, the U.S. side would for all practical purposes be the only loser. The USSR could use Urengoy gas to supply domestic needs or its European customers. It might not be able to ship LNG, since the tankers would be U.S. owned, but natural gas will remain in demand on the Eurasian continent, and within the USSR itself it could replace high-cost coal.

Another large LNG project proposal has grown out of Japan's desire to import natural gas. The most recent discussions pertain to the Vilyuy fields in the Yakutsk area of East Siberia. The gas would be piped to Nakhodka, the port on the coast of the Sea of Japan, where it would be liquefied and exported in LNG tankers to Japan and the U.S. West Coast. The project as a whole, including Japanese and U.S. LNG tankers, would require large investments by

Japan and private U.S. interests, as well as U.S. technology. The project is more problematical than the North Star project for two reasons: first, pipeline construction would have to overcome geological and climatic conditions worse than those between Urengoy and Murmansk (including seismic zones); second, it is by no means certain that enough gas is available at Vilyuy to justify the investment required.*

CONCLUSIONS

While the USSR during the remainder of the 1970s, at least, will only be a relatively minor participant in world trade in oil and gas, the potential exists for its becoming a significant energy supplier in world markets in the 1980s and beyond. The most likely significant increase in energy exports will be in natural gas for Western Europe; exporting West Siberian oil to Japan is also a possibility. The economics of a U.S.-Soviet LNG deal are not very attractive from the U.S. standpoint in that the U.S. East Coast price for Soviet gas would be substantially higher than the 1973 U.S. domestic price for natural gas. The decision will have to be made within the context of U.S. goals for political relations with the USSR.

Whether the USSR's energy export potential will be realized depends on a number of major considerations, including the willingness of the major non-Communist nations to make the necessary investments and to supply the necessary technology, and the willingness of the Soviet Union to export on a long-term basis vast quantities of nonrenewable fossil-fuel supplies. These considerations in turn will be shaped by the future evolution of Soviet relations with the non-Communist powers.

NOTES FOR CHAPTER 11

1. Unless otherwise indicated, data in this chapter are drawn from an unpublished background paper prepared for The Brookings Institution by Herbert Block, June 1973.

2. *Foreign Trade of the USSR*, and United Nations, Statistical Office, *World Energy Supplies*, Series J, Nos. 14-16 (New York, various years).

3. United Nations, *World Energy Supplies, 1961-1970*, No. 15, p. 78.

4. Herbert Block, "Value and Burden of Soviet Defense," U.S. Congress, Joint Economic Committee, *Soviet Economic Prospects for the*

*See Chapter 9 for further discussion of possible Soviet-Japanese deals for LNG.

Seventies, 93 Cong., 1st sess. (June 1973), p. 197. See also, U.S. Department of State, Bureau of Intelligence and Research, "The Planetary Product in 1973: Preliminary Report" (Washington, D.C., 1974).

5. U.S. Atomic Energy Commission, Forecasting Branch, *Nuclear Power 1973-2000* (Washington, D.C., 1972).

6. United Nations, *World Energy Supplies,* No. 16, p. 30.

7. V. V. Strichkov, G. Markova, and Z. E. Murphy, "Soviet Coal Productivity," in *Mining Engineering,* May 1973, p. 43.

8. United Nations, *World Energy Supplies,* No. 16, pp. 35 and 40.

9. A. Probst, *Puti Razvitiya Toplivnogo Khozyaistva SSSR* [Development paths of the fuel economy of the USSR] *Voprosy Ekonomiki,* No. 6 (1971), pp. 52-62.

10. U.S. Geological Survey, *Summary Petroleum and Selected Mineral Statistics for 120 Countries, Including Offshore Areas,* John P. Albers, et al., Professional Paper No. 817 (Washington, D.C.: Government Printing Office, 1973), pp. 142-43.

11. *Oil and Gas Journal,* worldwide issue, December 25, 1972, p. 5.

12. Robert E. Ebel, "Two Decades of Soviet Oil and Gas," *World Petroleum,* June 1971, p. 78.

13. Ibid., Table 1.

14. Probst, *Puti Razvitiya,* p. 52.

15. M. A. Adelman, *The World Petroleum Market* (Baltimore: The Johns Hopkins University Press, 1972), p. 76. (Adelman's cost estimate of $1.22 per barrel for the period 1960-63 was raised to $1.60 to reflect cost changes as of 1970.)

16. Discussed in Charles Issawi, *Oil, the Middle East and the World,* Washington Papers No. 4, Center for Strategic and International Studies (New York: Library Press, 1972), p. 24.

17. Adelman, *The World Petroleum Market,* p. 77.

18. Probst, *Puti Razvitiya,* p. 52.

19. "Vilaggzdasag" (Budapest), April 28, 1973.

20. United Nations, *World Energy Supplies,* Series J, various issues.

21. May 11, 1973, 16:00 GMT. The *Oil and Gas Journal* (December 25, 1972) also uses the official government reserves estimate.

22. Tass, May 10, 1973.

23. United Nations, *World Energy Supplies,* Series J, various issues.

24. Probst, *Puti Razvitiya,* pp. 52-53.

25. *Venshnyaya Torgovya* [Foreign trade], quoted in *Oil and Gas Journal,* August 20, 1973, p. 30.

26. Issawi, *Oil, The Middle East and The World,* p. 24.

27. "Imports of Soviet LNG Hinge on Policy Review," *Oil and Gas Journal,* January 22, 1973, p. 61; "Foreign Enterprises for Siberia," *Petroleum Press Service,* December 1972, p. 463.

Chapter Twelve

China

12. China

China has never played a major role in the world trade in energy and does not do so today.* Its modest dependence on imported oil ended in the mid-1960s, and now it neither imports nor exports significant quantities of energy materials. China's per capita consumption of energy is still very low, about that of Japan in 1938 or the Soviet Union in 1933. Nevertheless, China cannot be ignored in a study of international energy problems. Because of its sheer size, China today is probably the world's third largest consumer of energy, surpassed only by the United States and the Soviet Union. Also, China is industrializing rapidly, and its energy sector has considerable potential for growth. As will be brought out below, China could become an important exporter of oil in the early 1980s.

SOURCES OF ENERGY

China is well endowed with energy resources, which are widely, but unevenly, distributed and generally of good quality. Coal is still the leading source of energy but is rapidly giving way to oil and gas. (See Table 12-1.) Hydropower and fuelwood make small, but significant, contributions. China is not known to possess any nuclear power plants.

Coal

China's coal reserves are of the same order of magnitude as those of the United States. Only the Soviet Union has more coal. The

*In this study, China refers to the territory administered by the People's Republic of China; it does not include Taiwan.

Table 12–1.   Apparent Consumption of Commercial Energy in China by Source (trillion Btu)

Year	Coal	Hydro	Petro-leum	Wood	Gas	Total	Share of Coal (%)
1952	1,654	28	46	234	2	1,964	84.2
1953	1,759	32	68	239	4	2,102	83.7
1954	2,101	46	72	244	4	2,467	85.2
1955	2,468	49	114	250	4	2,885	85.5
1956	2,757	68	132	255	9	3,221	85.6
1957	3,282	71	155	270	12	3,790	86.6
1958	5,777	...	217	...	36	...	...
1959	7,537	...	304	...	54	...	...
1960	6,959	...	...	...	75	...	...
1961	4,286	...	...	...	108	...	...
1962	4,540	...	...	...	124	...	...
1963	4,794	...	...	...	215	...	...
1964	5,047	...	...	...	414	...	...
1965	5,543	208	234	317	430	6,732	82.3
1966	6,062	...	304	...	418	...	...
1967	4,794	...	...	...	398	...	...
1968	5,047	...	...	...	430	...	...
1969	6,309	...	610	...	1,131	...	...
1970	7,589	...	859	350	1,583	...	...
1971	8,191	268	1,098	357	1,979	11,893	68.9
1972	...	...	...	...	2,286	...	...

Source: Based on production estimates in Table 12–2.
Average annual growth rate (percent): 1952–71   9.9%
1957–71   8.5%
1965–71   10.0%

most frequently cited figures for Chinese coal reserves, 1.0 to 1.5 trillion metric tons, are based on Chinese statements of the mid-1950s.[1] Reserves today are almost certainly higher and should meet China's needs for the indefinite future.

China has not released any data on overall coal production since 1960. The most widely accepted estimates are those of R. M. Field of the Central Intelligence Agency, and they have been used in Tables 12-1 and 12-2.[2] As Table 12-1 indicates, coal still provides about two-thirds of China's energy. Coal output has expanded steadily over the past two decades, but at a declining rate. (See Table 12-2.) In the period 1965-71, coal production increased by only 6.7 percent annually, in contrast with the 30 percent and 29 percent annual growth rates estimated for oil and gas, respectively.

Coal output is not likely to grow rapidly in the future. Reserves impose no restrictions, nor do there appear to be obstacles to manufacturing or importing more mining machinery. The primary limitation will probably be the rising cost of coal relative to

Table 12-2.   Domestic Energy Production in China by Source

Year	Coal (million metric tons)	Electric Power Total (billion kwh)	Hydro	Petroleum (million bbl)	Fuelwood (million cubic meters)	Gas (100 million cubic meters)
1952	66	7.3	1.3	3.2	33.0	0.6
1953	70	9.2	1.5	4.6	33.7	1.0
1954	84	11.0	2.2	5.8	34.5	1.1
1955	98	12.3	2.4	7.2	35.3	1.1
1956	110	16.6	3.5	8.6	36.0	2.3
1957	130	19.3	3.5	10.8	38.0	3.3
1958	230	27.5	...	16.8	...	9.4
1959	300	41.5	...	27.4	...	14.2
1960	280	...	...	...	...	19.8
1961	170	...	...	...	...	28.3
1962	180	...	...	...	...	32.7
1963	190	...	...	...	...	56.6
1964	200	...	...	...	...	109.0
1965	220	52.0	16.5	40.0	44.6	113.2
1966	240	...	...	51.8	...	110.1
1967	190	...	...	...	...	104.7
1968	200	...	...	...	...	113.2
1969	250	...	...	105.2	...	297.7
1970	300	88.9	...	148.1	...	416.7
1971	324	106.2	29.8	188.1-190.4	50.2	520.9
1972	...	...	...	218.5-220.7	...	601.6

Average annual growth (percent):

Year	Coal (million metric tons)	Electric Power Total (billion kwh)	Hydro	Petroleum (million bbl)	Fuelwood (million cubic meters)	Gas (100 million cubic meters)
1952-57	14.5	21.0	23.0	27.0	...	41.0
1957-65	6.8	13.2	21.0	17.8	2.0	55.0
1965-71	6.7	12.6	10.4	30.0	2.0	29.0
1952-71	8.7	15.1	18.1	24.0	...	44.0

Source: Thomas G. Rawski, "The Role of China in the World Energy Situation," an unpublished background paper prepared for The Brookings Institution, June 1973. Rawski relied principally on official Chinese sources.

competing energy sources, particularly oil, for which domestic prices (and presumably production costs) have been declining over the past decade. The delivered cost of coal is apparently rising because the most easily accessible deposits are far from the areas of heaviest demand. Oil production, on the other hand, is shifting from remote northwestern areas to more accessible fields, especially Ta-ch'ing in Manchuria.

If, as appears to be feasible, the production of oil and gas continues to expand rapidly, there is little reason for the Chinese authorities to make a special effort to achieve higher rates of increase in coal output. The maximum growth rate for coal in the period 1971-85, therefore, should not go above 8 percent annually, just under the average rate for the previous two decades. For present purposes, the minimum rate will be set somewhat arbitrarily at 4 percent per year.

Oil

Earlier estimates of China's petroleum resources have been revised upward in recent years. No authoritative survey of China's oil reserves is available, however, and the most comprehensive study appears to be that of A. A. Meyerhoff,[3] whose data sources are not fully explained. Meyerhoff offers a figure of 19.6 billion barrels for proved plus probable reserves, and he describes this figure as "probably very conservative."

Meyerhoff's estimate does not include offshore resources. A team of geologists collected geophysical and geologic data in the East China and Yellow Seas in October and November 1968, and they concluded that the unexplored continental shelf between Taiwan and Japan "may be one of the most prolific oil reservoirs in the world." They also found that "a second favorable area for oil and gas is beneath the Yellow Sea."[4] At least part of these promising offshore areas, however, are subject to conflicting national claims.

The estimates of Chinese oil production presented in Table 12-2 indicate that output has grown at an average annual rate of 24 percent since 1952 and reached an impressive rate of 30 percent in the years 1965-71. In the 1950s, China imported small, but increasing, amounts of oil, principally from the Soviet Union. A peak of 28.7 million barrels was reached in 1959. This trend was reversed when the worsening Sino-Soviet dispute caused trade between the two great Communist powers to wither, and China successfully pursued a policy of self-sufficiency in energy. By 1967 imports from the USSR had fallen to negligible amounts.[5]

Limitations on the future growth of oil production are probably technical and economic, rather than natural. There should be no shortage of deposits to work. According to Meyerhoff, a number of China's known oil fields remain undeveloped. More important, many promising areas on land and offshore have apparently not yet been explored. While Chinese geologists and engineers have demonstrated the ability to find and develop oil resources on land, they have had almost no experience in working offshore. A common assumption is that China cannot engage in major exploitation of seabed oil resources without some form of technical assistance from foreign oil companies. Some caution in this regard may be wise, however, in view of China's clear desire to avoid dependence on foreigners and its past mastery of difficult technological problems, including the development of nuclear weapons.

Refining capacity may impose a barrier to the rapid growth of crude oil production for domestic consumption. Foreign observers have noted the apparent failure of refining capacity to keep pace with recent increases in the production of crude oil.[6] This lag may, however, merely reflect a temporary disruption of investment planning and construction during the Cultural Revolution (1966-68). If so, and if investment activity returned to normal after the Cultural Revolution, a sharp increase in refining capacity could appear in the near future. Nevertheless, continued very rapid rises in the consumption of petroleum products (say, 25 percent or more annually) could create requirements for new refineries beyond China's immediate capabilities. In that event, importation of refining equipment or reliance on the services of foreign refineries might become necessary.

Transportation is another difficulty that China must overcome in continuing the rapid expansion of oil output. Except for Ta-ch'ing (Manchuria) and Sheng-li (Shangtung Province), all of China's large oil fields are deep inland. If these interior resources are to be developed, a massive program of pipeline and railroad construction seems essential. China is certainly capable of such an effort, but the cost and the possible need to import steel or steel tubing may be unattractive to Chinese planners. Efforts to expand oil production may, therefore, be concentrated in coastal areas and offshore.

China's oil output could conceivably continue to expand at the recent very rapid rate of 30 percent annually,* but in view of the many imponderable factors involved, it is best to project a range of from 15 to 25 percent for the period 1971-85.

Natural Gas

Little is known about China's natural gas resources. Major deposits exist in the southwest and northwest, particularly in Szechuan Province. Gas has also recently been discovered in Chekiang Province south of Shanghai. Additional deposits may be found offshore in the Gulf of Po Hai, the Yellow Sea, and the East China Sea. Recent estimates place China's natural gas reserves (in the northwest and southwest only) at 500 to 600 billion cubic meters.[7]

The estimates of gas production presented in Table 12-2 were assembled from several sources. The 1952-54 figures are from C. S. Chen,[8] and the 1955-68 estimates are those given by A. A. Meyerhoff.[9] Figures for 1969-72 are largely extrapolations based on Chinese claims of percentage increases in production.[10] These

---

*Prime Minister Chou En-lai reportedly told Japanese Foreign Minister Ohira that China's oil production in 1973 was 50 million tons.[11] If so, 1973 output would have increased a staggering 67 percent above the approximately 30 million tons estimated to have been produced in 1972. Earlier reports had been that China's 1973 production was between 35 and 40 million tons, an increase of 17 to 33 percent over 1972.[12]

production estimates must be viewed with considerable reserve. Meyerhoff's figures are crucial, and they were presented with little documentation. There is a strong possibility that the estimates are high. Certainly, it is surprising that the high rate of growth shown (an average annual rate of 44 percent for the period 1952-71) could occur with little comment by the Chinese press. On the other hand, no available evidence contradicts Meyerhoff, and the Chinese have demonstrated their technical capability to develop new gas fields.

Obstacles to future increases in gas production may be less than in the case of oil. The same kind of transportation problems must be solved, since much of China's gas is also distant from major consuming centers. The difficulties in exploring and exploiting offshore resources are of course similar. The need to expand refining capacity, however, is not a constraint on gas production.

Even though gas production appears to have grown more rapidly than oil production in the past two decades, uncertainty concerning the available data requires a conservative approach to future projections. The same range (15 to 25 percent per year) will, therefore, be assumed in projecting future gas production as was adopted for oil.

Other Sources of Energy

The only other significant sources of energy in China at present are fuelwood and hydropower.

Estimates for fuelwood consumption between 1952 and 1957 are from Chinese sources cited by K. C. Yeh.[13] Figures for later years were derived by assuming constant per capita consumption. For the period 1958-72, population, and therefore fuelwood consumption, was assumed to have increased 2 percent annually. A similar growth rate will be used for the period 1972-85.

China's water power resources appear to be among the largest in the world. These resources have been estimated at 535 million kilowatts,[14] which far exceed similar estimates for the United States. Installed hydropower capacity (1971-72), however, is estimated at only 7.4 million kilowatts, or about 1.4 percent of potential capacity.

China's electric power industry continues to be dominated by coal-burning thermal plants. Hydropower accounted for only about 28 percent of the electricity produced in 1971. Hydropower grew at an average annual rate of 18 percent from 1952 to 1971, but it increased only about 10 percent per year in the period 1965-71.[15] Several possible explanations may be advanced for its relatively slow growth:

— Much of China's hydropower potential is in Yunnan, Tibet, and other remote areas.
— The Cultural Revolution, which interrupted long-term planning, may have curtailed or delayed large hydro projects.
— Some major hydro projects might interfere with water transport.
— Silting and seasonal variations in rainfall and water flow may pose technical problems.

A major expansion of hydropower facilities in the relatively near future is a distinct possibility. In the absence of concrete evidence of such plans, however, it seems best to project the recent 10 percent annual growth rate into the period 1972-85.

The development of nuclear power is another somewhat longer range possibility. Recent Chinese delegations to Japan and Canada have expressed interest in nuclear power stations.[16] China's achievements in the field of nuclear weapons suggest a capacity to develop nuclear power technology. Also, various sources report the existence of substantial deposits of uranium in China.[17]

Most of China, however, is probably not yet ready for nuclear power. To achieve low production costs, nuclear power plants must have large installed capacities and operate continuously at or near capacity. Nuclear power is therefore most suitable as baseload capacity in regions with heavy demand for electricity. These requirements limit the potential usefulness of nuclear power for China. At present, nuclear power could probably be considered only for the largest urban and industrial centers—e.g., Shanghai, where basic power facilities may be of pre-World War II vintage, the Mukden-Anshan-Luta industrial region in southern Liaoning, in which nuclear power might become attractive if coal costs rise markedly, or the Peking-Tientsin area.

Despite these considerations, China might decide to embark on a modest nuclear power program for reasons of prestige and to acquire knowledge of the technology involved. Lacking concrete evidence to the contrary, however, no development of nuclear power will be assumed in projecting energy output through 1985.

ENERGY REQUIREMENTS

China's consumption of energy appears to have grown at an average annual rate of about 10 percent over the past two decades. (See Table 12-1.) A rough sectoral breakdown of energy consumption for 1952, 1957, 1965, and 1971 is presented in Table 12-3. The major

Table 12-3.   Energy Consumption in China by Sector
(trillion Btu)

Sector	1952	1957	1965	1971
Nonindustrial uses:	173	234	500	726
Agriculture	1	34	50	143
Households	1,179	2,088	2,446	2,755
Nuclear weapons program	0	0	14	41
Total	1,353	2,356	3,010	3,665
Aggregate energy consumption	1,964	3,790	6,732	11,893
Industrial energy use (residual)	611	1,434	3,722	8,228

Source: Table 12-1 and Appendix B of Thomas G. Rawski, "The Role of China in the World Energy Situation," an unpublished background paper prepared for The Brookings Institution, June 1973.

shift revealed by these estimates is the replacement of households by industry as the major consuming sector.

In 1971, industry accounted for approximately 70 percent of total energy consumption. Projecting future energy requirements is therefore primarily a function of estimating industry's needs. Nonindustrial requirements (Table 12-4, col. 1) may be projected rather summarily as follows:

Transport: energy consumption rose at average annual rates of 6.2 percent during 1952-57, 10.0 percent in 1957-65, and 7.8 percent in 1965-71. Future increases of 8 percent per year are assumed in the energy requirements of this sector between 1971 and 1985.

Households: energy consumption was assumed to have grown 2 percent annually between 1957 and 1971, roughly keeping pace with population growth. For 1971-85, a 3 percent annual rate is assumed to allow for modest improvements in living standards.

Agriculture: consumption of commercial energy increased at rates of 11.7 percent during 1957-70 and 23.5 percent during 1965-70. (In 1952 consumption of commercial energy in agriculture was negligible.) An annual growth rate of 15 percent is assumed for 1971-85.

Nuclear Weapons: this sector can be ignored in projecting energy requirements since the amounts involved have no significant impact on China's energy balance.

Projecting industrial energy needs involves assumptions about both the growth of output and trends in unit energy requirements.* China's industrial output grew at average annual rates of over 19 percent during 1952-57, between 9.0 and 10.8 percent during 1957-65, and 9.9 percent during 1965-71. On general economic grounds, there is good reason to expect continued rapid growth in the future.[18] It therefore appears reasonable to use alternative growth rates of 8, 10, and 12 percent in Table 12-4.

No firm basis exists for estimating future unit energy consumption, but since the efficiency-inducing substitution of oil and gas for coal has only begun, unit energy requirements should decline only slightly, if at all, between now and 1985. Table 12-4, therefore, makes two assumptions concerning unit requirements; no change in the 1971 level (calculated by Rawski at 33,694 Btu per yuan) and a decline of 1 percent annually.

Table 12-4. Chinese Energy Requirements under Alternative Assumptions Concerning Industrial Growth, 1971-85 (trillion Btu)

| Year | Non-Industrial Requirements | Aggregate Energy Requirements if Industry's Growth Rate is: | | | | | |
| | | 8% | | 10% | | 12% | |
		(1)	*(2)*	*(1)*	*(2)*	*(1)*	*(2)*
1971	3,665	11,893		11,893		11,893	
1975	4,377	15,570	15,133	16,422	15,952	17,323	16,818
1980	5,624	22,073	20,664	25,024	23,362	28,441	26,487
1985	7,463	31,632	28,489	38,708	34,645	47,673	42,444
1971-85 Average annual growth		7.2%	6.4%	8.8%	7.9%	10.4%	9.5%

Notes: (1)   Assumes energy input per yuan of industrial gross output remains constant at the 1971 level (33,694 Btu per yuan).

(2)   Assumes energy input per yuan of industrial gross output declines 1 percent per year during 1971-85.

FUTURE PROSPECTS

Alternative Energy Balances

A major implication of Table 12-4 is that China can maintain a very respectable rate of industrial growth even if the energy sector expands more slowly than in recent years. Continua-

*The amount of energy required on the average to produce a yuan of output could vary with both the composition of the gross national product and the efficiency of energy use.

tion of past growth patterns in the energy sector will cause energy supplies to outrun domestic requirements at recent rates of economic growth. The resulting surplus could be used to support accelerated economic growth or increased energy consumption by the Chinese people. Or it could be exported.

These tentative judgments can be tested and refined by drawing up alternative energy balances for China in 1975, 1980, and 1985, combining in various illustrative ways the assumptions concerning energy production and requirements developed above. Table 12-5 provides alternative energy output figures. Table 12-6 uses the data in Tables 12-4 and 12-5 to derive a variety of possible energy balances.

Table 12-5.   Projected Output of China's Commercial Energy
(trillion Btu)

	Profile						
	1	*2a*	*2b*	*2c*	*2d*	*3a*	*3b*
1975	15,821	16,727	17,742	16,585	17,393	17,491	19,314
1980	23,652	28,482	35,132	25,846	28,398	30,856	39,878
1985	37,643	54,841	83,972	41,790	47,545	58,988	93,874
Underlying Annual Growth Rates (%)[a]							
Coal	4	4	4	6	8	6	8
Oil and gas	15	20	25	15	15	20	25

Source: Table 12-1 and text.

[a]In all profiles, output of fuelwood is assumed to increase 2 percent annually and that of hydroelectric power 10 percent.

Table 12-6 suggests a number of interesting possibilities. First, barring sharp and extremely improbable declines in the output growth rates of all major energy sources, domestic production will provide sufficient energy to support an 8 percent industrial growth rate to 1985 and beyond.

Second, if output of oil and gas grows in excess of 15 percent per year (*Profiles 2a-b and 3a-b*), projected domestic energy output will satisfy the needs generated by a 10 percent industrial growth rate with large surpluses appearing by the 1980s. If oil and gas output maintains past annual growth rates of 25 percent or higher (*Profiles 2b and 3b*), the energy surpluses generated after 1980 will be very large, even if the assumed rate of industrial growth is increased to 12 percent per year. Under any probable rate of industrial growth, expansion of oil and gas output in excess of 20 percent per year will provide China with an opportunity to become a major exporter of oil and gas in the 1980s.

Third, even if the growth rate of oil and gas output declines to 15 percent per year, energy requirements generated by a 10 percent rate of industrial growth can be satisfied if coal output grows at 6 percent per year. (*Profile 2c.*)

Fourth, if outputs of coal and oil and gas all grow more rapidly than is assumed in *Profile 1* (*Profiles 3a and 3b*), requirements generated by a 12 percent rate of industrial growth can be met

Table 12-6. China's Projected Energy Balance in 1975, 1980, and 1985 Under Various Assumptions Concerning Growth of Industry and of Energy Output: Domestic Production Minus Energy Requirements (trillion Btu)

Profile:	1	2a	2b	2c	2d	3a	3b
I. Industrial energy requirements per yuan of output assumed to be constant at the 1971 level							
8% growth for industry:							
1975	251	1,157	2,172	1,015	1,823	1,921	3,744
1980	1,579	6,409	20,877	3,773	6,325	8,783	17,805
1985	6,011	23,209	52,340	10,158	15,913	27,356	62,242
10% growth for industry:							
1975	-601	305	1,320	163	971	1,069	2,892
1980	-1,372	3,458	17,926	822	3,374	5,832	14,584
1985	-1,065	16,133	45,264	3,082	8,837	20,280	55,166
12% growth for industry:							
1975	-1,502	-596	-400	-1,557	-749	-651	1,172
1980	-6,755	41	14,509	-2,595	-43	2,415	11,437
1985	-10,030	7,168	36,299	-5,883	-128	11,315	46,201
II. Industrial energy requirements per yuan of output assumed to fall 1% per year from the 1971 level							
8% growth for industry:							
1975	688	1,594	2,609	1,452	2,260	2,358	4,181
1980	2,988	7,818	22,286	5,182	7,734	10,192	19,214
1985	9,154	26,352	55,483	13,301	19,056	30,499	65,385
10% growth for industry:							
1975	-131	775	1,790	633	1,441	1,539	3,362
1980	290	5,120	19,588	2,484	5,036	7,494	16,516
1985	2,998	20,196	49,327	7,145	12,900	24,343	59,229
12% growth for industry:							
1975	-997	-91	105	-1,052	-244	-146	1,677
1980	-2,835	1,995	16,463	-641	1,911	4,369	13,391
1985	-4,801	12,397	41,528	-654	5,101	16,544	51,430

Source: Energy balance is derived from figures for energy requirements (Table 12-4) and energy output (Table 12-5).

Note: (-) indicates a deficit.

Profile numbers correspond to profile numbers and underlying annual growth rate assumptions in Table 12-5.

beginning these exports in 1973. Some observers saw them as an effort to divert the Japanese from helping the Soviets build a pipeline along China's northern border from the Tyumen oil field to the Pacific.[23] Whatever their motives, however, if the Chinese decide to develop a large-scale oil export potential, they will find a ready market in Japan.

from domestic sources (despite a possible pinch in the 1970s) with an energy surplus which, in the case of *Profile 3b*, could still lead to major export potential after 1980.

The generally favorable picture presented in Table 12-6 would have to be modified, but not necessarily reversed, if the 1971 base estimate used in projecting gas production turns out to be too high.*

Export Possibilities

The energy surpluses shown in Table 12-6 should not be taken as quantitative predictions, but merely as illustrative calculations to show that, under a wide range of not unreasonable assumptions, China could become a significant exporter of energy.

In the mid-1960s, China apparently began to export oil to North Korea and North Vietnam.[19] No figures are available on this trade, but its volume is presumably not large. China also markets small quantities of diesel fuel in Hong Kong, and in early 1974 it reportedly agreed to sell 350,000 barrels of diesel fuel to Thailand and offered to sell an unspecified quantity of oil to the Philippines.[20]

Of much greater potential importance, China began to export oil to Japan in 1973. Exports in that year totaled 1 million tons, and the Chinese have reportedly assured the Japanese that exports in future years will be "not less than one million tons," or about 7.3 million barrels.[21] The target for 1974 is reportedly 1.5 million tons or about 11 million barrels.[22]

In the context of Japan's huge oil-import requirements, 1 million tons is an almost insignificant amount.** It is too early to judge prospects for substantially larger exports of oil from China to Japan. Some uncertainty exists concerning Chinese motives in

Conflicting Offshore Claims

China's ability to develop an energy surplus for export may depend on whether major oil and gas deposits in fact exist beneath the waters of the Yellow and East China Seas, and on when

---

*C. S. Chen and K. N. Au give an estimate of gas production in 1971 that is 56.7 percent of the Meyerhoff estimate used here.[24] They do not explain their estimate, however, other than to attribute it to "fragmentary information" collected by the Geographical Research Centre in the Graduate School of the Chinese University of Hong Kong. Dr. Chen is director of the center.

**It would have taken care of about 1.5 days' imports in 1972.

and how conflicting claims to those areas are resolved.[25] The positions of the various governments concerned may be described briefly as follows:

—   Tokyo follows the "median line" doctrine and claims roughly the eastern half of the East China Sea.
—   Seoul uses the median line doctrine in the Yellow Sea, but in the East China Sea applies the doctrine of "natural extension" of the shoreline out to a depth of 200 meters.
—   Taipei, speaking for China rather than only for Taiwan, argues that the area in question is "China's continental shelf" and claims all of the Yellow Sea and the East China Sea up to the Okinawa Trench, which runs close inshore along the western edge of the Ryukyus and Kyushu.
—   Peking has objected to the claims and offshore activities of the other parties, but it has apparently never clearly enunciated its own position.
—   Pyongyang's position is not known, but its claims may overlap with those of Peking.

The status of the various actual or potential disputes created by these conflicting claims differs widely:

—   The claims of Taipei and Peking, however formulated, can be assumed to be in total conflict, but this difference is submerged in the more fundamental disagreement over which claimant is the legal government of China. Taipei has not granted oil concessions in the Yellow Sea or the western half of the East China Sea, and offshore drilling under its authority has thus far been limited to the Taiwan side of the Taiwan Strait.
—   Tokyo's dispute with Peking and Taipei has been largely quiescent since the flare-up over ownership of the Senkaku Islands died down.* Tokyo and Taipei have in past years granted conflicting oil concessions, but neither seems to be pushing for

---

*In 1971, as reversion of the Ryukyus to Japan appeared imminent, the question of whether the Senkaku Islands west of the Ryukyus belong to China or Japan became a lively issue. At the time, it was widely believed that Japan needed to make good its claims to the Senkakus in order to get rights to part of the continental shelf. Taipei still argues that Japan "cannot jump over the Okinawa Trench," but Tokyo, using the precedent of Norway's doing precisely that in the North Sea, now confidently asserts the median line doctrine using the Ryukyus as a base line. The dispute over the Senkakus remains unresolved.

actual drilling in the contested areas. With respect to Peking, Tokyo seems most interested in promoting some kind of joint venture, but Peking has thus far been unreceptive.

— Seoul and Taipei have avoided pushing their overlapping claims. Seoul also appears willing to contract its claimed area to the south somewhat in order to open the way for a median line agreement with Peking. Peking has, however, denounced Seoul for "flagrantly and unilaterally" bringing foreign oil exploration companies into "Chinese coastal areas," areas that Seoul claims are on its side of the median line.[26]

— Seoul and Tokyo concluded an agreement in January 1974 for joint exploitation of disputed areas that could be a model for dealing with other similar problems.*[27]

— Pyongyang has apparently kept quiet about any offshore claims that it may have, possibly because it wants to avoid controversy with Peking.

The major uncertainty brought out by the above brief survey is the position of Peking. The Chinese leaders may face something of a dilemma. If they accept a median line settlement, which could easily be negotiated with Tokyo and Seoul, they would have to give up a large, promising area claimed for China by Taipei. If, on the other hand, they endorse Taipei's maximum claim, they would risk damaging relations with Japan (and possibly with the United States).** Even if they could make good a maximum claim, the only way to gain from it might be to employ the technical skills of the same companies that now hold concessions in the disputed area from Tokyo or Taipei.

CHINESE ENERGY POLICY

Current Policy

Current Chinese energy policy emphasizes diversification: regional diversification to increase energy production and consump-

---

*As of July 1974, this agreement had not been ratified by the Japanese Diet, primarily because of current strains between Seoul and Tokyo and secondarily because of the vigorous protest against the agreement by Peking. On February 4, 1974, the Chinese Ministry of Foreign Affairs denounced the agreement as "an infringement on China's sovereignty" and a violation of the principle that questions concerning the division of the continental shelf should be decided by the countries concerned through consultation.[28]

**The U.S. government has displayed nervousness over this very possibility and has warned U.S. companies that it would be inadvisable to explore for oil in the disputed area near the Senkakus.[29]

tion in the less industrialized inland areas and diversification among energy sources to lessen the dominance of coal in China's energy economy.

Regional diversification is desired for a number of reasons. Expansion of coal prospecting and production in coal-short areas, particularly those provinces south of the Yangtze River, promises to alleviate pressure on China's chronically overloaded railway systems, for which coal is the largest payload. Similarly, widespread construction of rural hydropower plants will permit extensive rural electrification without additional movement of coal. In addition to these economic benefits, regional diversification of energy production and consumption contributes to China's long-range political and military ambition of eliminating the distinction between developed coastal and backward interior regions.

Diversification among energy sources means the gradual substitution of oil and natural gas for coal as the leading source of energy. Continued commitment on the part of Chinese planners to increasing the role of oil in China's energy economy may be expected for several reasons. First, oil is virtually irreplaceable as the basis for a number of industries and activities that have formed a part of the modernization process elsewhere and may be expected to do so in China: motor vehicle transport, modern military forces, and the petrochemical industry, including chemical fertilizers. Second, petroleum products and gas have higher thermal efficiencies than other fuels in performing various tasks required for a modern economy. Finally, oil development is also essential to the expansion of road transport, which will facilitate the integration of remote rural areas into the mainstream of Chinese economic and political life.

To carry out these diversification policies, the Chinese government has relied primarily on its administrative control over resource allocation, supplemented by financial incentives. Recent reductions in prices of farm machinery, petroleum products, and electricity for farm production have encouraged agricultural mechanization and motor transport, while preferential interest rates stimulate rural electrification projects. Favorable publicity concerning regions and units that successfully implement desired policies offers further incentive for lower administrative levels to fall into step with official views.

Future Policy Problems

A basic policy decision that China must make is whether the energy sector should merely keep step with domestic requirements or produce a surplus that could be exported to finance imports needed to modernize other sectors of the economy. China's leaders have long displayed an ambivalent attitude toward the benefits of economic specialization as opposed to self-reliance. More recently, they have expressed a desire to stimulate China's technological development by the selective purchase of advanced Western capital goods. There is, therefore, a good possibility that China will try to expand energy exports, presumably in the form of oil, in order to pay for needed imports of capital goods.*

If China does adopt a policy of greatly expanding oil exports, a number of secondary policy problems will arise. The need for improved overland oil transport facilities has already been noted. Major Chinese oil exports will also require a program of harbor improvement or construction of offshore loading terminals. Ships of 30,000 tons are the largest that can dock at Chinese ports today.[30]

China must also decide whether to enlist the help of foreigners in the development of its energy resources, particularly those offshore. Thus far, China has shown a strong inclination to go it alone,[31] but this policy could conceivably change as the magnitude of the technological problems involved becomes more apparent.

China cannot indefinitely postpone dealing with the problem of conflicting claims to the seabed off its east coast. A clue to the approach to this problem that China may adopt is provided by a working paper submitted by the Chinese delegation in July 1973 to the UN Committee on the Peaceful Uses of the Sea-Bed and Ocean Floor Beyond the Limits of National Jurisdiction.[32] This paper states that "by virtue of the principle that the continental shelf is the natural prolongation of the continental territory, a coastal state may reasonably define . . . the limits of the continental shelf under its exclusive jurisdiction. . . ." The paper further declares that "states adjacent or opposite to each other, the continental shelves of which connect together, shall jointly determine the delimitation of the limits of jurisdiction of the continental shelves through consultations on an equal footing." How the Chinese would apply these general principles to their own situation is not clear. A negotiated settlement with Korea would appear to be consistent with the general Chinese position, but whether Peking is ready to deal with Seoul on this issue or would see much point in negotiating only with Pyongyang may be questioned. On the other hand, taken literally, the principles enunciated by the Chinese would confine the Japanese

*The energy sector itself may require substantial imports of advanced capital goods, notably: offshore drilling equipment, equipment for refineries, large-scale thermal generating equipment, and large-scale mining machinery.

to the area east of the Okinawa Trench and deny them any rights on the continental shelf. Whether China would adhere rigidly to this position, however, is uncertain.

Finally, as a participant in the world petroleum market, China must take a position on issues affecting other oil exporters. There is little doubt that it will give strong verbal support to efforts by the OPEC countries to maintain high prices and to reduce the power of the international companies. Such a policy would accord both with China's emerging interest as an oil exporter and with the Chinese government's desire to be accepted as a champion of the Third World in its struggle against Western imperialism. Whether China would cooperate in either a future Arab oil embargo, or in an effort by major oil-exporting countries to sustain prices by restricting production, is at least open to question. Certainly, China's oil export policy appears not to have been influenced by the Arab embargo and production cutback of 1973-74.

* * * *

If, as the above discussion suggests is possible, China does become an important exporter of oil in the late 1970s and 1980s, the most likely market for its oil would be Japan. The low cost of transporting oil to Japan would presumably enable China to command a premium price in much the same way as Libya does today from nearby European importers. And Japan, anxious to diversify its sources of energy, would welcome the emergence of a new source outside the politically troubled Middle East.

Although the United States might not receive any of China's oil exports, it would benefit along with other oil-importing countries if the emergence of China as a major exporter contributed to a general loosening of oil supplies internationally. Moreover, if U.S. oil companies became involved in the exploitation of China's offshore oil resources, a new, mutually beneficial element could be added to Sino-U.S. relations.

NOTES FOR CHAPTER 12

1. U.S. Congress, Joint Economic Committee, *An Economic Profile of Mainland China*, in two volumes, 90th Cong., 1st sess. (1967), I: 302; and Yuan-li Wu, *Economic Development and the Use of Energy Resources in Communist China*, Hoover Institution on War, Revolution and Peace (New York: Praeger Publishers, 1963), p. 35.

2. U.S. Congress, *Economic Profile*, I: 293; and U.S. Congress, Joint Economic Committee, *People's Republic of China: An Economic Assessment*, 92nd Cong., 2nd sess. (1972), p. 83.

3. A. A. Meyerhoff, "Developments in Mainland China," *The American Association of Petroleum Geologists Bulletin*, August 1970, pp. 1567-80.

4. John M. Wageman, Thomas W. C. Hilde, and K. O. Emery, "Structural Framework of East China Sea and Yellow Sea," *The American Association of Petroleum Geologists Bulletin*, September 1970, pp. 1611-43.

5. Meyerhoff, "Developments in Mainland China," p. 1569.

6. British Broadcasting Corporation, *Summary of World Broadcasts*, Part 3, "The Far East," W-706, A-7; *Current Scene* (Hong Kong), March 1973, p. 6.

7. Meyerhoff, "Developments in Mainland China," p. 1579; Ho Ko-jen, "Peiping's Petroleum Industry," *Issues and Studies* (Taipei), August 1968, pp. 22-25.

8. Cheng-siang Chen, *Petroleum Resources and their Development in China* (in Chinese) (Hong Kong, 1968).

9. Meyerhoff, "Developments in Mainland China," p. 1568.

10. British Broadcasting Corporation, *Summary of World Broadcasts*, Part 3, "The Far East," W-655, A-11 and W-693, A-11.

11. *New York Times*, January 9, 1974, p. 2.

12. *Ming Pao* (Hong Kong), November 15, 1973, p. 2B.

13. K. C. Yeh, *Communist China's Petroleum Situation* (Santa Monica, Calif.: RAND Corporation, 1962).

14. U.S. Congress, *An Economic Profile of Mainland China*, I: 301.

15. The estimates of total electric power output and the portion contributed by hydropower were derived by Thomas G. Rawski from a number of sources, as is explained in "The Role of China in the World Energy Situation," an unpublished background paper prepared by Rawski for The Brookings Institution, June 1973.

16. *Current Scene* (Hong Kong), March 1973, p. 7.

17. Republic of China, Ministry of Economic Affairs, *A General Survey of the Power Industry in Mainland China* (in Chinese) (Taipei, 1968), pp. 21-22; Yuan-li Wu, ed., *China: A Handbook* (New York: Praeger Publishers, 1973), p. 77.

18. Thomas G. Rawski, "Recent Trends in the Chinese Economy," *China Quarterly*, No. 53 (1973), pp. 1-34.

19. *Petroleum Press Service* (London), February 1973, p. 69.

20. *Washington Post*, January 10, 1974, p. A-14.

21. *Ming Pao* (Hong Kong), November 15, 1973, p. 2B.

22. *Oil and Gas Journal*, "Newsletter," February 11, 1974, p. 4.

23. *The Economist* (London), February 3, 1973, p. 64.

24. C. S. Chen and K. N. Au, "The Petroleum Industry of China," *Die Erde* (Berlin), 1972/3-4, p. 319.

25. This section is based in part on off-the-record talks by one of the authors with informed persons in Tokyo, Taipei, Hong Kong, and Seoul in February 1973. See also Choon-ho Park, "Oil Under Troubled Waters: The Northeast Asia Seabed Controversy," Harvard Law School, *Studies in East Asian Law* (China: No. 20, 1973), pp. 212-60.

26. *Far Eastern Economic Review* (Hong Kong), April 23, 1973, p. 36.

27. *Newsreview* (Seoul), May 5, 1973, p. 7; *Oil and Gas Journal*, February 11, 1974, p. 27.

28. Foreign Broadcast Information Service, *Daily Report, People's Republic of China*, No. 25, Vol. I (February 5, 1974), p. A4.

29. *Petroleum Press Service* (London), May 1971, p. 195.

30. *Ports of the World* (London: Benn Brothers, 1972), pp. 658-59.

31. In January 1973, Chinese Prime Minister Chou En-lai reportedly rejected a Japanese proposal for joint development of China's oil resources. *South China Morning Post* (Hong Kong), January 20, 1973.

32. United Nations, General Assembly, "Working Paper on General Principles for the International Sea Area," A/AC.138/SC.II/L.34 (August 1973), pp. 3-4.

Part Five

The World Oil Balance: Prospects and Implications

Chapter Thirteen

World Market Trends and Bargaining Leverage

by Edward R. Fried

Much of the current discussion of world energy problems proceeds on the assumption that the world oil market will be characterized by chronic supply shortages—whether for economic, technical, or political reasons—and that as a result the oil-exporting countries will continue to enjoy a dominant bargaining position for the indefinite future. Concern along these lines was already evident in 1972 and early 1973 as the oil-exporting countries showed steadily increasing strength in their negotiations with the international companies by imposing a succession of tax increases on each barrel of oil and gaining ownership shares in the concessions. The oil embargo of October 1973 and the subsequent quantum jump in tax levies transformed this concern into something close to despair.

This chapter examines the above assumption and outlook by exploring the major factors underlying the prospective world oil market and how they might be affected by the drastically changed conditions that now exist. The procedure will be to outline possible trends in world energy requirements and in world production of primary energy sources, to show how those trends could influence the size of the world oil trade and the ability of exporting countries to organize and control the world oil market, and on that basis to speculate about the direction of future oil prices and the evolution of marketing arrangements. Such projections are inevitably highly uncertain. Their value as an aid to policy formulation depends less on their precision and more on the degree to which they can isolate the major variables and show how those variables are likely to move as circumstances change.

PROJECTED WORLD ENERGY REQUIREMENTS

For present purposes, it is helpful to distinguish the pre-October 1973 situation from that which exists today. The oil embargo and the events that followed have substantially modified previous trends in total energy consumption and in the mix of fuels used and thus have fundamentally altered the world energy outlook. Using the pre-October 1973 situation as a starting point provides a convenient means of analyzing those changes and estimating their quantitative significance.*

Pre-October 1973 Projections

Before the October oil embargo, projections of world energy demand were fairly straightforward, deceptively so it turned out. For most of the postwar period, world energy demand, influenced principally by growth in GNP, increased by roughly 5 percent a year. Among industrializing economies, energy consumption rose by more than the growth in GNP; among industrially mature economies, the reverse was true. As to the mix of fuels, long-term trends showed a steady shift away from coal and toward the use of oil and natural gas.

The extraordinary cheapness of energy prices in general, and oil prices in particular, was a critical factor imbedded in these trends. In real terms, the price of energy fuels declined by approximately 30 percent between 1950 and 1970,[1] which encouraged energy consumption and reduced incentives for industrial investment in energy-saving products and methods of production. Prices also influenced the persistent preference for oil over coal. Throughout the period, oil prices were competitive with coal in terms of heat content and much cheaper when allowance was made for the advantages of oil over coal in convenience, efficiency, and flexibility of use.[2] Beginning in the mid-1960s, moreover, a growing emphasis on environmental considerations, particularly in the United States, gave oil a further competitive edge over coal. Consequently, despite the efforts of governments in Western Europe and Japan to subsidize and protect their coal industries, coal production steadily

*For convenience, the assessment of world market trends in this chapter excludes the USSR, Eastern Europe, and China on the assumption that those countries as a group will be self-sufficient in oil and therefore will not affect the world market. This is the minimum expectation and thus represents a conservative factor in these forecasts. If anything, this group of countries could prove to be a source of significant net exports, but there is very little data on which to estimate quantities. Where appropriate, such possibilities will be noted in the text without making specific allowances for them in the projections.

declined and oil accounted for fully two-thirds of the growth in energy consumption.

A projection of world energy requirements for 1980 and 1985, based principally on the continuation of these past trends, is shown in Table 13-1.[3] It rested on three critical assumptions: (1) in comparison with the 1960s, GNP in the United States would grow at a slightly higher rate, in Western Europe at about the same rate, in Japan at a substantially lower rate, and in developing countries at a significantly higher rate; (2) oil prices would stay about constant in real terms, that is, at about $3 a barrel (in 1973 dollars) for Saudi Arabian oil, f.o.b. the Persian Gulf; and (3) the world market would continue to be a reliable source of supply in the sense that the major exporters would not withhold oil from the market for either political or economic reasons.

On these assumptions, which it should be emphasized were made invalid by the October oil embargo, the non-Communist world's energy requirements over the period 1970-85 were projected to increase by 5.4 percent a year, or at about the same rate as in the 1960s. As to the mix of primary sources of energy to meet these requirements, the projections in Table 13-1 highlight the following trends:

— Oil would maintain its favored position as a source of energy. Between 1970 and 1980, the proportion of oil in total world energy supplies would continue to rise, from 52 percent to 58 percent, and decline only moderately thereafter.
— Natural gas would provide roughly one-fifth of total requirements, the proportion dropping slightly over the period. Its declining relative importance in the United States would be almost offset by its rapidly growing use elsewhere, particularly in Western Europe.
— The most striking changes were projected to take place in coal and nuclear power. A continued sharp decline in the world's relative dependence on coal was expected to be more than offset by the rapidly rising use of nuclear power. By 1985 nuclear power would supply almost as much energy as coal.

Thus, for most of the 1970s, oil would again supply a more than proportionate share of the growth in world energy requirements, as was true during the 1960s. In the 1980s, however, nuclear power would take over as the most dynamic element on the supply side.

Several features of this projection relating to the changing geographic pattern of energy consumption also merit attention. (See

Table 13-1.   Pre-October 1973 Projection: Energy Requirements of Non-Communist Nations, 1970-85

	Requirements (in million barrels per day of oil equivalent)			Projected Annual Growth (Percent)		
	Actual 1970	1980	1985	1960-70	1970-80	1980-85
United States:	*32.7*	*49.9*	*60.8*	*4.2*	*4.3*	*4.0*
Oil	13.7	24.5	26.7	4.0	6.0	1.7
Gas	11.3	12.2	13.9	5.7	0.8	2.6
Coal	6.3	8.8	10.1	2.4	2.9	2.7
Nuclear	.1	2.7	8.2		39.0	24.9
Hydro	1.3	1.5	1.6	5.0	1.4	1.1
Geothermal	neg	.2	.3			
Western Europe:	*21.0*	*35.2*	*44.5*	*5.5*	*5.3*	*4.8*
Oil	12.5	23.3	28.7	12.0	6.4	4.3
Gas	1.4	4.5	6.0	20.3	12.3	5.9
Coal	6.2	4.8	4.3	-0.2	-2.6	-2.2
Nuclear	.2	1.5	4.3		22.3	23.4
Hydro	.7	1.0	1.2	5.2	3.6	3.6
Japan:	*5.4*	*11.9*	*16.0*	*11.9*	*8.2*	*6.1*
Oil	3.8	9.0	11.2	19.7	9.0	4.5
Gas	.1	.7	1.2	15.2	21.0	11.2
Coal	1.3	1.3	1.5	2.7	1.0	1.0
Nuclear	neg	.7	1.8			20.8
Hydro	.2	.2	.3	3.8	2.8	2.8
Other Industrial:	*4.5*	*8.1*	*10.6*	*6.3*	*6.1*	*5.9*
Oil	1.9	3.7	4.4	6.9	6.9	3.5
Gas	.8	1.6	2.2	14.4	7.1	6.6
Coal	1.5	2.1	2.5	3.4	3.4	3.4
Nuclear	neg	.2	.7			28.5
Hydro	.3	.5	.8	5.8	5.8	5.8
Developing Countries:	*8.7*	*18.9*	*27.5*	*6.8*	*8.1*	*7.8*
Oil — domestic	4.7	10.5	14.9	7.0	8.4	7.3
Oil — bunkers	1.0	2.0	2.8	6.6	7.5	7.5
Gas	1.2	3.1	5.0	11.8	10.0	10.0
Coal	1.6	2.6	3.3	3.4	5.0	5.0
Nuclear	neg	.3	.8			21.7
Hydro	.2	.4	.7	9.2	9.2	9.2
Total:	*72.3*	*124.0*	*159.4*	*5.4*	*5.5*	*5.2*
Oil	37.6	73.1	88.7	7.6	6.9	3.9
Gas	14.8	22.1	28.3	7.0	4.1	5.1
Coal	16.9	19.6	21.7	1.0	1.5	2.0
Nuclear	.3	5.4	15.8		33.5	24.0
Hydro, Geothermal	2.7	3.8	4.9	5.6	3.5	5.2

Sources and Notes: *United States.* Supply projections are based on Case III, National Petroleum Council, *U.S. Energy Outlook* (Washington, D.C., 1972), p. 32. This lower of the two intermediate alternatives assumes moderately rising domestic production of oil, declining production of natural gas, and substantially increasing imports of both oil and gas. The price assumptions are ambiguous; they are projected to double between 1970 and 1985, principally on the assumption of low or moderate drilling and finding rates, but they are not related to market conditions or to the prices of competitive fuels. In comparison to the 1960s, energy demand is assumed to grow somewhat more rapidly between 1970 and 1980 and somewhat less rapidly beyond 1980. The assumption of a higher GNP growth rate accounts for the former, and the influence of the long-term trend toward declining energy use in relation to GNP for the later. Energy growth rates for 1960-70 are from Walter G. Dupree and James A. West, *United States Energy Through the Year 2000*, U.S. Department of the Interior (Washington, D.C., 1972). (Also see note on nuclear energy.)

Western Europe. Projections for 1980 are based on Organisation for Economic Co-operation and Development (OECD), *Oil: The Present Situation and Future Prospects* (Paris, 1973), pp. 41-42. For 1985, projected growth rates for total energy consumption and for coal and gas are based on a background paper on *Western European Energy Policy* prepared for The Brookings Institution by the European Community Institute for University Studies (1973). (Also see note on nuclear energy.)

Japan. Projected growth rates are from a background paper prepared for The Brookings Institution by the Japan Economic Research Center and the Institute of Energy Economics (Tokyo). Growth rates for 1960-70 and figures for 1970 are from OECD, *Oil: The Present Situation and Future Prospects.* (Also see note on nuclear energy.)

Other Industrial Countries. This category consists of Canada, Australia, New Zealand, and South Africa. Figures for 1970 and the growth rate for 1960-70 are from United Nations, Department of Economic and Social Affairs, *World Energy Supplies, 1961-1970,* Series J, No. 16 (New York, 1973). Projections for 1985 were based on maintenance of the 1960-70 growth rate in GNP and a moderate decline in the ratio of energy consumption to GNP. (For nuclear energy projections, see note below.)

Developing Countries. Figures for 1970 and growth rates for 1960-70 are from United Nations, *World Energy Supplies, 1961-1970.* They include oil-exporting and oil-importing developing countries. Projections assume a GNP growth rate of 6.3 (the 1968-72 rate) and a ratio of energy consumption to GNP of 1.3 (slightly higher than for 1960-70). The rate of coal production is assumed to rise because India, the principal coal producer in this category, would be under severe balance-of-payments pressure to conserve oil imports, even at the pre-October 1973 price for oil. (Also see note on nuclear energy.)

Oil for bunkers supplied in developing countries is shown separately because projected requirements are based on trade volume rather than GNP. Bunkers supplied for outgoing transportation are treated as a consumption requirement on the assumption they are roughly equal to bunkers supplied elsewhere for incoming transportation. Since exports of developing countries are high in volume in relation to value, this simplifying procedure moderately exaggerates the oil-consumption requirements of developing countries. Figures for 1970 are from United Nations, *World Energy Supplies,* 1961-70. (Projections are based on current growth rates in exports of developing countries.)

Nuclear Energy. Standard projections before the oil embargo were based on the worldwide forecast made by the U.S. Atomic Energy Commission (AEC), "Nuclear Power, 1973-2000" (Washington, D.C., 1972), projections for the European Community prepared by the Commission, and projections for Japan made by various Japanese agencies and institutions. All proved to be substantially too high in light of trends prevailing before October 1973. A more

accurate reflection of the situation at that time is the low forecast of the most recent AEC projection, "Nuclear Power Growth, 1974-2000," which is based on information available as of February 1974. The low forecast essentially shows what would have happened had there been no change in emphasis on nuclear power or on procedures in constructing nuclear plants. Thus it assumes for the United States that licensing and technical difficulties would result in an interval of ten years between the planning and completion of a nuclear generating plant, and similar delays abroad in bringing nuclear plants on line. This low projection therefore has been used to show the most likely trend before the October embargo. Specifically the figures shown in Table 13-1 assume an installed capacity in the United States of 85 million kilowatts in 1980 and 231 million kilowatts in 1985, and abroad (non-Communist countries only) of 87 million kilowatts in 1980 and 214 million kilowatts in 1985. The combined figures for all non-Communist countries for 1985 represent a reduction of 15 percent from what the AEC had believed two years earlier to be the most likely forecast, which reflects the growing problems that had been developing in the licensing and construction of nuclear generating plants.
Figures may not add due to rounding.

Table 13-2.) One is the declining U.S. share in total energy consumption of non-Communist nations and the markedly rising shares of Japan and, particularly, the developing countries. By 1985, the developing countries were projected to consume about as much energy as Western Europe and Japan together consumed in 1970. (Table 13-1.) Even so, the differences among nations in per capita energy consumption would continue to be strikingly and to some extent inexplicably large. Thus Americans, who can be expected by 1985 to have per capita incomes not substantially higher than those of West Europeans or Japanese, would be consuming about twice as much energy on a per capita basis. This in itself dramatizes the

Table 13-2. Geographic Shares and Per Capita Energy Consumption: Non-Communist Nations, 1970-85 (Pre-October 1973 Projection)

	Share of Total Energy Consumption (percent)		Per Capita Energy Consumption (in barrels per year oil equivalent)	
	1970	1985	1970	1985
United States	45.2	38.5	57.9	87.8
Western Europe	29.0	28.2	21.4	40.4
Japan	7.5	10.1	19.0	48.1
Other industrial countries	6.2	6.8	35.9	48.3
Developing countries	12.0	17.4	1.6	3.6
	100.0	100.0		

Sources: Consumption data as projected in Table 13-1. Population projections shown in United Nations, Department of Economic and Social Affairs, Population Division, *Total Population Estimates for World, Regions and Countries Each Year, 1950-1985* (mimeographed), ESA/P/WP.34 (New York, 1970).
Figures may not add due to rounding.

uniquely energy-intensive character of the American society, a fact that may now be in process of substantial change. On the other hand, the gap in per capita energy consumption between the developing countries and either Western Europe or Japan, while enormous, is largely explainable by differences in per capita GNP.

The Effect of the October Oil Embargo

Events following the October war decisively changed the foregoing outlook. For one thing, cheap oil is clearly a phenomenon of the past. Even before the October crisis, the process of reversing the long-term trend in oil prices was well under way. As a result of the Tehran and Tripoli negotiations and the Geneva agreements, the export price of oil by June 1973 had doubled over the low point of June 1970.* Then, in the final three months of the year, the producers imposed a threefold price increase on the market. (See Table 13-3.) So large a jump in price will necessarily reduce consumption and weaken oil's competitive position in relation to other primary sources of energy.

Table 13-3.   Posted and Estimated Market Prices for Oil
(U.S. $ per barrel)

Date	Posted Price	Estimated Market Price in Current Dollars	Estimated Market Price 1973 Dollars
January 1955	1.93	1.93	3.26
September 1960	1.80	1.45	2.16
January 1970	1.80	1.30	1.48
February 15, 1971	2.18	1.65	1.80
January 20, 1972	2.48	1.85	1.96
January 1, 1973	2.59	2.20	2.27
April 1, 1973	2.76	2.30	2.32
June 1, 1973	2.90	2.70	2.67
October 16, 1973	5.12	3.65	3.54
January 1, 1974	11.65	8.45	8.20
May 1974	11.65	9.55	9.05

Sources and Notes: *Market price in current dollars*—from January 1955 through October 16, 1973 —are from a World Bank Staff Memorandum (informal). The market price for January 1974 is estimated on the assumption that 75 percent of exports was equity oil (tax paid cost of $7.00 plus $.65 for production costs and company profits) and 25 percent was participation oil sold at 93 percent of posted price. For May 1975, equity oil is assumed to be 40 percent and participation oil 60 percent.

Prices in 1973 dollars are calculated by use of the implicit price deflators on GNP, U.S. Department of Commerce Series.

Note: Prices are for Saudi Arabian light oil 34°, f.o.b. Ras Tanura.

*See discussion of these negotiations and agreements in Chapter 2.

Perhaps equally important, the traumatic impact of the embargo set governments in the importing countries on a course to achieve greater self-sufficiency in energy. The United States is wrestling with a suitable framework for its Project Independence in light of the differences in financial and other costs that would be required to meet alternative goals; the European Community is seeking to work out a new long-term energy strategy aimed at greater self-sufficiency; and Japan is concentrating on energy conservation. These programs are not working blueprints, but rather first approximations of goals that will be reformulated as better information on the cost of alternative policies becomes available. The main point is that initial commitments have been made to new energy policies aimed at greater security of supply. A fall in oil prices probably would modify these commitments and slow down the enactment of necessary measures, but concern over excessive dependence on unreliable sources of energy imports is now too well entrenched to permit the trend to be completely reversed.

In effect, therefore, the embargo and its aftermath have brought about the end of oil's age of innocence and have made obsolete the assumption of an ever-expanding world oil market. How then will higher oil prices and government policies to achieve greater security of supply change the world energy outlook from that shown in the pre-October 1973 or base forecast? Specifically, by how much will world oil requirements be reduced from projected levels? Answers to these questions must necessarily be quantitatively approximate and in some respects oversimplified, since the world has had virtually no experience with high-priced and uncertain oil supplies. The approach adopted here is to outline some working assumptions that seem reasonable in light of present information about the effect of the new situation on future energy consumption and on the future production and use of non-oil primary sources of energy, and then to trace their effect on the pre-October 1973 projection. (Possible increases in oil production in the importing countries as a result of higher oil prices will be noted in the following section as part of the discussion of the prospective world market for oil.)

For this exercise, world oil prices are assumed to be pegged to a U.S. price of $7.50 a barrel, *in constant dollars of 1973 purchasing power.* This would be approximately equivalent to a price of $6.25 a barrel, f.o.b. Persian Gulf, for Saudi Arabian light oil, or roughly one-fourth lower than market prices at the beginning of 1974. The U.S. price has been chosen for this purpose because the United States is the largest consumer of energy and potentially the largest source of incremental energy supplies among the importing countries; hence the U.S. price would be the most important single determinant of the size of the world oil market. The long-term price of $7.50 a barrel is used initially because a number of studies suggest that this price (and its competitive equivalent for other primary sources of energy) would be sufficient to bring forth substantial additional U.S. production of conventional oil, natural gas, and coal, and also to encourage investments in production of synthetic oil from shale and possibly coal.[4] Moreover, a price of $7.50 a barrel could be expected to have a significant impact on consumption in all industrial countries, since it would represent at least a doubling of the crude oil prices assumed in the pre-October 1973 forecast of world energy demand.

World energy consumption is the most important variable. A U.S. price of $7.50 a barrel, and its equivalent elsewhere, combined with national conservation measures in all importing countries, is assumed to reduce the projected growth of world energy consumption by about 1 percentage point a year—that is from 5.4 percent to 4.5 percent a year. In aggregate terms, this would mean a decline of 8 percent in 1980 and 13 percent in 1985 from the consumption levels shown in Table 13-1.

Unfortunately, it is difficult to test this assumption on the basis of past economic studies of the relationship between price and the demand for energy, again because those studies suffer from the lack of experience and data derived from a world of high oil prices. The work now going on, however, notably that of Houthakker, suggests that the response of energy demand to price is substantially greater than is here assumed. A critical conclusion in these recent investigations is that energy consumption and oil use grew rapidly in the past because prices were falling and that "the abrupt recent turnaround in oil prices will sharply curtail demand growth just as previous price weakness encouraged it."[5]

The assumption of a 1 percentage point reduction in the rate of growth in energy consumption is in line with present government forecasts and other indications, although in these instances as well it may be conservative. In the United States, for example, the Federal Energy Office looks toward a reduction of 1.5 to 2.5 percentage points in U.S. energy consumption.[6] In Western Europe, an indication of prospects can be gained from the new proposals of the Commission of the European Communities, which envisage by 1985 a reduction of 10 percent in energy consumption from the amount estimated in policy statements drawn up before the October crisis.[7] In Japan and the developing countries, the growth in energy consumption has been much higher than the world average; hence a reduction of 1 percentage point in the rate of growth would be proportionately smaller there than in the United States or Western Europe. Also, in both areas the pressure for energy conservation will be very great; in Japan, because of the nation's almost complete dependence on imported supplies of primary energy fuels; and in the developing countries, because of balance-of-payments constraints.

Industry could be expected to account for a more than proportionate share of reductions in energy consumption resulting from high prices and conservation. It is the largest consumer of energy (38 percent in the United States, 45 percent in Western Europe, and 55 percent in Japan) and the sector in which consumption is likely to be most responsive to changes in the price of crude oil.* The economics of energy saving in industry become exceedingly attractive as the price of fuels rises. At $7.50 a barrel for crude oil (and nearly comparable increases in the price of natural gas and coal), a reduction of 1 percentage point in the rate of growth of industrial energy use would mean a potential saving for U.S. industry of $5 billion a year by 1980. This would provide a very large incentive for investments in energy-saving equipment and industrial processes. These considerations apply equally to the industrial sectors in Western Europe and Japan.**

---

*At the assumed price of oil, a reduction of 1 percentage point in the rate of growth in industrial demand for energy would imply a long-range price elasticity of only –.2, which seems inordinately low, thus suggesting that industrial demand for energy would be cut by more than the projected average. This implicit price elasticity is based on the following calculation. Compared to the pre-embargo situation, a price of $6.25 a barrel f.o.b. the Persian Gulf, would mean an average increase of approximately 100 percent in the price of crude oil in all OECD countries. Because of the virtual absence of internal taxes on residual and distillate fuel oil, an increase in the price of oil results in a relatively large increase in the price of oil products used by industry. Taking into account existing margins for refining and distribution costs and for profits, it is estimated that a doubling of the price of crude oil would result in an increase of more than 50 percent in the price of oil products used by industry. Prices of other primary sources of energy presumably would rise by somewhat less than 50 percent if they are to improve their competitive position against oil. In all, therefore, the projected increase in the price of crude oil might be said to cause an average increase of 50 percent in the price of all forms of energy used by industry. In response to this price increase of 50 percent, the assumed reduction of 1 percentage point in the rate of growth of energy demand works out to be roughly equivalent to a reduction of 10 percent in projected consumption.

**Initial reactions in the United States suggest that this process is under way. Former Federal Energy Administrator William Simon, in testimony before the House Committee on Government Operations, April 9, 1974, said that "both industry and government were quick to take advantage of the dollar savings that accompanied energy economies. Energy audits prompted many reports of 15 to 30 percent savings." The Dupont Company, on the basis of its

A similar conclusion, not so sharply drawn, would apply to commercial and residential use of energy, which account for approximately 30 percent of energy consumption in OECD countries.

On the other hand, energy growth rates in the transportation sector could well decline more slowly than is here assumed if higher crude oil prices are the only restraining influence. This is particularly significant for present purposes because energy requirements for transportation are uniquely dependent on oil. A major reason for the difference in the response of demand to higher crude oil prices is that consumer taxes on oil, which are extremely heavy in Western Europe and larger in Japan than in the United States, are concentrated on gasoline and diesel oil. Consequently, a doubling of the price of crude oil would mean a much smaller percentage increase in the price of gasoline—about 25 percent in the United States, 18 percent in Japan, and only 11 percent in Western Europe.*

For this and other reasons, the possibilities for fuel saving in the transportation sector differ substantially among the three areas. In Western Europe and Japan, where transportation accounts for an average of 25 percent of oil requirements, fuel savings are likely to be relatively small because of the more limited effect of higher crude oil prices on product prices, because small energy-saving vehicles are already the standard, and because mass transit is relatively well developed. Conversely, in the United States, where transportation accounts for more than half of total oil consumption, energy savings are likely to be greater because higher crude oil prices will significantly affect gasoline prices and accelerate the trend toward smaller cars, and because energy-saving considerations are causing the government to invest more heavily in mass transit facilities.

On the supply side, higher oil prices (and therefore higher prices for competing sources of energy), together with governmental measures to achieve greater self-sufficiency in energy, could bring about significant increases in the production and use of natural gas, coal, and nuclear power in place of oil. What might be the

---

own experience both inside and outside the company, has estimated that a significant conservation effort on an operating industrial plant will normally yield a 7 to 15 percent reduction in energy demand within a reasonably short time.[8]

*These estimates are based on data supplied by the Office of Energy Research, Central Intelligence Agency. Taxes per barrel of gasoline are estimated to be $5 in the United States, $14 in Japan, and $28 in Western Europe. The margin for refining, distribution, and profits is estimated to be $6.50 in the United States, $5.50 in Japan, and $6 a barrel in Western Europe.

consequences for production and use of each of these non-oil sources of energy compared with the base forecast projected in Table 13-1?

For natural gas, increased production in the United States and Western Europe presents the major possibilities.

As far as the United States is concerned, a number of studies suggest that a wellhead price (in 1973 dollars) in the range of 55 to 95 cents per 1,000 cubic feet (28 cubic meters)* could, by 1985, increase natural gas production by the oil equivalent of approximately 5 million barrels a day. At 95 cents per 1,000 cubic feet, the upper end of the range,** natural gas (in Btu equivalents) would still be one-third cheaper than crude oil priced at $7.50 a barrel, to say nothing of the fact that it is an environmentally preferred and more convenient fuel for heating purposes. To be conservative, it is assumed for present purposes that higher prices will increase U.S. natural gas production by half the amount suggested in these studies, that is by the oil equivalent of 1 million barrels a day by 1980, and by 2.5 million barrels a day in 1985.

In Western Europe, the European Commission's proposals for greater energy self-sufficiency envisage an increase in natural gas production of over 2 million barrels a day in oil equivalent by 1985. This is to be achieved by higher drilling rates and financial subsidies. It seems evident that higher oil prices alone will increase drilling rates in the North Sea at least and that this will result in increased natural gas production, if only as a joint product with oil. Again, to be conservative, it will be assumed that natural gas production in Western Europe will increase by half the target amount, that is, by the oil equivalent of 0.5 million barrels by 1980 and 1 million barrels by 1985.

Soviet natural gas could also become an extremely important source of energy supply in Western Europe. Presently planned levels of Soviet gas exports to Western Europe are not significant. However, the possible construction of a 2.5 meter pipeline from eastern Siberian fields to West Europe—which is now under West German-USSR technical study—would be a different matter since the pipeline could carry amounts of natural gas

---

*In mid-1974, the average wellhead price for gas moving across state lines was about 30 cents per 1,000 cubic feet. During the first half of 1974, however, the Federal Power Commission approved a number of new contracts calling for prices in the 40 to 50 cents range. Prices for gas produced and sold within the same state are much higher.

**The range of estimates for 1985 is shown and analyzed in Milton Searl, *Towards Self-Sufficiency in Energy Supply*. Searl also points out that the natural gas resource base in the United States appears adequate to support growing production through the year 2000.

equivalent to 3 million barrels a day of oil. Soviet natural gas reserves are more than ample, and higher prices of oil would greatly increase the economic incentive. However, as pointed out in Chapter 8, such a development is at best a prospect for the more distant future. Thus, no allowance for it will be made in the current projection.

Coal will have a substantially larger role in the post-embargo energy world, both for use directly and as a source of synthetic gas and oil. Only its direct use will be considered here; the possibilities for producing oil from coal, which potentially are more important as an incremental source of usable energy in the more distant future, will be noted in the discussion on prospective oil supplies.

In the United States, coal resources are huge, but coal now provides less than one-fourth of total U.S. energy requirements. Manpower, financial, and environmental problems stand in the way of larger production. Nevertheless the recent spurt in coal prices should be a powerful incentive. Projections by the Federal Energy Office suggest that, technically, coal production could be increased 50 percent by 1980 and 100 percent by 1985,[9] which in comparison with the base forecast would mean additional supplies equivalent to 0.5 million barrels of oil a day by 1980 and 2 million barrels a day by 1985. Even more substantial increases would evidently be possible in light of the extensive resource base. To use additional coal in place of oil on a substantial scale in the United States, however, would require difficult technical solutions to environmental problems or a relaxation of environmental standards. Perhaps equally important, changes in rate-making rules would be required to provide utilities with assurances that they would obtain economic advantages from, or at least an adequate return on, the substantial investments necessary to convert generating capacity from the use of oil to coal.

To err on the side of caution, it will be assumed in drawing up the current projection that no additional coal will be used in the United States in 1980 and 1985 over the amounts previously shown for the pre-October 1973 projection. This assumption is excessively pessimistic but it has the advantage of providing a cushion for two unfavorable contingencies: (1) that the unique characteristics of the U.S. energy mix will impose technical constraints on the extent to which additional production of other primary energy sources can substitute for oil; and (2) possible shortfalls in production increases of natural gas or nuclear generating capacity or in the assumed rate of savings in energy consumption.

In Western Europe, conversion problems do not arise to any sizable degree because the objective need not be to increase the use of coal but rather to halt or moderate its decline. The European Commission's proposals for a new energy strategy would seek to maintain future production at 1973 levels, principally through rationalizing production, granting financial subsidies, and instituting new manpower training programs. French officials have termed this objective overly optimistic and the analysis in Chapter 8 suggests a similar conclusion in view of the strong movement of labor out of coal mining in both Germany and the United Kingdom, the two principal coal-producing countries in the European Community. On the other hand, subsidies and substantially higher market prices for coal could change the situation. Furthermore, in Turkey, the other major coal producer in OECD Europe, the new energy situation is certain to bring about increased coal output. In light of these considerations, projecting the decline in coal production in Western Europe at half the rate shown in the base forecast would seem to be a reasonable adjustment. This would mean an addition to supplies equivalent to 0.7 million barrels a day of oil in 1980 and 0.8 million barrels a day in 1985.

Among the developing countries, the possibilities for sizable increases in coal production exist principally in India, which produces about three-quarters of the total coal output in developing countries, and in Indonesia, Korea, and several African countries. Between 1968 and 1971, India's coal production actually declined because of the lack of investment and adequate transportation facilities. The availability of cheap oil was a major reason for these deficiencies. Conversely, the fourfold rise in oil prices from that period and the increased balance-of-payments difficulties that now plague India because of higher oil-import costs are powerful incentives for sharply reversing this trend in coal production. Time and substantial amounts of capital will be required. Principally on the basis of probable developments in India, coal production in developing countries is projected to increase by 7 percent a year, about double the rate of growth in the 1960s. In comparison with the base forecast, this would mean an addition to coal supplies in the developing countries equivalent to 0.8 million barrels of oil a day in 1980 and 1.1 million barrels a day in 1985.

In the case of nuclear power, the AEC's middle range or most likely forecast, revised as of February 1974, has been used to reflect the new energy situation. Before the October embargo, time lags between the proposal and completion of nuclear generating plants had been lengthening to intervals of 8 to 10 years. As a result, it appeared likely that nuclear generating capacity in 1980 and 1985, both in the United States and abroad, would be substantially below previously formulated plans and projections, as explained in the notes to Table 13-1. Because of high oil prices and aggravated concern about security of energy supply, a significant speed up in the number of nuclear generating plants under construction is now likely.

In the United States, the AEC has proposed administrative changes that are designed to reduce the time between proposal and completion from 10 years to 7½ years, and the administration has submitted legislative proposals to further shorten this interval to 6 years.[10] Reaching these goals, however, depends not only on altering AEC licensing procedures but also on actions by industry to standardize reactor designs and to overcome safety and technical problems. Taking these difficulties into account, the AEC in its most recent forecast now projects that nuclear power capacity in the United States by 1985 will amount to 260 million kilowatts.

For all other non-Communist countries, the AEC now projects that nuclear power capacity will be 114 million kilowatts in 1980 and 310 million kilowatts in 1985. This could turn out to be a conservative projection. In Western Europe, which accounts for about half the capacity outside the United States, the European Commission's proposals look toward an increase in capacity of more than 50 percent by 1985 over previous objectives, and the French in particular recently have contracted for a large number of new plants.

In comparison with the base projection for nuclear generating capacity shown in Table 13-1, these adjustments would mean an increase in capacity for all non-Communist countries equivalent to 1.4 million barrels of oil a day in 1980 and 4.6 million barrels a day in 1985.

In sum, it is evident that the combined effect of all these changes would profoundly alter the size and composition of future world energy requirements. How, in particular, would they affect world demand for oil?

To answer this question, oil is taken to be the residual source of primary energy supply and thus the balancing element between total world energy requirements and world production of other primary sources of energy—natural gas, coal, and nuclear generating capacity. This means that within the projected margins of change increased production of other primary sources of energy would be fully substitutable for oil and that a reduction in total energy consumption would reduce oil requirements by an equal amount.

Application of this assumption to Japan and Western Europe poses no substantial technical difficulty, since in both areas oil accounts for so large a proportion of total energy requirements (75 percent for Japan and 60 percent for Western Europe). Furthermore, as noted earlier, gasoline and other fuel requirements for transportation, which are uniquely dependent on oil, take up a relatively small proportion of total oil consumption (22 percent in Japan and 29 percent in Western Europe). Evidently there would be a wide margin in industrial, commercial, and residential uses of energy for substituting other fuels for oil—certainly more than enough to manage the reduction in Japanese and West European energy consumption and the increases in production of non-oil fuels described above.

In the United States, the situation is very different. Oil accounts for a smaller proportion of total energy requirements (45 percent), and gasoline and other fuels for transportation account for over half of total oil consumption. Compared with Japan and Western Europe, there is a much smaller margin in the United States for substituting other forms of energy for oil. Moreover, depending on the size of the changes, such substitution in the United States would require more time to work out the necessary technical adjustments than would be the case for Western Europe and Japan. On the other hand, the projections for 1980 and certainly for 1985 provide substantial time for adjustments, such as converting present oil-burning industrial and generating plants to, and basing new capacity on, gas, coal, or nuclear power. In addition, much wider possibilities exist in the United States than in other countries to reduce energy consumption in the transportation sector, which would reduce oil requirements directly.

Based on the working hypothesis that oil will be the residual source of primary energy supply and on the specific changes outlined earlier, a current projection of world energy requirements in 1980 and 1985 is shown in Table 13-4. It should be recalled that this projection assumes (1) a long-term crude oil price of $7.50 a barrel in the United States (in constant 1973 dollars) and $6.25 a barrel f.o.b. Persian Gulf; and (2) government policies in the importing countries designed to achieve greater energy self-sufficiency. The components of the projection have been calculated conservatively, that is, biased on the side of smaller rather than larger changes when the available information suggested room for choice.

Even so, the combined effect of these changes on future oil requirements is strikingly large. Compared with the pre-October 1973 projection, oil requirements are down by 12 million barrels a day for 1980 and by almost 30 million barrels a day in 1985. Reduced energy consumption accounts for about two-thirds of the

Table 13-4. Current Projection and Change from Pre-October 1973 Projection (Base Forecast): Energy Requirements of Non-Communist Countries, 1980 and 1985 (in million barrels per day oil equivalent)

	Current Projection		Change from Oct. 1973 Projection		Growth Rates 1970-1985 (percent)	
					Pre-Oct. 1973 Projection	Current Forecast
	1980	*1985*	*1980*	*1985*		
United States:	*45.5*	*52.7*	*-4.4*	*-8.1*	*4.2*	*3.2*
Oil	18.6	15.0	-5.9	-11.7	4.5	0.6
Gas	13.2	16.4	+1.0	+2.5	1.4	2.5
Coal	8.8	10.1	...	...	3.2	3.2
Nuclear	3.2	9.3	+0.5	+1.1	34.1	35.3
Hydro	1.5	1.6	...	...	1.4	1.4
Geothermal	0.2	0.3	...	...		
Western Europe:	*32.1*	*38.6*	*-3.1*	*-5.9*	*5.1*	*4.1*
Oil	18.5	19.7	-4.8	-9.0	5.7	3.1
Gas	5.0	7.0	+0.5	+1.0	10.2	11.3
Coal	5.5	5.0	+0.7	+0.7	-2.5	-1.4
Nuclear	2.1	5.7	+0.6	+1.4	22.7	25.0
Hydro	1.0	1.2	...	...	3.7	3.7
Japan:	*10.8*	*13.9*	*-1.1*	*-2.1*	*7.5*	*6.5*
Oil	7.7	8.3	-1.3	-2.9	7.5	5.3
Gas	0.7	1.2	...	...	18.0	18.0
Coal	1.3	1.5	...	...	1.0	1.0
Nuclear	0.9	2.6	+0.2	+0.8		
Hydro	0.2	0.3	...	...	2.7	2.7
Other Industrial:	*7.4*	*9.5*	*-0.7*	*-1.1*	*5.9*	*5.1*
Oil	2.6	2.0	-1.1	-2.4	5.8	0.3
Gas	1.6	2.2	...	...	7.0	7.0
Coal	2.4	3.1	+0.3	+0.6	3.5	5.0
Nuclear	0.3	1.4	+0.1	+0.7		
Hydro	0.5	0.8	...	...	6.8	6.8
Developing Countries	*17.9*	*25.3*	*-1.0*	*-2.2*	*8.0*	*7.4*
Oil — Domestic	8.7	11.0	-1.8	-3.9	8.0	5.8
Oil — Bunkers	2.0	2.8	...	...	7.1	7.1
Gas	3.1	5.0	...	...	10.1	10.0
Coal	3.4	4.4	+0.8	+1.1	5.0	7.0
Nuclear	0.3	1.4	...	+0.6		
Hydro	0.4	0.7	...	...	8.7	8.7
Total:	*113.7*	*140.0*	*-10.3*	*-19.4*	*5.4*	*4.5*
Oil	58.1	58.8	-15.0	-29.9	5.9	3.0
Gas	23.6	31.8	+1.5	+3.5	4.4	5.2
Coal	21.4	24.1	+1.8	+2.4	1.7	2.4
Nuclear	6.8	20.4	+1.4	+4.6	30.2	32.5
Hydro, Geothermal	3.8	4.9	...	...	4.1	4.1

Sources and Notes: Pre-October 1973 projection is shown in Table 13-1 with sources noted in the footnotes. This projection is used as the base forecast from which changes are projected in the light of the new energy situation resulting from the October oil embargo and the subsequent rise in oil prices.

Current projection reflects the changes from the base forecast described in the text. The price of crude oil is assumed to be $7.50 in the United States and $6.25, f.o.b. the Persian Gulf. Prices of other sources of energy are assumed to rise less rapidly so that the demand for energy is reduced but, at the same time, the difference in relative price changes is sufficient to encourage the substitution of coal, gas, and nuclear power for oil. In addition, the projection assumes that governments will take specific measures to reduce requirements for oil (such as mandatory speed limits) to encourage the use of coal and nuclear energy in new generating plants and to provide financial subsidies for non-oil energy production.

For nuclear energy, the projections are the middle range or most likely forecasts shown in the AEC's *Nuclear Power Growth, 1974-2000* (Washington, D.C., February 1974). The breakdown among countries and regions outside the United States is based on information supplied by the Commission.

The rate of growth in energy consumption is assumed to decline by 1 percentage point in the oil-importing developing countries and to remain unchanged in the oil-exporting developing countries. Consequently, the average decline in the growth of energy consumption for all developing countries is less than 1 percentage point.

savings; increased supplies of non-oil energy sources for the rest. In general, the significance of the latter grows through time, reflecting the substantial lags involved in increasing production of primary sources of energy.

The United States accounts for somewhat less than half the total savings in oil requirements and Western Europe for about one-third. Japan's savings are relatively small since its lack of resources provides little scope for increased production of non-oil sources of energy; virtually all of its reduction in oil requirements must come from energy conservation.

Generally, oil plays a substantially smaller role in world energy under the current projection than in the pre-October 1973 outlook. This can be seen in Table 13-5. Instead of supplying a steadily increasing proportion of total world energy requirements, as had earlier been forecast, oil's relative importance would decline from 52 percent in 1970 to 42 percent in 1985. In the new situation, oil provides about one-third of the absolute increase in energy requirements compared with two-thirds in the pre-October 1973 projection. And finally, the growth of oil consumption declines to 3.0 percent a year—about half the rate that might have been expected in the era of low prices and more reliable oil supplies. Indeed, by 1980 world oil consumption begins to level off in absolute terms.

In view of its importance in the projection, the hypothesis that additional production of other fuels could substitute for oil (and that savings in the use of any form of energy would reduce oil requirements) deserves testing. One means of doing so is to allocate

Table 13-5. Composition of World Energy Requirements, 1970,
and Pre-October 1973 and Current Projections (Percentage of Total)

	Actual *1970*	*Projected:* *Pre-October 1973*		*Projected:* *Current*	
		1980	*1985*	*1980*	*1985*
Oil	52	59	56	51	42
Gas	20	18	18	21	23
Coal	23	16	14	19	17
Nuclear	—	4	10	6	15
Hydro	4	3	3	3	4
Total	100	100	100	100	100

Sources: Tables 13-1 and 13-4.
Note: Figures may not add due to rounding. Does not include Communist countries.

the projected level of oil consumption in 1980 and 1985 between transportation and all other uses so as to determine whether minimum requirements in each sector could be met. Since substitution of other fuels for oil, for all practical purposes, is not possible in transportation, the amount available for other uses would be the residual left after transportation requirements had been fulfilled. The question then is whether the projection provides for a sufficient amount of oil in the non-transportation sectors.

These calculations, broken down for each of the major industrial countries or regions, are shown in Table 13-6. They are based on the rather special assumption that transportation's share of total energy consumption will remain constant, which would mean that the rate of growth of energy consumption in the transportation sector would decline by 1 percentage point, or by the rate of decline projected for total energy consumption. As pointed out earlier, however, the rate of fuel savings in the transportation sector is likely to be smaller in Western Europe and Japan and larger in the United States. How much of a difference would this make?

As far as Western Europe and Japan are concerned, the calculations in Table 13-6 support the belief that the margin for substituting other forms of energy for oil is very wide. In the substantially changed energy environment reflected in the new projections for 1980 and 1985, at least two-thirds of the projected oil supply would still be available for non-transportation purposes. Even if fuel savings in transportation were much smaller than the average, there should be no technical obstacles in Western Europe and Japan to substituting the projected additional supply of other primary sources of energy for oil.

Table 13-6.   The Substitution Problem: Availability of Oil for
Transportation and for Other Uses, 1971, and Projections for 1980
and 1985 (in million barrels per day)

	1971 (Actual)		1980 (Projected)		1985 (Projected)	
	Trans- portation	Other Uses	Trans- portation	Other Uses	Trans- portation	Other Uses
United States	8.2	7.3	10.9	7.7	12.5	2.5
Western Europe	3.9	9.6	5.7	12.8	6.9	12.8
Japan	.9	3.4	1.3	6.4	2.3	6.0

Sources: Data for 1971 and distribution of oil requirements by sector are from
Organisation for Economic Co-operation and Development, *Economic Outlook*,
December 1973.

Note: Calculations assume that transportation in the future will account for the
same proportion of total energy consumption as in 1971, and that oil will supply
the same proportion (95 percent or more) of fuel requirements in the
transportation sector as in 1971. To calculate transportation requirements for
oil, these percentages were applied to the energy requirement projections for
1980 and 1985 shown in Table 13-4. To obtain the amount of oil available for
other uses, transportation requirements were deducted from projected total oil
requirements shown in Table 13-4.

As expected, a much tighter situation would exist in the
United States. Assuming a reduction of 1 percentage point in the
growth of energy consumption in transportation—from 4.1 percent
to 3.1 percent a year—the amount of oil available for other than
transportation uses in 1980 would be almost 8 million barrels a day,
about the same absolute amount as in 1970. This would seem to be
tolerable. By 1985, however, the amount of oil available outside the
transportation sector would be only 2.5 million barrels a day, which
clearly would be inadequate.*

Thus, in the United States fuel savings in the transportation sector would have to be much greater than average if there was
to be sufficient leeway to substitute additional production of other
primary sources of energy for oil. What might be possible? The rate
of energy growth in transportation could be reduced from 4.1
percent a year to less than 2 percent per year if by 1985 the average
miles per gallon ratio for U.S. cars was raised from 13.5 to the goal
of 17 suggested by the Federal Energy Office. In that event, oil
required for transportation uses would be approximately 10.5

*The projections of Dupree and West show that by 1985 the United
States would require 2.5 million barrels per day of oil just for non-fuel uses in
the industrial, household, and commercial sectors.[11] This projection was
based on comparatively low oil prices. Even at the prices assumed in this
chapter, however, oil requirements for non-fuel uses in 1985 probably would not
be much below 2 million barrels a day.

million barrels per day and oil available for other uses would increase to 5 million barrels a day. Additional fuel savings might be achieved through intensified conservation measures. However, rather than count on attainable but perhaps overly optimistic targets, it will be assumed, somewhat arbitrarily, that technical constraints on substituting other fuels for oil will increase U.S. oil requirements in 1985 by 4 million barrels a day above the amount shown in Table 13-4.*

Two conclusions emerge from this examination of the substitution problem. For Western Europe and Japan, the reductions in oil requirements envisaged in the current projection are well within technically feasible limits. In fact, oil requirements could readily be reduced further by larger than expected imports of coal from the United States or the Soviet Union and of natural gas from the USSR and the Middle East, or by additional reductions in total energy consumption. For the United States, on the other hand, a substantial reduction in long-term reliance on oil for meeting total energy requirements will depend critically on large fuel savings in the transportation sector, not simply on energy savings in general or on additional production of non-oil primary sources of energy. This highlights the importance of policy measures and market pressures that could accelerate the trend toward smaller cars, more energy-efficient driving habits, and more extensive use of mass transit facilities.

THE PROSPECTIVE WORLD OIL MARKET: A TRIAL BALANCE

Even with the lower projection for world oil requirements, the demand for oil would rise from 37 million barrels per day in 1970 to 59 million barrels per day in 1985, or by about 60 percent. So large an increase immediately raises the question of whether world oil reserves are sufficient to support the required level of production. For this period, at least, most experts contend they are. The OECD Oil Committee, for example, concluded in 1972 that "world reserves of crude oil are more than ample to satisfy projected needs for the

---

*In drawing up the projections in Table 13-4, no allowance was made for the possibility of substituting for oil additional coal production equivalent to 4 million barrels per day. In effect, therefore, with this additional allowance of 4 million barrels a day, it is being assumed that U.S. oil requirements in 1985 will be increased by 4 million barrels a day because of an inability to reduce sufficiently the rate of growth of energy consumption in the transportation sector or to convert sufficient energy creating capacity in other sectors from oil to other energy sources, or because of a combination of both.

foreseeable future."[12] At the reduced level of oil requirements that now may be contemplated, that conclusion presumably could be reached with even greater assurance. The difficulty, of course, is that so large a portion of these reserves is concentrated in a relatively few Middle Eastern countries. Hence, as the events of 1973 dramatically demonstrated, uncertainty exists about how much oil the key producing countries will choose to supply, under what conditions, and at what price, rather than about the physical sufficiency of reserves.

As against projected requirements, how much oil would theoretically be available? A first approximation, shown in Table 13-7, assumes once more a long-term U.S. crude oil price of

Table 13-7.   World Crude Oil Production, 1972 and 1973, and Potential Production, 1980 and 1985 (in million barrels per day)

	1972 (Actual)	1973 (Estimated)	Potential Production 1980	Potential Production 1985
Oil-Importing Countries:	13.7	13.4	20.5	24.0
United States	11.6	11.0	13.0	14.5
Western Europe	.4	.5	4.0	5.0
Japan	—	—	—	—
Other Industrial	.3	.4	1.0	1.0
Developing Countries	1.4	1.5	2.5	3.5
Exporting Countries:				
Canada	1.5	1.8	3.0	4.0
Latin America	3.9	4.2	5.0	5.5
Venezuela	3.4	3.5	3.5	3.5
Other	.5	.7	1.5	2.0
Africa	5.5	5.7	9.0	11.0
Algeria	1.0	1.0	1.5	2.0
Libya	2.1	2.0	3.0	3.0
Nigeria	1.8	2.0	3.0	4.0
Other	.6	.7	1.5	2.0
Far East	1.3	1.7	3.5	4.5
Indonesia	1.1	1.4	2.5	3.0
Other	.2	.3	1.0	1.5
Middle East	18.0	21.4	38.0	47.0
Saudi Arabia	6.0	7.8	15.0	20.0
Iran	5.0	5.9	9.5	10.0
Iraq	1.5	2.0	5.0	6.0
Kuwait	3.3	3.1	3.0	3.0
United Arab Emirates	1.2	1.4	4.0	6.0
Other	1.0	1.2	1.5	2.0
Total	43.9	48.2	79.0	96.0

Sources and Notes: Data for actual production in 1972 and estimated production for 1973 are from United Nations, *Monthly Bulletin of Statistics*, March 1974. The 1973 estimates do not reflect the Arab cutbacks in the fourth quarter.

Net Importing Countries:

United States—National Petroleum Council (NPC), *U.S. Energy Outlook* (Washington D.C., 1972). Figures used are essentially NPC supply case I-A, which assumes a high drilling rate, which would be consistent with a price of $7.50 a barrel, but a low finding rate. Production of oil from coal by 1985, however, is assumed to be negligible, in contrast to an allowance of .7 million barrels per day from this source in the NPC projection. Figures include natural gas liquids.

Western Europe—Based on a background paper on *Western European Energy Policy* prepared for The Brookings Institution by the European Community Institute for University Studies (1973). Projections were increased by 1 million barrels per day in 1980 and 1985 to allow for effects of high oil prices.

Other industrial countries—Consist of Australia, New Zealand, and South Africa. Production estimates are for Australia.

Developing Countries—Oil production in these countries increased by 6 percent a year between 1961 and 1970. Since higher oil prices will stimulate oil exploration and development, production is projected to increase by 7.5 percent a year between 1973 and 1980, and by 9 percent a year, 1980 and 1985.

Net Exporting Countries:

Canada—Government of Canada, *An Energy Policy for Canada—Phase 1* (Ottawa, 1973), pp. 101-104.

Latin America—Venezuela: Industry and Venezuelan government sources. *Other:* Derived from regional estimates in NPC study cited above. Ecuador and Peru are included in this category.

Africa:

Algeria—1980 estimates from "Land, Man and Development in Algeria, Part III: The Four Year Plan," by John Waterburg. American University Field Staff Reports (Washington, D.C., March 1973), p. 13. 1985 figure assumes discovery of additional reserves.

Libya—Assumes government policy will permit moderate increase in production over current levels but will continue to keep production below 1970 level of 3.3 million b/d.

Nigeria—See Chapter 7.

Other—Derived from regional totals shown in NPC study, cited above. The resulting figures, however, were too large to be explained by specific country data. Consequently, the projection for "other countries" in Africa derived from the NPC regional total was reduced by 2 million barrels per day in 1980 and 1985.

Far East:

Indonesia—Industry sources. Also, in an interview on November 1, 1973, Lt. General Ibnu Sutowo, president-director of the state-owned oil company, Pertamina, said that Indonesia could double its production of 1.3 million barrels a day "in the next few years." *New York Times*, November 7, 1973.

Other—Derived from regional totals shown in NPC study, cited above.

Middle East:

Iran—The basic figure is the Shah's agreement with the consortium on production of 8 million b/d in 1976. Additional production is assumed to come from development of offshore sources.

Iraq—The Iraqi Under Secretary for Oil Affairs has stated that 1974 production would be "doubled in four years." *New York Times,* March 26, 1974, p. 59. Furthermore, the *Middle East Economic Survey* on May 17, 1974 (Vol. 17, No. 13), carried a report from Baghdad stating that the Iraq National Oil Company had raised its production target for 1981 to 6 million barrels per day.

United Arab Emirates—For Abu Dhabi, based on government statements. (See *Petroleum Intelligence Weekly,* February 5, 1973.) Also includes production forecasts for Dubai and Sharja.

Kuwait—Present government policy, which permits no significant production increases, is assumed to continue. Includes half the production of the Neutral Zone.

Saudi Arabia—The basic figure is the government goal of 20 million b/d by 1979-80 announced by Sheik Yamani in late 1972. (*Petroleum Press Service,* November 1972.) It is assumed conservatively that this goal would not be reached until 1985. Includes half the production of the Neutral Zone.

Other—Individual country estimates for the smaller producers in the region.

$7.50 a barrel ($6.25 f.o.b. Persian Gulf), takes into account the standard information on world oil reserves, and ascribes policies of maximizing energy production to the net oil-importing countries and, generally speaking, the continuation of pre-embargo production policies to the exporting countries. Under these conditions, increases in world oil production by 1985 could be twice as large as projected increases in oil requirements.

Obviously this is not a forecast, since in such circumstances neither the price, the production policies, nor the demand assumptions would hold. The figures in Table 13-7 are best described as showing the amount of production that would be available on the basis of a mixed set of production policies concerned largely with resource constraints and a fairly cautious view among the major exporting countries of feasible rates of resource exploitation. Concern for market consequences is virtually ruled out. The figures, thus, may be taken as a useful indication on the supply side of the framework in which exporting countries will have to develop their market policies. To understand the nature of these initial supply projections, it is useful to explain the key components.

Among the importing countries, the key components are as follows:

— *In the United States,* high drilling and low finding rates lead to an increase in production to approximately 14 million barrels a day by 1985. Production of oil from shale by 1985 is assumed to be less than 1 million barrels a day and no allowance is made for production of oil from coal. This level of production would be near the lower end of a range of estimates corresponding to the assumed price; some of these estimates suggest that U.S. production could be 2 or 3 million barrels a day more.*

— *In Western Europe,* production from the North Sea is about 1 million barrels a day below the more optimistic forecasts. (See Chapter 8.)

— *In the oil-importing developing countries,* the rate of increase in oil production is about one-third above that actually experienced in the 1960s. This would reflect the impact of higher prices on new discoveries and on the pace of developing present oil-producing regions.

Among exporting countries, the following are the key components of the projection:

— *In Saudi Arabia,* production expands fairly rapidly, but by considerably less than the maximum amount that reserves can support.

— *In Iran,* expansion continues to the full extent permitted by economic exploitation of available reserves.

— *In Iraq,* production increases sharply in consonance with the rapid expansion in 1973 and with recent statements by Iraqi officials. Iraq's 1973 settlement with the oil consortium should set the stage for new investment and exploration and development of its reserves, which are widely believed to be very extensive.

— *In Libya and Kuwait,* conservationist policies are continued.

— *In Nigeria and Indonesia,* exploration, development, and production are accelerated as a result of higher prices.

— *In Canada,* production rises at about the rate of the past ten years. This would mean that the effect of higher prices would be offset by possible limitations in reserves. Production of oil from tar sands during this period is assumed to be minor.

— *In Venezuela,* production remains flat principally because of limitations in reserves.

*Searl analyzes a range of estimates, including some that suggest production of 17 million barrels a day at a price considerably lower than that assumed here. Figures include natural gas liquids. The estimate used in Table 13-7 is essentially the total shown in the National Petroleum Council's Case 1-A, but without an allowance for the production of oil from coal. In this latter respect, some optimistic forecasts exist. For example, Federal Energy Office Administrator Sawhill, in an address to the National Petroleum Council, April 22, 1974, said that "by 1985 we could build plants that would convert coal into 2 million barrels of oil per day." Searl's analysis indicates this timetable would be overly optimistic.[14]

— *In other oil-producing countries*, output increases rapidly among newly growing producers (e.g., Peru, Ecuador, Angola, and Gabon) and moderately elsewhere as a result of higher prices and normal expansion.

As the conventional wisdom suggests, these theoretical production schedules emphasize the dominance of the Middle East—and of Saudi Arabia in particular—as the major source of additions to world supplies. Of the potential increase in production of 48 million barrels a day between 1973 and 1985, roughly half would come from this area. Equally significant, however, and less well recognized, is the large number of other sources of additional production. By 1985, production outside the Middle East could increase by 23 million barrels a day; this would equal almost two-thirds of present world consumption and would alone be about enough to cover the currently projected increase in world oil requirements. Furthermore, so large a dispersion of additional production, in itself, would provide some assurance about security of supply and would hamper producing countries in their attempts to organize and control the world market.

Indeed, the principal import of these production possibilities lies in the question they raise about the future world oil market. It will not be a sellers' market automatically. To the contrary, combining these production possibilities with currently projected oil requirements indicates that potential production could exceed import requirements by about 25 percent in 1976, 60 percent in 1980, and more than 100 percent in 1985. This is shown in Table 13-8.

Why does this tendency toward growing surpluses appear and what does it mean? The reasons of course are inherent in the estimated effect of higher prices (and the drive toward greater energy self-sufficiency) on the importing countries: in reducing their energy consumption, expanding their production of non-oil sources of energy, and increasing their production of oil itself. All three have the effect of reducing oil imports. In earlier years, say through 1976, the impact on oil-import requirements would depend principally on the size of the reduction in energy consumption, since the effect of price on the supply side tends to be limited in the short run. Furthermore, oil production from the North Sea and from Alaska will not come to the market in sizable volume until near the end of the decade. By 1985, however, the consequences could be strikingly large; compared with pre-October 1973 trends, the three factors together could reduce the projected size of world trade in oil by fully one-half.

Table 13-8. Trade Positions in Oil: Actual 1972, Estimated 1973, and a Trial Balance for 1976, 1980, and 1985 (in million barrels per day)

	1972 (Actual)	1973 (Estimated)	Prospective Positions 1976	1980	1985
Import Requirements of Net Importing Countries:					
United States[a]	4.8	6.0	7.5	5.5	4.5
Western Europe	14.4	15.0	16.0	14.5	14.5
Japan	4.8	5.5	6.0	8.0	9.0
Other Industrial[b]	.5	.5	.5	—	—
Developing Countries[c]	4.0	4.5	5.0	5.0	5.0
Total Net Imports	28.5	31.5	35.0	33.0	32.0
Potential Export Availabilities of Exporting Countries:					
Canada	.1	.2	.5	1.0	1.5
Venezuela	3.2	3.3	3.0	3.0	3.0
Indonesia	1.0	1.2	1.5	2.0	2.5
Algeria	1.0	1.0	1.5	1.5	2.0
Nigeria	.8	2.0	2.5	3.0	4.0
Iraq	1.5	2.0	3.5	5.0	6.0
Iran	4.8	5.6	7.5	9.0	9.0
Other Africa, Latin America, and Asia	1.5	2.0	2.5	3.0	4.0
Subtotal	14.8	17.3	22.5	27.5	32.0
United Arab Emirates[d]	1.2	1.4	3.0	4.0	6.0
Kuwait	3.3	3.1	3.0	3.0	3.0
Libya	2.1	2.0	2.5	3.0	3.0
Qatar and Oman	1.0	1.0	1.5	1.5	1.5
Saudi Arabia	6.0	7.8	11.0	15.0	20.0
Subtotal	13.6	15.5	21.0	26.5	33.5
Total Export Availabilities	28.5	32.6	43.5	54.0	65.5
Potential Surplus of Availabilities over Requirements			8.5	21.0	33.5

Source: Author's estimates derived from projections in Tables 13-4 and 13-7. To determine export availabilities, production data in Table 13-7 have been adjusted where appropriate to take into account domestic requirements in exporting countries. Available data for 1973 are approximate; no attempt has been made to balance imports and exports or to estimate changes in stocks.

[a]Includes allowance of 4 mbd in 1985 because of assumed limits on substitution of additional production of other primary sources of energy for oil, as discussed in the text.

[b]Australia, New Zealand, and South Africa.

[c]Includes requirements to supply bunkers estimated at .7 mbd in 1972; .8 mbd in 1976; 1.3 mbd in 1980, and 1.9 mbd in 1985.

[d]United Arab Emirates includes production of Abu Dhabi and estimated production for Dubai and Sharja combined of .5 mbd in 1976 and 1 mbd in 1980 and 1985.

In sum, a very different picture emerges of the world oil market and of the relative bargaining position of exporters and importers over the next ten years or so than is implied by the reactions to the October oil embargo and its immediate aftermath. Most of the bloom has been taken off the world oil market. A steady increase in oil consumption is no longer an assured prospect, and import requirements by 1985 could be no higher than they were in 1972. The industrial countries will continue to be heavily dependent on oil imports, although in varying degree: Japan entirely so; Western Europe for about three-fourths of its requirements; and the United States by rising and then sharply falling amounts—nearly reaching self-sufficiency by 1985. Traditionally, there has been nothing inherently disturbing about this type of dependence. Japan and Western Europe have always been in this position in oil. In other commodities, tin for example, all industrial countries have accepted with equanimity the fact that they are entirely dependent on imports. What makes the current situation different is that the world market can no longer be taken for granted as a source of oil needs.

The exporting countries clearly have a strong initial bargaining position. Energy is critical to industrial economies, and oil in important respects is a unique form of energy for which alternative sources of supply are not readily available in the short or even medium term. Hence, the fact that import requirements are large and will continue to grow for a while, combined with fears of future supply interruptions aroused by the embargo, has made the industrial countries extremely nervous about the world oil market. Consequently, the exporting countries were able to impose unprecedentedly large price increases on the market in 1973, and prices in 1974, although weakening in real terms, are still very high.

That bargaining position, however, contains fundamental weaknesses. Potential increases in world oil production are so much greater than prospective demand that the exporting countries will be faced with a chronic need to limit supply to prevent the erosion of prices. Their ability to do this will be complicated by the fact that, while total oil demand is growing, a significant portion of the potential incremental requirements of oil-importing countries between now and 1985 will be satisfied internally rather than from imports—that is by North Sea, Alaskan, and other oil production, by increased natural gas production in Western Europe and the United States, and by the more extensive production and use of coal in many areas. Most important, nuclear power will be the most rapidly growing source of energy and will take up a portion of demand, particularly in Western Europe and Japan, that otherwise would have been supplied by oil. Finally, this market assessment does not take into account the possibility of significant exports of oil or gas from either the USSR or China, or of unusually large new oil discoveries, or of an accelerated time schedule for producing substantial quantities of oil from shale or coal, or of technological break-throughs during this period in the production of energy from other sources. Any of these eventualities would further reduce the world oil market available for the non-Communist oil-exporting nations.

CAN AN EXPORTERS' CARTEL WORK?

In these circumstances, what marketing options are open to the exporting countries and what would be their consequences? As a first step toward examining this question, the underlying market situation shown in Table 13-8 is assumed to prevail during the period 1972-85. This would mean that in order to sustain prices at a level of $6.25 a barrel, f.o.b. the Persian Gulf, exporting countries as a group would have to freeze the volume of oil exports at more or less the 1973 volume (some 30 million barrels per day) throughout the period. Thus, the volume of exports would have to be cut substantially from the amounts that the exporting countries would be able and might be planning to produce. How could production be controlled and which countries would accept the major burdens?

This statement of the problem, of course, rules out a purely competitive solution. A modified version could eventually occur, but not in present circumstances: exporters are already organized and have had the heady experience of exercising unprec-edented market power; the key governments are in a strong financial position to withhold supplies; the international oil companies tend to mitigate the effects of competition and thus help to provide a floor to the marketing structure; and the industrial countries, anxious about the future, are prepared to compete among themselves in the hope of obtaining assured sources of supply. For these reasons the exporters—certainly the major suppliers—currently view their marketing position in quasi-monopolistic rather than competitive terms, and they will try to develop their marketing strategy accordingly.

Nevertheless, they are, at bottom, a disparate rather than a cohesive group. As higher prices and energy conservation narrow the market, the divergences in their interests will become evident. Price cutting is likely to occur, temporarily at first and in selected markets, as individual exporters seek to improve their market share in the belief that marginal additions to supply will not affect prices in general. Eventually, this could cause a breakdown in cooperation among suppliers and an erosion of the market as a whole. The experience of some other commodity cartels or cartel-like arrangements is worth bearing in mind.

First, there are the examples of individual countries, or groups within a country, that sought to support the price of a commodity by their own actions alone. The United States for many years maintained an umbrella over the world market for cotton by unilaterally restricting its cotton exports to the amount necessary to avoid a fall in its world price. That U.S. restraint, however, merely enabled and indeed encouraged other countries to increase production and exports, eventually causing the world price to fall and forcing the United States to abandon the policy. At various times Brazil adopted a similar policy in coffee, with similar results. On the other hand, a relatively high domestic price for oil was maintained in the United States for over a decade by production restrictions in Texas and Louisiana, the main domestic producing areas, and by quotas on imports. In this case, a price-maintenance policy proved feasible because the total market grew steadily and production did not get out of hand in those parts of the United States where it remained unrestricted.

International commodity agreements also have become ineffective when the efforts of the principal exporters to withhold supplies from the market were frustrated by substantially increased sales by smaller producers. Both the International Wheat Agreement and its successor, the International Grain Agreement, are examples. These price maintenance schemes broke down when the world production surplus became substantially greater than the amount that the United States and Canada were willing to keep off the market and store.

On the other hand, the International Coffee Agreement performed reasonably well as a means of preventing a sharp fall in the price of coffee despite the existence of huge Brazilian stocks hanging over the market. In this case as well, prices were under pressure because small producers chose to cut prices secretly and export more than their assigned quotas whenever their production was in surplus. Brazil, the dominant supplier, had most to lose from price cutting and to sustain the market it had to sell less than its quota even though it was the lowest-cost producer. Brazil found it advantageous to do this only because the cooperation of consuming countries in enforcing the agreement kept quota violations within manageable limits and because the agreement as a whole tended to strengthen the marketing position of the exporting countries and to maintain fairly remunerative world prices.

Arrangements for sugar provide still another model. Many importing countries obtain part of their requirements by protecting or subsidizing domestic production. The United States, the United Kingdom, and the Soviet Union acquire the balance of their needs by special import arrangements with favored suppliers, under which they pay higher than world market prices and receive assurances about supply. The so-called world market, which consists of that portion of trade not covered by special arrangements, is relatively small and frequently volatile. Generally countries that rely on it—Japan is an example—have obtained their sugar at bargain prices, although they have had to pay premiums during periods of shortage. International commodity agreements for sugar, which have been in operation at various times since the 1930s, in effect sought to stabilize prices for that portion of the trade not covered by special arrangements, that is, the residual world market.

What does this experience suggest about oil cartels, and specifically about the ability of oil-exporting countries to maintain high prices by sharp limitations on potential production?

Efforts by the oil exporters to exercise such discipline would be facilitated by three factors. First, the key exporters now have, or soon will have, the financial resources to shock the market by reducing or even halting sales for a considerable period of time.* Second, surplus stocks, as distinct from surplus capacity, are not in existence and therefore are not a burden on the market in the sense that was true for coffee, wheat, or cotton when attempts were made to bolster prices of those commodities. Instead, production is under the control of governments, which can enforce production limitations, rather than the responsibility of numerous individual producers, some of whom might be in a financially weak position to withhold supplies. Third, some of the major exporters have very large current revenues in relation to absorptive capacity and therefore might tend to view foregone production as a means of deferring income rather than as lost income.

The obstacles, on the other hand, stem from the need to impose very sharp limits on future production increases. This will be difficult because, as in other commodity agreements, the problem of controlling the market is complicated by rising production from new

---

*The existence of a number of financially strong exporters is unusual in commodity agreements but not unique. The United States was a "strong seller" in cotton and wheat but nonetheless was eventually forced to give up the game.

exporting countries, whose governments would refuse to accept market-sharing arrangements based on the status quo. Furthermore, most of the governments in producing countries need expanding oil revenues to carry out development plans; while they might recognize the need to restrict production to sustain prices, their natural instincts could be to argue that others were in a better position to carry the burden and to refuse to join in a coordinated scheme to restrict exports.

Oil exporters in these circumstances could, in theory, approach the problem of controlling the market in four alternative ways: (1) by relying on strong sellers, such as Saudi Arabia, to assume the residual suppliers' role; (2) by establishing an OPEC-administered producers' cartel in which oil exporters on a negotiated basis share responsibility for restricting exports; (3) by making special deals for part or all of their production with particular consuming countries; and (4) by entering into a commodity arrangement with importers. What are the prospects for each?

As an aid to the analysis, an assumed allocation of the production restrictions required to achieve a rough balance between supply and projected demand at high prices is shown in Table 13-9. This allocation is based on a grouping of exporting countries into four categories: (1) Canada is shown separately as a relatively small exporter which would receive higher than world market prices in the U.S. market because of transportation-cost advantages and whose export policy would be determined independently of the producer cartel; (2) countries not likely to restrict production, at least not for a number of years, either because they have urgent revenue needs or because their production is only beginning to become significant; (3) countries already restricting production and not planning future increases; and (4) potential residual suppliers, that is countries that would have to accept more than proportionate market-control responsibilities either because of the size of their present, and potential increases in, production, or because of their strong financial position, or both.

The data in Table 13-9 suggest that production restrictions among the potential residual suppliers alone probably could maintain the projected price of $6.25 a barrel over the next few years, but with increasing difficulty beyond then. In any circumstances, maintaining the $6.25 price would not be assured. While some exporters would be increasing exports and revenues, a number would have to act with unusual self-sacrifice—principally, Saudi Arabia and Iran, by freezing exports at about the 1973 level, as well as Iraq and

Table 13-9. Export Restrictions Needed to Maintain a Tight World Oil Market and High Prices, 1976, 1980, and 1985 (in million barrels per day)

Sources of Supply:	1973 (Estimated)	Projected Exports		
		1976	1980	1985
Canada	.2	.5 (.5)	1.0 (1.0)	1.5 (1.5)
Countries that are unlikely to restrict output initially:	6.2	8.0 (8.0)	8.5 (9.5)	9.5 (12.5)
Algeria	1.0	1.5 (1.5)	1.5 (1.5)	1.5 (2.0)
Indonesia	1.2	1.5 (1.5)	1.5 (2.0)	2.0 (2.5)
Nigeria	2.0	2.5 (2.5)	2.5 (3.0)	3.0 (4.0)
Other	2.0	2.5 (2.5)	3.0 (3.0)	3.0 (4.0)
Countries already restricting output:	8.4	7.0 (8.5)	6.5 (9.0)	5.0 (9.0)
Kuwait	3.1	2.5 (3.0)	2.0 (3.0)	1.5 (3.0)
Libya	2.0	2.0 (2.5)	2.0 (3.0)	1.5 (3.0)
Venezuela	3.3	2.5 (3.0)	2.5 (3.0)	2.0 (3.0)
Potential residual suppliers:	18.0	18.5 (26.5)	17.0 (34.5)	14.5 (42.5)
Iran	5.6	6.0 (7.5)	5.5 (9.0)	5.0 (9.0)
Iraq	2.0	3.0 (3.5)	3.0 (5.0)	4.0 (6.0)
Saudi Arabia	7.8	7.0 (11.0)	6.0 (15.0)	5.5 (20.0)
United Arab Emirates	1.4	1.5 (3.0)	1.5 (4.0)	1.5 (6.0)
Qatar, Oman, and Other Middle East	1.2	1.0 (1.5)	1.0 (1.5)	1.0 (1.5)
Total Exports	32.8	34.0 (43.5)	33.0 (54.0)	30.0 (65.5)
Import Requirements	31.5	34.0	33.0	32.0

Sources: Authors' estimates. Import requirements and export availabilities as shown in Table 13-8. Allocation of production and export restrictions as described in text.

Notes: Numbers in parentheses show potential exports as defined in Table 13-8. Import requirements correspond to prices of $6.25 a barrel for Saudi Arabian oil, f.o.b. the Persian Gulf, and normal market differentials for oil from other sources. (All prices in 1973 dollars.)

the United Arab Emirates, by sharply cutting planned production increases. Either of the first two countries or the second two countries combined could break the market by refusing to cooperate and increasing production. Also Kuwait, Libya, and Venezuela might have to reduce exports below what they may have planned, even though they have already been carrying out restrictionist policies. By 1980, practically all exporting countries would have to restrain production increases or actually reduce production. Subsequently, an increasing number would have to reduce exports to keep supplies reasonably tight. All this would be a great deal to expect without some formal system of apportioning burdens through a negotiated agreement.

Thus, the first option eventually merges into the second— an OPEC-administered system of export quotas. This would not be easy to bring into being. The marketing successes experienced by the producers in 1973 obscure the fact that up to now OPEC has been notably unsuccessful in getting its members to agree on market shares as a means of supporting the world price. When this was tried in 1965, Saudi Arabia, Libya, and Iran exceeded their allotments and the attempt failed. In the following year, a number of countries simply opted out of the negotiations to establish a basis for allocating export quotas among OPEC members. And through most of the 1960s, some of the major producers, notably Iran, constantly pressed for more export volume and larger market shares and were singularly unconcerned about the effect on world prices.

Attitudes could be different in today's circumstances, since a fall in prices might galvanize the major exporters to take quick action to shore up the market and then try to negotiate an agreement on export quotas covering all exporters. The basis for such an agreement, however, would be difficult to work out. One obstacle is the traditional economic rivalry among the major Middle Eastern exporters, to say nothing of their sharp political differences, which in some cases have involved and continue to involve armed conflict. Furthermore, the smaller exporters that are heavily dependent on oil revenue to meet current expenditure requirements probably would be strongly opposed to accepting other than token cuts in their production plans; if they did accept substantial cuts nominally, their surplus productive capacity would be a potential source of illicit, extra-quota sales that would weaken market prices and cause the producers' cartel to come apart. And finally, exporting countries with large reserves, such as Saudi Arabia, the United Arab Emirates, and Iraq, have different economic interests than other exporters.

They have to think about the price of oil in the more distant future and therefore are more concerned about actions that would reduce oil consumption or encourage eventual large-scale production of other primary sources of energy. Hence their optimal long-term price of oil is lower than that in countries with more limited reserves, like Libya, Algeria, Venezuela, and even Iran.

What about bilateral arrangements covering part of the oil trade and a world market for the remainder, as in sugar? For example, the United States might aim at producing three-fourths of its oil requirements domestically, set the domestic price at the level necessary to achieve this target, and establish individual country quotas to import its remaining needs. These quotas, say with Canada, Venezuela, Iran, and Saudi Arabia, might be for specified quantities over a period of three or five years at a price equal to or somewhat below the U.S. domestic price. The West European countries might make similar arrangements with producing countries. Japan, which does not have domestic production to protect, might decide either to enter into long-term contracts at a fixed price, or rely on the residual world market. A similar choice would face many developing countries.

These special arrangements could lead to a number of outcomes, depending partly on the extent to which long-term supply contracts provided importing countries with sufficient assurances and economic incentives to cause them to limit domestic production of primary energy sources. For the producing countries, a different calculation would be involved. Those whose capacity was largely used to fill long-term contracts could decide to withhold the remainder or dispose of the oil on the world market at such prices as it might command. Those whose exports were not covered by bilateral contracts, or covered only to a small degree, would be forced to sell on the world market. In all likelihood, world trade in oil would be characterized by two prices: generally higher prices for oil under long-term contract, and generally lower prices for oil sold on the residual world market. This system persisted for a long time in sugar, but it seems unlikely to be durable for oil. For one thing, a significant price difference in oil would involve enormous costs for those importing countries relying on fixed contracts, much more so than in sugar. Furthermore, the political considerations that help to perpetuate bilateral arrangements in sugar—for example, relations between the United States and Latin America, the Soviet Union and Cuba, and the United Kingdom and the Commonwealth—would not apply as strongly to oil.

Consequently, bilateral arrangements in oil, while they might seem attractive initially, are inherently unstable. This would probably be true even if they included reciprocal undertakings in which the oil-exporting countries agreed to purchase specified amounts of manufactured goods in the oil-importing countries. In that event, the possibility of using artificially high prices for manufactured goods to bring the contract price of oil closer to the price in the world market might be a means of prolonging the arrangements. Soon, however, these arrangements would degenerate into the bilateralism that characterizes Eastern Europe's trade, without the state trading institutions and the political pressures required to perpetuate that system. Furthermore, special bilateral trade arrangements in oil could lead to political ties and responsibilities that neither party really wanted to accept. All this confirms what the history of world trade has amply demonstrated: bilateral trade arrangements would be inferior to multilateral trade on both economic and political grounds.

The fourth approach, an international commodity agreement involving both exporters and importers, could solve some of the difficulties inherent in each of the other three approaches but would involve severe negotiating difficulties of its own. The two groups of countries would have to agree on a price range and both would have to accept market obligations to support it—the exporters by assuring that there would be sufficient supplies on the market to prevent prices from exceeding the ceiling, and the importers by helping to enforce assigned export quotas so as to prevent prices from dropping below the floor. In theory, benefits could be sizable for both groups of countries. Exporters could be confident about a quantitatively larger export market at a remunerative price, without as much risk of price erosion and price breaks as would be involved in a producers' cartel or in bilateral arrangements. Importers, if they could be sufficiently assured about the reliability of future imports, could reduce their commitments to invest in sources of energy that would cost more than the agreed international price of oil and also would entail potentially large environmental losses. In practice, the negotiating problems would be formidable, including the problems involved in agreeing on price, on the allocation of export quotas, on provision for new exporters, and on enforcement measures. The difficulties would be all the greater in the present atmosphere because of the political factors that surround the oil market, because the exporting countries have exaggerated views of their bargaining leverage, and because of suspicions among the industrialized countries about each other's motives.

EFFECT OF STRUCTURAL CHANGES IN THE INTERNATIONAL OIL INDUSTRY

Before drawing conclusions from the above analysis of market forces, it is necessary to ask what effect current changes in the structure of the international oil industry may have on the bargaining leverage of importers and exporters and, therefore, on market trends. Chapter 2 described the recent events that have shaken the industry: the appearance of the independents, the rise of OPEC, the emergence of a sellers' market and producer-determined prices, the imposition of participation agreements, and the quickening trend toward nationalization of company properties in the wake of the October war. These changes are shifting power from the seven or eight major international companies that long dominated the world oil industry to the governments of the producing countries. How far will this trend go and to what practical consequence for the distribution of bargaining leverage?

It is useful to bear in mind that the influence of the majors has been declining for some time—in earlier periods, because of the competition of the independents and of the state-owned companies in both importing and exporting countries. The majors' share of crude oil production outside the United States dropped from a near monopoly in 1950 to 75 percent in 1970, and their share of refining capacity and distribution is now down to between 50 and 60 percent of the total market.

This trend is likely to continue and, in light of the redistribution of power over the production and export of crude oil, accelerate. The pace of acceleration, however, will be moderated by the fact that the majors will continue to possess substantial competitive advantages at all stages of the industry.

In exploration and production activities, the majors possess experience, skilled personnel, and financial resources that cannot be matched by their competitors and that will continue to be attractive to governments seeking to develop their oil-production potentialities. Moreover, because of the worldwide scope of their operations, the majors can afford to take risks in a single country that would be too great for independent companies. If recent moves to create a large German national oil company succeed, or if Japan decides to launch such a company, an important new element could be introduced into the world oil industry. The potential financial resources of German and Japanese national companies would give them a risk-taking capability comparable to that of the majors. In general, however, the national companies of the importing countries, because of their late entry into the field, suffer from a relative lack of experience and skilled personnel, which will take time to overcome.

In transportation, marketing, and refining, the majors also will continue to have substantial competitive assets. New facilities can only be built slowly, and they require enormous financial investments. Such investments will of course be made by the new actors entering the arena. For some time to come, however, the majors will have a large share of available capacity and that capacity will have to be used, whether on the basis of their "own" oil, or "buy-back" participation oil, or simply oil purchased from wholly nationalized concessions that the majors operate on a service contract.

The combined effect of the structural changes in the industry will probably still leave the majors responsible for finding, producing, transporting, refining, and marketing a large part of the world's oil. Their exploration and production functions will, however, increasingly be carried out under service contracts, rather than the traditional concession agreements. The form of their participation, combined with other changes taking place in the industry, will significantly affect the functioning of the international oil market in future years. How much so will depend on the way that market is organized.

It is evident that the producing country governments now have the decisive voice in determining prices and production levels and this is likely to persist as long as the market is tight. In the past, the companies were able to prevent sharp fluctuations in prices by bargaining with these governments and by adjusting their worldwide operations to support the negotiated level of prices. Now they simply pass on price increases that the governments of the producing countries impose.

However, if supply threatens to outrun demand, as the analyses in the preceding sections suggest will happen, the role of the companies could once again become important. Should the producing countries seek to control the market on their own, they would have a continuing stake in using the major international companies to the maximum degree, either by permitting them to have a large equity share in output or by giving them access to a large portion of production at favored prices. As long as these companies constitute a major factor in marketing oil, they can in a soft market exert a stabilizing influence on prices or more accurately be a force for moderating a fall in price.* Thus it could be contrary to the interests of the producing countries to market too large a proportion of their oil through their national companies or the independents. Conversely, the interests of the importing countries would lie in adopting measures of enforcing trade practices that would require or encourage greater competition in the purchase of crude oil. This would reduce the ability of the major international companies to support the attempts of a formal or informal producers' cartel to shore up a weakening market.

If crude oil prices are determined either through bilateral government-to-government arrangements or under the terms of an international agreement including governments of both importing and exporting countries, the operations of the international companies would matter very little insofar as the market is concerned. The companies might be useful as a means of enforcing the terms of intergovernmental agreements, but crucial decisions would be made by governments.

CONCLUSIONS: MARKET OPTIONS AND FUTURE PRICES

The assumption that the world oil trade has permanently shifted from a buyers' to a sellers' market is not supported by the evidence. To the contrary, underlying trends point to a large potential surplus in oil-producing capacity. For example, maintaining over the long term a price of $6.25 a barrel for Saudi Arabian oil, f.o.b. the Persian Gulf (which after transportation costs would be approximately equivalent to the present blend price of $7.50 a barrel in the United States), would require the oil-exporting countries as a group to freeze exports at about 1973 levels until 1985. This would be increasingly difficult to do as total surplus capacity grew and particularly as the smaller exporters became able to expand production and exports. Consequently, present prices of about $9.00 a barrel, which would lead to even smaller markets and require even larger production cutbacks, are likely to fall substantially.

On the other hand, a return to the soft markets of the 1960s, when oil prices averaged about $2 a barrel (in 1973 dollars), seems equally improbable. Importing countries, for one thing, will

*In a soft market in which the cartel was coming apart, the large international oil companies might be able to restrain and informally allocate production increases among the exporting countries, somewhat along the lines of their operations during the oil surplus situation of the 1960s.

rely to a greater extent than heretofore on high-cost, protected, domestic sources of energy. Further, the oil-exporting countries now have an enhanced view of their bargaining power and are in a financially stronger position to exercise it by withholding supplies from the market. As a result, meeting growing world energy needs is likely to require considerably higher real prices for all sources of primary energy than prevailed during the 1960s.

Where the price of oil will move or settle within this broad range will not simply, or even principally, be a function of the course of oil supplies in the producing countries. Rather it will depend much more heavily on what happens to the production and consumption of all forms of energy everywhere in the world. Because, in present circumstances, primary sources of energy can be substituted for each other over a wide area, the market for oil is inherently part of a world market for energy.

This seemingly obvious conclusion is frequently lost to sight in current discussions of the oil problem. Yet it has important implications for assessing how bargaining leverage between oil exporters and oil importers may shift in the future and how that shift may be influenced by policy measures. In this connection, a number of conclusions that stem from the projections in this chapter are worth special emphasis:

— Energy consumption is the critical variable determining the future course of the oil market. In the short or medium term, a reduction in the growth rate of energy consumption is virtually the only means by which importing countries can improve their bargaining leverage. Even over the longer term, changes in energy demand probably have much greater potential for affecting the oil market than changes in supply. Small changes in the rate of growth in world energy demand can decisively alter the long-term world oil equation—probably well beyond the practical capacity of oil exporters to offset.

— On the supply side, actions anywhere in the world to increase the production of primary sources of energy are equally effective in putting downward pressure on the world price of oil. Thus, an increase of coal production in India, or of natural gas production in the Netherlands, or of Japanese investments in, and purchases of, Soviet coal, gas, and oil would have about the same effect on the world oil market as an equivalent increase in the production of oil in the United States.

— The bargaining leverage of importers in the world oil market will depend most importantly on what happens to energy consumption in the United States. The United States is the world's predominant consumer of energy, not only because its economy is large, but because it is so much more energy-intensive than other industrialized countries. A significantly slower rate of growth in U.S. energy consumption, with more than proportionate savings achieved in U.S. gasoline consumption, would move the United States a long way toward energy self-sufficiency. This alone would take the buoyancy out of market prospects for oil. By the same token, a failure of the growth rate for energy consumption in the United States to moderate probably could not be fully made up by an increase in domestic production of primary energy sources. This would mean larger imports and a greater likelihood that oil prices would be high.

Thus, developments in the importing rather than the exporting countries are likely to have the greater influence on the future prospects of oil. The analysis in this chapter suggests that the combination of a $6.25 price of crude oil at the Persian Gulf and of government policies aimed at greater energy self-sufficiency would so affect the consumption and production of energy in the importing countries as to create a situation of growing potential over-supply in the world oil market. This in turn would eventually cause prices to fall and consumption to rise. Where might equilibrium be reached and through what means: the free play of competitive forces or cartel-like arrangements?

In a deteriorating market, the major Middle Eastern producing countries, whose production costs are a small fraction of present prices, could resort to price cutting as a means of forcing higher-cost producers to assume the entire burden of restricting production. Oil consumption and the size of the world market would increase. But the low-cost producers, through price-cutting measures, could no longer hope to affect substantially domestic oil production in the United States, Western Europe, and other importing countries, which on present trends seem likely to account for about one-third of total world oil requirements. That would have been possible before October 1973. Now, however, the importing countries are likely to go to considerable cost to protect domestic production as a secure and therefore preferred source of supply. Consequently, if price cutting was relied upon as the general solution to achieving market equilibrium, its effects would have to be concentrated on production in the exporting countries alone and would involve very sharp price reductions and huge revenue losses, even for the Middle Eastern countries that increased their market shares. In all likelihood, therefore, the major exporters would stop short of a purely competitive solution of the problem. Price wars could well occur, but more as a temporary test of bargaining strength among exporters in the maneuvering to form a cartel than as a permanent way of life.

The success of a cartel or cartel-like arrangements would depend on the market being large enough to accommodate a set of export quotas that could be negotiated among a sufficient number of major producers and to which they would adhere. Only then might supply be kept in balance with demand at something substantially higher than a competitively determined price. For the next few years, the major Middle Eastern producers with growth potential—Saudi Arabia, Iran, Iraq, and the United Arab Emirates—conceivably could take upon themselves the role of residual suppliers and seek to moderate the erosion of present high prices by restricting or reducing their own production even if other exporters refused to cooperate. Or Saudi Arabia, with or without others, might find it preferable to reduce prices more quickly and move toward a somewhat larger world oil market, thus establishing a sound base for negotiating an agreement in which most or all of the producers would share responsibility for restricting production.

To illustrate the factors involved, the possible effect on world import requirements of a price of $5 a barrel is compared below with projections of import requirements corresponding to prices of $3 and $6.25 a barrel:

Price of Saudi Arabian Crude Oil, f.o.b. Persian Gulf (in 1973 dollars)	*World Oil Import Requirements (in million barrels per day)*		
	1973	*Projected*	
	Actual	*1980*	*1985*
$3.00	32	48	60
$5.00	32	39	43
$6.25	32	33	32

No predictive significance should be attached to this projection, although in light of the limited evidence available it does not appear to be unreasonable. The market corresponding to a price of $5 a barrel has been calculated within a framework set by the two projections examined in this chapter—the pre-October 1973 projection, which was based on a Persian Gulf price of crude oil of approximately $3 a barrel and a willingness of governments to rely increasingly on imported oil, and the current projection shown in Table 13-8, which is based on a Persian Gulf price of $6.25 a barrel and on continuing government measures to achieve greater self-sufficiency in energy. The calculations rest on two propositions: (1) that the response of energy consumption to a reduction in price would be somewhat greater at the higher than at the lower end of the range; and (2) that a reduction in the price of imported oil would have a much less than proportionate effect on domestic production of primary sources of energy in oil-importing countries because those countries will now provide higher levels of protection or subsidy to this production.* In the short run, exporters probably gain substantially from present high prices. From an intermediate or longer range point of view, these calculations suggest they would be better off with lower prices. As a group, they might be able to maximize their revenues at a price somewhere between $6.25 and $3.00 a barrel—probably somewhat above the middle of the range.

Would the exporting countries be able to agree on arrangements to share an oil market corresponding to a price of about $5 a barrel and exercise the necessary discipline to maintain that price? The answer is not self-evident. The prospects certainly would be considerably better than in the case of the virtually stagnating oil market projected to correspond to a price of $6.25 a barrel, to say nothing of a steadily declining oil market that would correspond to maintenance of the present price of about $9 a barrel. While market growth at $5 a barrel would be slow, it could be sufficient to leave a margin for new exporters to participate, for smaller exporters to expand, and for the major exporters to share a small incremental residual over the next decade. Obviously, the negotiating difficulties and the risks of failure would be considerable, but this course is certainly possible in light of the very large financial costs that the major exporters would incur if they slid into the price-cutting alternative.

An international arrangement for oil involving importers and exporters would, of course, greatly enhance the prospects for achieving market stability within such a price range. Whether, in light of what has happened since October 1973, each group can obtain the assurances it would require from an international commodity agreement for oil is at the moment questionable. In time and within limits, the possibility that they might choose to rely somewhat more on each other through such an institutional arrangement is likely to improve. Certainly in economic terms, the potential gains from cooperation provide a wide area of negotiation, more than sufficient for both groups to benefit substantially and for the world to avoid an economically wasteful and environmentally costly diversion of resources.

*Specifically, in comparison with the projection corresponding to a price of $6.25 a barrel (Table 13-8), a price of $5 a barrel is assumed to reduce the saving in energy consumption by one-half and to reduce the increase in domestic production of primary sources of energy, including oil, by one-fourth.

NOTES FOR CHAPTER 13

1. United Nations, *Monthly Bulletin of Statistics*, June 1960, June 1965, and June 1970.

2. Organisation for Economic Co-operation and Development, *Energy Policy* (Paris, 1966), as quoted in Joel Darmstadter, Perry D. Teitelbaum, and Jaroslav G. Polach, *Energy in the World Economy*, Resources for the Future (Baltimore: Johns Hopkins University Press, 1971), p. 18.

3. This projection relies primarily on the following forecasts: Organisation for Economic Co-operation and Development, *Oil: The Present Situation and Future Prospects* (Paris, 1973); National Petroleum Council, *U.S. Energy Outlook* (Washington, D.C., 1972); assessments made in background papers on energy requirements and supplies in Western Europe and Japan prepared for The Brookings Institution; and specific adjustments as specified in the footnotes to Table 13-1.

4. These studies are summarized in Milton Searl, *Towards Self-Sufficiency in Energy Supply*, Resources for the Future, a study prepared for the Ford Foundation Energy Policy Project and scheduled to be published in 1974 by Ballinger Publishing Company.

5. Hendrick Houthakker and Michael Kennedy, *Demand for Energy as a Function of Price.* (Processed.) Specific estimates cited in this paper for average long-run elasticities for OECD countries are: gasoline −.82; kerosene −2.0; distillate fuel oil −.76; and residual fuel oil −1.58.

6. For example, in an address at American University (Washington, D.C.) April 18, 1974, John C. Sawhill, then deputy administrator of the Federal Energy Office, said, "Our target in the conservation area is to reduce our current 4-5 percent annual energy growth rate to 2 to 3 percent."

7. The proposals of the Commission are summarized in *L'Europe*, No. 1483 (new series), March 22, 1974.

8. "Energy Management in the Industrial Community," an address presented at National Energy Forum II, Washington, D.C., 1973.

9. Federal Energy Administration. *U.S. Energy Self-Sufficiency: An Assessment of Technological Potential*, preliminary draft, February 6, 1974. This projection of coal production is from the "intermediate scenario" outlined in the study.

10. *Wall Street Journal*, March 12, 1974.

11. Walter G. Dupree and James A. West, *United States Energy Through the Year 2000*, U.S. Department of the Interior (Washington, D.C., 1972). Organisation for Economic Co-operation and Development, *Oil: The Present Situation*, p. 14. See also Jack Hartshorn, *Europe and World Energy*, manuscript (London: Chatham House, August 1973 revised), Chapter II. Looking to 1980, Hartshorn suggests that confidence about the sufficiency of reserves could be maintained at about present levels if during the 1970s additions to proved reserves, either through new discoveries or uprating, were about equal to the additions made in the 1960s, or 30 to 50 billion tons.

This would maintain reserves at the level equivalent to 15 years of consumption. This calculation is based on the pre-October projection of oil requirements. At currently projected requirement levels, the same additions to reserves would provide a reserve-to-consumption ratio closer to 20.

12. Searl, *Towards Self-Sufficiency in Energy Supply*, p. 30.

Chapter Fourteen

Financial Implications

by Edward R. Fried

Oil has been the most important primary commodity and a rapidly growing factor in world trade for some time. The value of oil exports (in 1973 dollars) rose from $7 billion in 1962 to $20 billion in 1972, primarily because of an expansion in volume. During that period, the share of oil in world exports grew by one-half, from 3.2 percent in 1962 to 4.8 percent in 1972. This substantial growth posed no special problems for the world economy, and indeed went largely unnoticed.

When prices were escalated in 1973, the situation changed drastically. The value of oil exports increased to approximately $32 billion in 1973, or 7 percent of world exports, and could amount (in 1973 dollars) to perhaps $95 billion in 1974, or about 17 percent of world exports. This trend obviously cannot continue for long; in fact, the analysis in Chapter 13 suggests that it may soon be reversed. Even so, the financial flows generated by the world trade in oil will have substantial consequences for individual countries and will severely test the capacity of the international economic system to adjust to change.

This chapter explores some of these international adjustment problems: first by estimating the size and disposition of oil revenues over the period through 1985; and then by examining the impact of those revenues on the international financial system, on the U.S. balance of payments, and on the flow of development assistance. While the focus of this study is on longer term effects, the suddenness of the jump in prices will in some instances create special and particularly difficult adjustment problems for the next year or two. These will be noted where appropriate.

277

POSSIBLE SIZE AND DISPOSITION OF
OIL REVENUES

Estimates of the oil revenues of exporting countries over the period 1974-85 depend on assumptions about export prices, export volume, and the production costs and profits of the producing companies. These are specified below. Throughout the discussion in this chapter, prices, costs, and revenues will be stated in 1973 dollars, thus abstracting as much as possible from the effects of inflation on relationships among factors of production.

Production costs and profits are a relatively small element in the oil equation as the world market now stands and for convenience they will be discussed first. Production costs are assumed to remain constant at estimated present levels—say 15 cents a barrel for Persian Gulf countries and 40 cents a barrel elsewhere. The profit margin of the producing companies over the period through 1985 is projected to be fixed at an average of 35 cents a barrel for all oil production, whether earned through the sale of "equity oil" or "buy-back" participation oil, or from service contracts. This is less than the companies realized in 1973 and less than they will realize in 1974. As the market eases, however, and as the companies' equity share in the concessions declines, their margin from the production and marketing of crude oil is apt to be lower.

For present purposes, export prices and volume are the key assumptions and, as discussed in Chapter 13, they are closely related. Before the October 1973 embargo, the exporting countries might be said to have had a choice between two broad policy alternatives. They could increase real prices steadily but moderately (as suggested by the tactics they employed in the Tehran and Geneva agreements), thereby attaching importance to a continued increase in the size of the market. This in turn would make it easier for them to continue to share and control the market by price-fixing arrangements among themselves. Or, they could take full bargaining advantage of the short-term situation—in which neither quick reductions in consumption nor quick increases in supply are feasible—and sharply raise prices from the outset. This would quickly strengthen their financial position and enable them to shore-up prices by withholding supplies from the market whenever a condition of excess supply developed. This tactic, however, would risk triggering reactions in the importing countries that eventually could reduce the size of the market and thereby compound the difficulties of negotiating export quotas among producers and maintaining an effective cartel.

As the events following the embargo demonstrated, the exporting countries adopted the second alternative, partly by accident and partly by design. They now face the task of controlling a market in which, as a consequence of higher prices, export capacity will rise more rapidly than import requirements.

What export price and volume trends might seem reasonable for the next ten years or so in these circumstances? The analysis in Chapter 13 indicated that: (1) prices can be expected to fall substantially from the present level of approximately $9 a barrel; (2) a price of $6.25 a barrel (equivalent on a delivered basis to the present blend price of $7.50 a barrel in the United States) would mean a virtually stagnant if not declining export market and probably would not be sustainable for more than a few years; and (3) a price in the area of $5 a barrel might be an equilibrium price for the period through 1985 in that it would permit a steady growth in exports, provide a better base for negotiating export quotas among producers or an arrangement between producers and consumers, and tend to maximize producer revenues for the period. (These prices are for Saudi Arabian light oil, f.o.b., the Persian Gulf, in 1973 dollars.)

Prices might decline from present levels in two alternative ways. As one possibility, the major Persian Gulf producers, by the combined exercise of export restraint, might be able over the next few years to keep supplies close to reasonable balance with demand. Based on the gradual growth of import requirements projected in Table 13-9, this could require that Saudi Arabia restrict exports to about 7 million barrels a day (almost 2 million barrels a day below the May 1974 level), that Iran freeze exports at 6 million barrels a day, and that Iraq and the United Arab Emirates increase exports by much less than they now plan. If these restraints were exercised, the present price of $9 a barrel might erode gradually over the next few years, as other producers increased their output, averaging say between $6 and $7 a barrel. World import requirements might then increase slowly, from perhaps 30 million barrels a day in 1974 (excluding increases in stocks) to 34 million barrels a day in 1976. Thereafter, the market might level off or become smaller as coal, gas, and oil production in the importing countries began to increase significantly in response to higher prices. The market situation then might be beyond the capacity of the major producers to control—or, more accurately, beyond their willingness to accept the necessary export restraints. Prices could be expected to fall further and, after a period of testing, an equilibrium might be reached—perhaps at a price around $5 a barrel and at a level of exports averaging between 40 and 45 million barrels a day. This equilibrium might be the result of an export-quota arrangement negotiated solely among producers, or of an international agreement that included the cooperation of the importing countries.

Alternatively, Saudi Arabia, acting alone or with the cooperation of others, might seek to reduce prices much more quicky at the outset in the hope of blunting the drive toward greater energy self-sufficiency in the importing countries. Lower prices would mean a larger export market and would place Saudi Arabia in a position in which it was under somewhat less pressure to restrict its own exports and at the same time in a better position to control the market over the future. Acting along these lines would be consistent not only with repeated official Saudi statements favoring a reduction in oil prices but, more importantly, with Saudi interests, which differ in a fundamental respect from most other oil exporters. Whereas countries with more limited oil reserves might be interested in maximizing revenues over a relatively short period of time, Saudi Arabia, with its huge reserves, would tend to look more at how present actions might affect the oil market well into the future and therefore would be more worried about prematurely encouraging production of competitive sources of energy.

Either hypothesis about how the world oil market might develop would indicate a substantial fall in prices from present levels and an increase in prospective export volume. The difference between them would largely be a matter of timing. For purposes of calculating prospective oil revenues over the period 1974-85, this difference can be disregarded. For the period as a whole, it is assumed that (1) the price of Saudi Arabian light oil, f.o.b. the Persian Gulf, will average between $5 and $6 a barrel, with appropriate differentials for prices in other exporting countries where oil-quality characteristics and transportation advantages are factors; (2) world oil exports will average 38 million barrels a day; and (3) the major Persian Gulf oil producers will assume primary but not sole responsibility for keeping the supply of oil in balance with demand at the assumed price. On these assumptions, the governments of the oil-exporting countries* will receive revenues averaging $71 billion a year, or about one-third less than they may realize in 1974, and almost three times more than they realized in 1973. The projected revenues of each of the major exporting countries is shown in Table 14-1.

*This estimate includes only the revenues of the developing-country oil exporters and therefore excludes those of Canada. However, Canadian oil exports are included in projections of the world market; they are estimated to average 1 million barrels a day over the period 1974-85.

Table 14-1. Projected Annual Average Oil-Export Revenues,
1974-85

	Export Volume[a] (in mbd)	Export Price[b] ($ per barrel)	Government Revenues[c] ($ per barrel)	Annual Govt. Revenues[d] (in $ billion)
Countries in Continuing Need of Development Finance:				
Algeria	1.5	6.50	5.75	3.1
Indonesia	2.0	6.50	5.75	4.2
Iran	6.5	5.50	5.00	11.9
Iraq	3.5	5.50	5.00	6.4
Nigeria	2.5	6.25	5.50	5.0
Venezuela	2.5	6.25	5.50	5.0
Other	3.0	6.25	5.50	6.0
Total	21.5			41.6
Countries in Financial Surplus:				
Kuwait	2.5	5.50	5.00	4.6
Libya	2.0	5.50	6.00	4.4
Qatar & Oman	1.0	6.00	5.50	2.0
Saudi Arabia	8.0	5.50	5.00	14.6
United Arab Emirates	2.0	6.00	5.50	4.0
Total	15.5			29.6
Total	37.0[e]			71.2

Source: Author's estimates.

Note: All dollar figures are in 1973 U.S. dollars.

[a]Allocation of exports among countries is based on Table 13-9 and discussion in text.

[b]Differentials over base price for Saudi Arabian light oil are approximately equal to those prevailing during first six months of 1973. Differentials were much larger following the October 1973 embargo, but they are likely to return to more normal levels as the market eases.

[c]Government revenues per barrel equal export price less adjustment for production costs and company margins, as discussed in text.

[d]Does not include Canadian oil revenues.

[e]Does not include Canadian oil exports, projected at 1 million barrels per day.

These are very large incremental revenues. How might they be used?

Approximately $42 billion, or 60 percent of the total, will go to relatively poor countries that have sizable-to-large populations and comparatively diversified economies in which oil is an important but subsidiary economic sector. These countries should have ample

domestic investment opportunities to use additional oil revenues.*
This is evident from the data in Table 14-2.

Indonesia and Nigeria are very poor countries that have
large populations and development needs that are much greater than
could be financed by their oil revenues alone. Indeed, since about
1970, the imports of both countries have been growing by 20
percent a year.

Iran and Algeria have in common a development strategy
aimed at maintaining a growth rate of 10 percent a year, which, in
turn, would require rapidly increasing imports. At this rate of
growth, Algeria probably would continue to be a significant net
importer of capital. Iran's net, long-term capital position might seem
to be subject to more doubt in view of the huge size of its projected
oil revenues. Large as they are, most of these oil revenues are likely
to be needed to finance development-related imports, even if some
decline in the growth rate of these imports takes place over the next
decade. Iran's foreign exchange expenditures to upgrade its military
equipment will also be large; military equipment now on order or
reportedly being considered for purchase is estimated to cost almost
$4 billion.[1] A modernized military force, moreover, will require
spares, replacements, and maintenance services on a continuing basis,
most of which will have to be imported.

Iraq's development programs over much of the past two
decades have been frustrated by political turmoil and instability.
Present prospects are more promising and may set the stage for
accelerated development of the economy as a whole. The current
development program (1970-75) is targeted to achieve a growth rate
of 7 percent a year. A prospective rise in import requirements for
development, plus Iraq's substantial expenditures on military equipment, should leave little room for financial surpluses to emerge from
projected oil revenues.

Venezuela has had the longest experience among the major
oil-producing countries in using oil revenues to diversify and expand
its economy. If economic growth continues at the recent rate of 4 to
5 percent a year, financial surpluses could emerge. Here again,
however, a moderate increase in the development of the non-oil
sector, which accounts for 75 percent of the economy, could result
in substantial additions to import requirements. Furthermore,
Venezuela may invest heavily in domestic oil-refining capacity, which
would require large-scale imports of equipment and technical
services.

*See Part Two for a full discussion of the development needs of
these countries.

Table 14-2.   Present and Projected Importance of Oil in the Economies of the Oil-Exporting Countries

	Est. 1973 GNP per Capita (1973 U.S. $)	Oil Revenues as Percent of GNP (Est. 1973)	Oil Exports as Percent of Total Exports, 1972	Projected Population 1980 (millions)	Projected 1980 Per Capita Oil Revenues (1973 U.S. $)
Countries in Continuing Need of Development Finance:					
Algeria	425	17	75	23.9	121
Indonesia	90	12	46	183.8	21
Iran	575	24	86	38.8	276
Iraq	440	35	91	13.9	410
Nigeria	170	21	81	72.8	63
Venezuela	1,200	24	93	15.0	307
Other	...	...	...	...	...
Countries in Financial Surplus:					
Kuwait	4,300	55	95	1.6	2,560
Libya	2,000	65	98	3.1	1,290
Qatar & Oman	1,000	95	99	1.0	1,800
Saudi Arabia	950	80	99	10.5	1,250
United Arab Emirates	11,000	95	99	0.2	17,500

Sources and Notes: Estimated 1973 GNP per capita based on data for 1971 in International Bank for Reconstruction and Development World Bank Atlas (Washington, D.C., 1973), and adjusted for recent growth rates in GNP and increases in oil revenues. Oil revenues for 1973 calculated by using export volume from Table 13-8, average price for Saudi Arabian oil from Table 13-3, price differentials in Table 14-1, company-profit margin of 35 cents per barrel, and production costs of 15 cents a barrel for Persian Gulf Oil and 40 cents a barrel for other countries. Figures for 1973 oil and total exports—International Monetary Fund, International Financial Statistics; per capita oil revenues for 1980 assume a $5 price for oil and a world oil market of 39 mbd. United Nations, Economic and Social Affairs Department, World Population Prospects as Assessed in 1968 (New York, 1973). Figures represent the "medium variant." Projected oil revenues per capita were derived by dividing the projected average oil revenues, 1974-85, shown in Table 14-1 by the projected population for 1980.

As shown in Table 14-2, on a per capita basis, the oil revenues of this group of countries will not be overwhelming—ranging in 1980 from only $25 in Indonesia to $500 in Iraq. Nor will the GNP of these countries, even with the stimulus of oil-financed economic expansion, put them in the category of net long-term capital exporters. Only Venezuela at present has a per capita income of over $1000; Iran and Algeria, the most rapidly growing economies in this group, probably will not reach even that point until the end of this decade. Some may invest in tankers or in oil-refining and marketing facilities abroad, as, for example, Iran and Iraq seem to be doing; such investments, however, would be part of their oil-marketing strategies rather than an indication that their financial capacity had become surplus to domestic needs. Most of these countries, moveover, will have temporary financial surpluses, simply because their oil revenues will be so large and so suddenly accumulated that a lag will necessarily exist between the time the revenues are received and the time at which they can be spent for additional imported goods and services. These factors aside, the countries in this group can be expected to have a continuing need to use oil income to finance growing import requirements generated by economic expansion, rising consumption, and in some instances, military investment.

The economic base of the second group of countries in Table 14-1 is too narrow to absorb their large projected oil revenues at home—at least not during the period through 1985. Consequently, the oil revenues of the countries in this group, averaging approximately $30 billion a year, will give rise to persistent financial surpluses, which will have to be invested abroad either through the purchase of financial debt or equity instruments or through direct investment in property, plant, and equipment. The main oil-exporting countries in this group are Kuwait, Libya, Saudi Arabia, and the United Arab Emirates.

In speculating about the size of these surpluses, two caveats are worth emphasizing. Both stem from a situation in which countries characterized by small populations and oil sectors that dominate their economies will be able to formulate their domestic programs virtually free of financial constraints. This will mean, first, that their spending will not necessarily be limited by their capacity to absorb capital investment; they can subsidize consumption directly through the import of consumer goods. Second, their economic and social programs will necessarily have an unusually large import component since they will have to rely on foreign sources not only for capital equipment but for a wide range of other goods, as well as technical services.

The countries in this group presently differ in the stage of their economic development, the extent of their present commitment to economic growth, the scope of their social programs, and their view of military requirements. Future policies in these matters will substantially determine their capacity to use future oil revenues at home—that is, to finance imports of goods and services. The following policy forecasts have been used to estimate import requirements for each of these countries.[2]

Libya's imports grew rapidly—at a rate of approximately 15 percent a year—during the period 1965-71. Much of this increase was due to the explosive expansion of its oil sector. This phase of its development is now completed, the more so because Libya has adopted a conservationist oil-marketing policy. In 1972, however, Libya launched a three-year, $3 billion development program. If the country should sustain this degree of emphasis on internal development, despite the uncertainties of Libyan politics, the non-oil sector might grow rapidly. Accelerated expansion of this sector, which accounts for approximately one-third of Libya's GNP, could easily require a growth in imports of perhaps 10 percent a year. To allow for a possible shortfall in the development program, no allowance is made in this projection for imports of military equipment, which in recent years have been substantial.

Kuwait has been rapidly increasing its expenditures on economic development for more than a decade, mostly for roads, ports, schools, housing, and medical facilities. During the initial phase of the program, imports grew by 15 percent a year. More recently, the rate of increase has declined to 8 percent a year. The already high level of social-development expenditures would indicate a further slackening in the growth of imports. On the other hand, population is increasing and the government is buying much larger quantities of military equipment. Taking these factors into account, import requirements are projected to continue to grow at the rate of 8 percent a year.

The United Arab Emirates (UAE)—a federation of ministates whose economy is based on the oil production of Abu Dhabi, Dubai, and Sharja—has a population of approximately 200,000. It combines the highest per capita income in the world with a tribal society having few modern institutions. At present the UAE is embarked on a path of establishing a paternalistic welfare state similar to that of Kuwait. If that policy continues to be followed, import requirements are rather arbitrarily projected to increase at 15 percent a year—or the rate experienced in Kuwait during the initial phase of its social and economic development program. The lack of skilled people and the need to build a social and economic infrastructure could be limiting factors. On the other hand, in view of the enormous size of the UAE's per capita oil revenues, the means for directly increasing consumption could be purchased from abroad on a continuing basis. If so, imports would rise rapidly, quite apart from the progress made in social and economic development.

Saudi Arabia is by far the largest potential holder of surplus financial resources. To put these holdings in perspective, however, it should be noted that even if its huge projected oil revenues should underwrite a GNP growth rate of 10 percent a year, Saudi Arabia's per capita income in 1985 would be little more than one-half that in the United States today.

For the future, Saudi officials have indicated that high priority will be assigned to accelerating economic and social development. A base for this has been established; over the past ten years Saudi Arabia has made substantial progress in opening up its tradition-dominated society and creating a new middle class of professionally trained people. While Saudi Arabia's leadership group will continue to be concerned about the preservation of traditional values, the movement toward modernization probably is no longer reversible given the recent quantum jump in oil revenues.[3]

Government expenditures, even with much smaller oil revenues than are now anticipated, grew by 15 percent a year during the period 1967-72, and this rate of growth is projected to continue over the next decade.[4] So massive an increase in expenditures on social and economic development might seem unrealistic considering the present limited size of the non-oil sector of the economy. However, present development projects are heavily capital intensive (e.g., steel, petrochemicals, and dock facilities), a large-scale effort in health, education, housing, and rural electrification seems to be under way, and military expenditures will certainly grow.[5] Against this background, the projection of government expenditures for the period through 1985 is broken down as follows: (1) an increase of 15 percent a year in development outlays; (2) an increase of 18 percent a year in expenditures on health and education, which would bring those expenditures to about $200 per capita in 1985, or roughly the 1973 Kuwait level; and (3) an increase of 13 percent a year in other government expenditures, of which about half (or an average of $1.5 billion a year) would be used to finance the substantial military buildup now in progress.

These projected domestic expenditures would have a high import content for both goods and labor-embodied services. Saudi Arabia's narrow non-oil sector could not supply the investment goods and could produce only a small portion of the requirements for consumer goods, including food. In addition, a large number of foreign technicians would be needed. In time, this dependence on imports could be expected to diminish. Imported goods and services are projected to amount to three-fourths of total expenditures for the period through 1980, and for two-thirds of the total for the following five years. This would mean that Saudi Arabia's imports of goods and services would grow from $1.6 billion in 1972 to $8.4 billion in 1985, a rate of increase of almost 14 percent a year.

In sum, these policies would mean that the financial-surplus countries as a group would increase their imports of goods and services from $5 billion in 1973 to $20 billion in 1985, or by 13 percent a year. The magnitudes are impressive, but not unrealistic in view of the dramatic growth in the oil income of these countries, their prospective interest in social and economic development, and their need to import a large proportion of required goods and labor-embodied services. Deducting these import requirements from projected oil revenues indicates a potential financial surplus of approximately $19 billion in 1976, $14 billion in 1980, and $8 billion in 1985. The figures for individual countries are shown in Table 14-3. (For 1974-75, see Table 14-5.)

To arrive at the estimated financial surplus that would be available for foreign investment, an allowance should be made for bilateral economic aid to developing countries and for military aid to other Arab governments. In the past, these amounts have been relatively small. Under arrangements made in Khartoum following the 1967 war, Saudi Arabia, Kuwait, and Libya gave the UAR and Jordan somewhat less than $300 million a year to offset foreign exchange losses from the closure of the Suez Canal and the occupation of Jordan's West Bank. Kuwait has had an economic development fund for some years, and Abu Dhabi established a similar one early in 1973. Disbursements from both funds were probably no more than $100 million a year. The October war and the subsequent high oil prices have led to a substantial increase in these activities. Saudi Arabia reportedly financed large purchases of military equipment for Egypt and presumably will continue to do so. Libya's actions are unpredictable, but Kuwait and the United Arab Emirates probably are helping to finance Egyptian and Syrian arms purchases and are rapidly expanding their development lending plans.[6] In addition, these countries are or probably will be supplying some oil to developing countries on terms that include partial financing at concessional rates of interest. Somewhat arbitrarily, a

Table 14-3.   Projected Import Requirements and Potential Financial Surpluses of Selected Oil-Exporting Countries (in billions of 1973 U.S. dollars)

Countries	1976 Import Require- ments	1976 Potential Financial Surplus	1980 Import Require- ments	1980 Potential Financial Surplus	1985 Import Require- ments	1985 Potential Financial Surplus
Kuwait	1.4	3.2	1.9	2.2	2.8	1.3
Libya	1.9	2.5	2.7	1.3	4.4	—
Saudi Arabia	3.1	10.6	5.2	7.9	10.1	4.7
United Arab Emirates	0.5	2.5	0.8	2.9	1.6	2.1
Total	6.9	18.8	10.6	14.3	18.9	8.1

Source: Author's estimates.

Notes: Import requirements include all goods and services except those associated with oil production. The latter are assumed to be financed from production costs and profits of the oil companies, not from government oil revenues.

To show the trend of potential financial surpluses over the period, oil exports (in mbd), based on a price after 1976 of $5 a barrel, were projected in Chapter 13 to be 34 in 1976, 39 in 1980, and 43 in 1985. Exports of the four financial surplus countries (in million barrels per day) were projected as follows: Kuwait 2.5 throughout the period; Libya 2 throughout the period; Saudi Arabia 7.5 in 1976, 8 in 1980, and 9 in 1985; and the UAE 1.5 in 1976 and 2.0 in 1980 and 1985. These figures are consistent with the allocation of average exports for the period 1974-85 shown in Table 14-1.

figure of $2 billion a year—perhaps five times the early 1973 level—will be assumed for the bilateral economic and military aid of these four financial-surplus countries.

These conjectures suggest that over the period 1974-85 the financial-surplus countries would have excess oil revenues averaging approximately $14 billion a year that would be available for foreign investments or for the purchase of primary reserve assets, such as gold or Special Drawing Rights (SDRs). Foreign investments might include short- or long-term foreign debt instruments, equities, or direct investments. In addition, for the period as a whole, reinvested earnings could average over $4 billion a year. As Table 14-4 shows, these investments, including reinvestment of earnings, might total $136 billion by 1980 and $211 billion by 1985, of which more than half would accrue to Saudi Arabia.

Estimates of financial surpluses are sensitive to assumptions about the price of oil, but perhaps less and somewhat differently than might be expected. The financial surpluses shown in Table 14-4 were based on a price after 1976 of $5 a barrel (in 1973 dollars). At higher prices, say the price of $6.25 a barrel assumed for the market analysis in Chapter 13, total oil exports by 1985 might be reduced by one-third. Saudi Arabia and the UAE would probably have to accept a disproportionately large reduction in their exports to sustain the price at that level; consequently, their export revenues and surpluses would be lower than the amounts shown in Table 14-4. Iran and Iraq might also be worse off, depending on the outcome of their bargaining with Saudi Arabia on how export restraints were to be shared. Other exporting countries, however, would be better off. Conversely, at lower prices—say $4 a barrel—and a larger total export market, Saudi Arabia and the UAE could relax their production constraints and would therefore enjoy a more than proportionate increase in their exports; as a consequence they would have larger total revenues and financial surpluses. Iran and Iraq might experience little change in their position, but other exporting countries would not be as well off, since they would be receiving a lower price for about the same volume of exports. This line of analysis emphasizes once again the difference in interest among the exporting countries on the price issue—a difference that stems from the probability that some would almost certainly have to bear much larger responsibilities than others in restricting exports to sustain a cartel-fixed price.

Before considering the implications of these financial surpluses for the operation of the international economic system, it may be useful to address the question of whether these countries would find it advantageous to become capital exporters on so large a scale as implied by Table 14-4 in preference to cutting back their production and, so to speak, leaving their wealth underground. In a sense, the answer to this question is implicit in the foregoing examination of market trends and prospects. If that analysis is valid, the financial-surplus countries, as primary residual suppliers, would already have curtailed their exports by the amounts necessary to maximize their revenues. A further reduction of exports, while tightening supplies and increasing prices in the short run, would reduce the demand for oil-exports over the longer term and worsen the financial position of the residual suppliers.

Much the same answer is derived from making the calculation in terms that take into account the revenues that are foregone because of a preference for keeping wealth underground. In other words, considering the case for Saudi Arabia, how high would the price have to be in the distant future to have made it worthwhile for Saudi Arabia to keep its oil underground rather than selling it during the present period at the projected average price of $5 a barrel? Saudi Arabian reserves are so large that production curtailed now would not be needed until at least the beginning of the next

Table 14–4.   Projected Investments of Financial Surplus Oil-
Exporting Countries (in billions of 1973 U.S. dollars)

Countries	Surplus of Oil Revenues over Import Requirements (Annual Average)[a]	Bilateral Economic and Military Aid (Annual Average)[b]	Oil Surplus Available for Foreign Investment (Annual Average)[b]	Reinvested Earnings (Annual Average)[b]	Total Investment 1980	1985
Kuwait	2.4	0.5	2.0	0.7	20	31
Libya	1.6		1.7	0.6	21	27
Saudi Arabia	8.8	1.0	7.9	2.6	80	124
United Arab Emirates	2.4	0.5	1.9	0.5	15	29
Total	15.3	2.0	13.5	4.4	136	211

Source: Author's estimates.

Note: These projections are based on a price of oil over the period 1974-85 averaging $5-$6 a barrel for Saudi Arabian light oil, f.o.b. the Persian Gulf, in 1973 dollars, and total world exports increasing from 32 million barrels per day in 1974 to 43 million barrels per day in 1984.

[a]Derived from Table 14-3 and Table 14-5.

[b]Assumes a 4 percent real rate of return expressed in 1973 dollars.

century.* Meanwhile, however, the revenue foregone by cutting back production could have been invested in earning assets with a return of perhaps 4 percent a year in real terms. For Saudi Arabia to break even, then, the export price of oil in the year 2000 would have to reach $13 a barrel on the Persian Gulf.** At prices lower than $13 a barrel in that year, which seems much more likely in light of the market analysis in Chapter 13, Saudi Arabia would sustain substantial losses from this type of production limitation. It would seem to be strongly against Saudi Arabia's economic interest, therefore, to base its production decision on a preference for wealth underground rather than in the form of earning financial assets.

*Saudi Arabia's proved reserves are placed at 132 billion barrels. Most experts believe that by 1985 additional reserves at least equal to that amount will be found or upgraded. Suppose it is assumed, however, that (1) only one-half this amount is added to proved reserves by the end of the century; and (b) production averages 8 million barrels a day from 1974-85 and 12 million barrels a day from 1985-2000. In that event, the ratio of reserves to production at the beginning of the next century would still be better than 20-1, or at the upper end of the range generally considered to be economic for production in the Middle East.

**If the calculation is based on the March 1974 price of $9.50 a barrel instead of $5 a barrel, the break-even price in the year 2000 would be $26 a barrel, f.o.b. the Persian Gulf (all figures in 1973 dollars).

IMPLICATIONS FOR THE INTERNATIONAL ECONOMIC SYSTEM

Higher oil prices will require the importing countries to make a number of difficult domestic and international economic adjustments. The domestic adjustments, which inherently are the most burdensome, are merely noted here by way of introduction to the international issues.

Domestically, it must be recognized that the importing countries will have to accommodate to a substantial increase in real energy costs, which sooner or later will have to be borne by consumers. The amounts are substantial. At a price averaging between $5 and $6 a barrel, for example, oil imports over the period 1974-85 could cost the importing countries almost $40 billion a year more than they would at the average 1973 price of $2.75; these costs eventually would have to be paid for in real resources, that is, in exports of goods and services to the oil-producing countries. The resource cost of producing domestic sources of energy will also rise.* In all, the increase in the real cost of energy could amount, on the average, to approximately 1 percent of the total GNP of the non-Communist oil-importing countries.

The importing countries also face serious domestic economic dislocations (e.g., lower output and employment in the U.S. automobile industry as it shifts to production of smaller cars) and difficult problems of demand management, foremost among them, the determination of a fiscal and monetary policy mix that can contain the inflationary impact of higher energy costs at minimum sacrifice of employment. These latter problems are transitional in the

---

*The resource costs of increased domestic production of energy are difficult to assess, but they are likely to be much smaller than the incremental cost of imports. Some rough indication might be gained from the following line of reasoning. Over the period 1974-85 domestically produced primary sources of energy are projected to constitute about two-thirds of total energy requirements in the non-Communist world. The price of these energy sources will rise but, if substitution is to take place, not by as much as imported oil. For the base quantity of production, so to speak, price increases would represent a transfer of income within each country, rather than a resource cost. Production above this amount, however, would require a rising input of capital and labor. Table 13-4 suggests that such incremental production might grow, on an oil equivalent basis, from relatively small quantities in the first few years to roughly 3 million barrels a day in 1980 and 9 million barrels a day in 1985. For the period as a whole, incremental domestic energy production might average 10 percent of the volume of imported oil. Additional resource costs might then average something under $3 billion a year—less at the beginning and much more toward the end of the period.

sense that they will disappear as countries complete the adjustment to a world of higher energy costs. If they are not successfully managed, however, the consequences for employment and income could involve even greater costs than those that consumers will have to pay because of the rise in the real cost of energy.

As for the international economic system, higher oil prices raise a somewhat different set of adjustment problems. These problems stem principally from the fact that the oil-exporting countries, both those in continuing need of development finance as well as those in a position of quasi-permanent financial surplus, cannot spend their oil revenues on imported goods and services as rapidly as they receive them. Consequently, the importing countries will necessarily defer the transfer of part of the real resources that are being exchanged for oil imports. In turn, the exporting countries will necessarily take part of their payment for oil exports in the form of investments (financial or direct) in the oil-importing countries. Thus, in transactions between the two groups of countries, each taken as a whole, capital transactions in the balance of payments will have to offset current transactions. Can the importing countries make the necessary policy adjustments in this balance-of-payments environment?

A second difficulty arises from the fact that the oil-importing countries will not fare equally in receiving increased export orders or investments from the exporting countries. Some (such as the United States and Germany) may receive proportionately more, others (such as most developing countries and some financially troubled industrial countries, like Italy) may receive proportionately less. There is of course nothing unusual about such bilateral imbalances in a multilateral trade and payments system. The point simply is that the massive size of the oil transactions and the sudden financial shifts they will set in motion will require unusually large second-stage adjustments through which funds can be recycled among importing countries so as to restore a working equilibrium.

In tracing possible international consequences, it is useful to distinguish between requirements during the next few years and those for the period through 1985 as a whole. Higher oil prices have imposed a severe shock on the world economy, as much because they were sudden as because they were so sharp. Adjustment requirements in almost any conceivable circumstances will be heavy in the initial period. Subsequently, they can be expected to diminish and take on characteristics that are more usual in international economic relations.

The Transitional Period, 1974-76

The adjustments required during this period will depend principally on the way in which prices move toward a longer term equilibrium level, and on the length of the lag between the receipt and expenditure of oil revenues. The analysis earlier in this chapter suggested that: surplus oil-export capacity already exists and is likely to grow, prices are pointed downward, and the primary residual suppliers could decide either to reduce prices fairly quickly so as to seek from the outset a sustainable balance between import requirements and a negotiated set of export quotas, or delay the pace of price erosion as long as possible by accepting the necessary export restraints themselves. There is no basis on which to forecast which course they are likely to take since their interests differ and very substantial market uncertainties exist.

For present purposes, the following assumptions will be used:

— Export prices in real terms will go down gradually beginning in 1974. As a result, government revenues per barrel (in 1973 dollars) will be $8.50 in 1974, $7.00 in 1975, and $5.50 in 1976.
— Total exports will be 32 million barrels per day in 1974, 33 million barrels per day in 1975, and 34 million barrels per day in 1976 (which allows for some buildup of stocks).
— Expenditures of oil revenues will differ between the two groups of exporting countries. Those countries in continuing need of development finance will spend: (a) amounts equal to 1973 oil revenues on a current basis; and (b) revenues over that amount on the following schedule: 25 percent during the year in which they are received and 25 percent during each of the following three years. Financial-surplus countries will increase their imports of goods and services at the rate projected in Table 14-3 and in addition will spend a total of $2 billion a year on bilateral military and financial assistance.

On the basis of these assumptions, oil-exporting countries would have a total current account surplus of $64 billion in 1974, $39 billion in 1975, and $11 billion in 1976, and oil-importing countries, as a group, would have equivalent current account deficits. About two-fifths of the surpluses accumulated during this period would be temporary in the sense that they would accrue to low-income exporting countries and would disappear (and past surpluses be drawn down) as disbursements by those countries caught up to their expenditure commitments. The derivation of these figures is shown in Table 14-5. No claim to precision is made for these figures; they should be viewed simply as a means of emphasizing that the oil trade in almost any circumstances will require very large offsetting capital account flows in 1974, but that these requirements will decline rapidly thereafter, both because of sagging oil prices and rising expenditures by the oil-exporting countries.

If events take this course, what problems will arise for the international economic system? Adjustment issues can be distinguished in three closely related areas: financing arrangements, trade policy, and exchange rate policy.

As is now generally understood, financing between the two groups of countries will take place automatically in that oil-exporting countries must necessarily invest their surpluses in financial or other claims on the oil-importing countries. Investment inflows will not of course match current account deficits in each importing country. Thus the policy problem comes down to the question of whether the importing countries can recycle these funds among themselves so as to maintain confidence in the system as a whole and to reduce the risk that some will feel compelled to take destabilizing trade or exchange-rate actions in the face of rapidly rising oil-import costs.

The recycling problem is complicated by the unprecedentedly large size of the offsetting capital flows required in 1974—perhaps three times as much as the financing required in 1972 by all countries with current account deficits from their entire trade. On the other hand, the possible ways in which financial claims can be transferred between importing and exporting countries and recycled among importing countries are extensive and, more to the point, highly elastic. They include the use of the large and flexible Euro-currency markets, the now unrestricted U.S. capital market, the credit and intermediation facilities of the International Monetary Fund, bilateral credit facilities among central banks (the swap network), and the borrowing and lending capacity of the World Bank. In addition, of course, the importing countries can draw down and the exporting countries can build up their financial reserves.

The above facilities, while more than adequate technically, will have to be used in non-conventional ways, however, if they are to meet the unique characteristics of the oil-financing problem.[7] For one thing, borrowing short and lending long will have to be a regular practice during the transitional period. The reason is that oil-

Table 14-5.   Projected Current Account Surplus of Oil-Exporting Countries

	1974	1975	1976
Govt. revenues per barrel (in 1973 dollars)	8.50	7.00	5.50
Volume (in million barrels per day)	32	33	34
	(in billions of 1973 U.S. dollars)		
Total revenues	99	84	68
Countries in continuing need of development finance:			
Oil revenues	60	51	41
Expenditures	27	36	42
1973 level	16	16	16
From incremental revenues	11	20	26
Current Account Surplus	33	15	−1
Financial Surplus Countries:			
Oil Revenues	39	33	27
Expenditures	8	9	10
Imports of goods & services	6	7	8
Military & Economic Aid	2	2	2
Current Account Surplus	31	24	17
Total Current Account Surplus	64	39	16

Sources: Assumptions as stated in text. Countries included in each category are shown in Table 14-1. Export volume for each group of countries projected on the basis of Table 13-9.

Notes: For the purpose of this calculation, the current account is defined to include oil-exporting country bilateral grants and credits for military and economic aid.

Revenues per barrel represent weighted averages for all oil-producing countries. For Saudi Arabia and other countries producing equivalent quality oil, revenue per barrel would be 50 cents less than the amount shown.

exporting countries are likely, at least initially, to place a large portion of their surpluses on short-term deposit in private capital markets, while borrowing countries will need funds for the duration of the adjustment process. Thus, even if the loans are short-term in form, borrowing countries will need assurance that the loans can be renewed for as long as necessary. Conventional standards of credit worthiness similarly will have to be altered since the oil-importing countries as a group will not be able to repay their obligations until the exporting countries are in a position to accept payment in the form of goods and services. In this respect, many developing countries face particularly difficult problems since doubts about their ultimate repayment capability severely detract from their present borrowing capacity.* Taken together, these operating requirements suggest that private commercial banks will not be willing, on their own, to handle a large portion of the financial load on a continuing basis. Therefore, the effectiveness of the financing network as a whole will depend on the extension or contingent availability of very large-scale intergovernmental credits; these credits could be provided bilaterally or through international financial institutions and would have to be offered on a quasi-automatic basis. In effect, the complex of intergovernmental credits will have to serve internationally the lender-of-last-resort function that central bank credits serve domestically.

An effective financing network would mean that most, if not all, industrialized countries could have current account deficits—some of very substantial proportions—even in a situation in which they were in balance-of-payments equilibrium. If they accept such deficits as being inevitable in the unique circumstances surrounding the adjustment of the world economy to higher oil prices, trade problems need not arise. However, if neo-mercantalistic attitudes toward the trade account gain ascendancy, those deficits could contribute to protectionist tendencies or touch off attempts to subsidize exports. A few instances of unilateral action already have occurred, notably the adoption by Italy and Denmark of measures to curb imports. Fortunately, however, the danger that these practices will spread has been lessened by a common awareness of the nature of the problem and of the costs to all countries of a policy failure in this area. As a result, the OECD countries pledged in May 1974 to avoid protectionist trade measures and to abstain from competitive export subsidies for one year.[8] Nonetheless, large trade deficits are a powerful force for restrictionism; it should be stressed, therefore, that a code of good conduct will be durable only if the facilities to finance those deficits are adequate.

This raises the thorny question of appropriate guidelines for managing exchange rates during the transitional period: at what point does reliance on financing end and reliance on exchange-rate adjustment begin? It is evident that oil-importing countries cannot improve their current account by depreciating their exchange rates as a group in relation to the oil-exporting countries; that would only worsen their deficits. However, exchange-rate adjustments in response to a deterioration or strengthening of one oil-importing country's competitive position toward the others would still be necessary. How can such adjustments be accomplished in mutually

*See Chapter 10 and the subsequent section on development assistance in this chapter.

acceptable ways when market signals are obstructed or distorted by the overwhelming impact of the oil trade on both the current and capital accounts? In other words, how and to what extent should oil-related financial flows be eliminated from the calculation of appropriate exchange-rate adjustments among the industrialized countries?

Here again, financing is critical, specifically the procedures used for redistributing among importing countries the initial investment of the surplus revenues of the oil-exporting countries. If these surpluses were recycled in proportion to the increase in oil-import costs,* the resulting payments position of each country would more nearly represent its underlying position, specifically its performance in non-oil transactions, its progress in moving toward greater self-sufficiency in energy, and its ability to compete for incremental exports of goods and services to the oil-producing countries. Changes in financial reserves and market pressures on exchange rates could then be useful signals to guide government intervention in foreign exchange markets, as well as international surveillance of such intervention.

The additional concern that the Arab countries might wish to use their financial surpluses to destabilize or destroy the international monetary system can virtually be rejected out of hand. In the first place, about half the oil revenues will go to non-Arab countries. Furthermore, among the Arab countries only a few will accumulate large surpluses. The major countries—Saudi Arabia, the United Arab Emirates, and Kuwait—are essentially conservative in outlook. If they wanted to achieve militant policy objectives they would use their oil exports, not their financial investment policies. As far as financial surpluses are concerned, their interests lie in finding attractive investment opportunities, and achieving that goal would be enhanced by stability and diminished by instability in the exchange markets.

To be sure, as pointed out earlier, a sizable portion of these surplus funds will be held in short-term deposits, and this can pose a problem for the system quite apart from the question of

*A numerical example using round numbers might be useful. Suppose, for example, that the increase in oil revenues in 1974 compared with 1973 amounted to $60 billion, and that the oil-exporting countries in 1974 spent one-fourth of that amount, or $15 billion, on increased imports of goods and services. Suppose further that each importing country received as oil-surplus investments—either directly or through recycling—an amount equal to three-fourths of the increase in its oil-import bill. The resulting surpluses and deficits of each importing country, then net of the phenomenon of surplus oil revenues, could provide a basis for guiding exchange rate adjustments in a world of managed floating.

whether the holders of the funds have predatory intentions. The existence of these large balances will add to the need for the international financial system to bring countervailing forces into play during periods of uncertainty in financial markets. Maintenance of flexible exchange rates, which is one means of creating such forces, will therefore be essential. Very large intergovernmental swap lines of credit which, in any event, will be necessary as part of an effective financing network, can also be used when market forces are pushing heavily in directions that are contrary to underlying economic trends. And the availability of suitable financial instruments, such as the International Monetary Fund's special oil facility, which might encourage the oil-exporting countries to shift assets from short to longer term holdings, would also have a stabilizing influence on the system as a whole.

The Period Through 1985

Once the transitional adjustments have been completed, the impact of the new oil situation on the world economy should substantially diminish. Oil prices, while much higher in real terms than in the 1960s, can be expected to decline from 1974 levels; the growth in oil consumption and imports also can be expected to moderate as a result of relatively high prices and the increase in domestic production of primary energy sources; and the oil-exporting countries, in time, will expand current imports of goods and services more nearly to the level their oil revenues can support.

Consequently, neither the ongoing financial surpluses nor their accumulated amount should cause unmanageable problems for the world economy. These surpluses will of course be large. As projected in Table 14-4, they would average about $18 billion a year for the period as a whole (including reinvested earnings) and accumulate to approximately $135 billion by 1980 and $210 billion by 1985. Thus, by the end of the period, the international investment position of the financial surplus oil-exporting countries might be about equal in size to that of the United States today. (In 1972, the addition to U.S. direct investment abroad, including reinvested earnings, amounted to $8 billion and the total value of private, non-liquid U.S. investments abroad was $144 billion.[9]) On the other hand, these investment flows from oil must be seen in the perspective of a large and growing world economy. Over the period 1974-85, the combined GNP of the United States, Western Europe, and Japan might increase from approximately $2.8 trillion to $4.3 trillion (1973 dollars). Financial transactions, both domestic and international, will be similarly large. Former Secretary of the

Treasury George P. Schultz has estimated that in 1980 annual capital formation in the industrial countries might be $700 billion, new issues of stocks and bonds might be perhaps $250 billion, and the total value of stocks and bonds in the world might exceed $3 trillion.[10] A world economy of that size should be able to absorb fairly smoothly the projected savings of the oil-exporting countries.

Could special problems arise because these surpluses and the corresponding investments will be owned by governments? In other words, if a large portion is used for direct investment, will the host countries countenance the potential influence that foreign governments might thus gain in their affairs? One can of course construct a pattern of investment so heavily concentrated in one country or in one economic sector as to be politically troublesome, but such a pattern would be rather improbable. First, the investing governments would be chary about assuming the obvious risks. Second, they do not have the necessary managerial experience or expertise to operate or control a network of foreign enterprises. Third, and most important, their interests lie in the development of their own countries, not in managing enterprises abroad. In the words of Sheik Yamani, they tend to view their financial holdings as temporary surpluses that eventually "must be converted to industries that can provide them with their livelihood after their oil is depleted."[11] Consequently, the financial decisions of the oil-exporting countries are likely to show a preference for portfolio over direct investments and to stress protection against the erosion of capital, safety of yield, and diversification—rather than maximum return or managerial control.

Finally, the existence of substantial financial surpluses in a few oil-producing countries is a factor that might have to be taken into account in determining the need for the creation of additional international liquidity. Part of the settlement of oil payments might be made, in effect, through the drawing down of international financial reserves in the importing countries and their accumulation by the exporting countries. This trend would not soon be reversed and would not be subject to the normal operation of the adjustment process. Consequently, for the importing countries, who account for the predominant portion of world trade and payments, a continuation of this process could have the same effect as an extinction of reserves for the system as a whole. Unless reserves had been excess to needs, a failure to allow for this phenomenon in decisions on reserve creation could be a source of future disequilibrium in the system.

In sum, the most dangerous pressures on the international economic system from higher oil prices will arise during the transitional period rather than over the longer term. The initial shock will be severe because the first-year financial changes will have been so sudden and so large. In succeeding years, the adjustment burden should be smaller and the policy problems less demanding. With close and quickly achieved cooperation, the transitional adjustments are manageable at sustainable cost. All the necessary institutional machinery is in place, but it will have to be applied flexibly and pushed to its limits. This is evident from the nature of the requirements: nonconventional ground rules for extending inter-governmental credits on a massive scale; special arrangements to ensure that developing countries are not disastrously penalized because of their relatively weak credit position; avoiding trade restrictionism in the face of large trade deficits; and guidelines for making mutually acceptable exchange-rate adjustments in unusual balance-of-payments circumstances.

It follows that a failure to achieve such cooperation would lead the major industrial countries to seek solutions to their problems by unilateral means, that is, by intensifying deflationary domestic policies and pursuing restrictionist foreign economic policies. The actions of each country would cause reactions in others and in the end would be self-defeating. Together, this would result in overcompensation for the effects of higher oil prices and an unnecessarily large loss in world income and employment.

How will this new situation affect ongoing international economic negotiations? International monetary reform obviously will not take the form of an early agreement on a new system to be put into operation on a fixed date. Rather, it will have to proceed piecemeal as part of a continuing attempt to protect world income and employment while the adjustments to higher oil prices are worked out. This changes matters very little since an evolutionary approach to the problem of constructing a more flexible inter-national monetary system was in any event necessary and a great deal of progress has already been made. As for trade negotiations, higher oil prices have, if anything, made them more urgent, among other reasons, to forestall retrogression in trade policy, to help combat world inflation, and to provide the developing countries with an opportunity to solve their oil problems through their own efforts.

IMPLICATIONS FOR THE U.S. BALANCE OF PAYMENTS

Foreign exchange outlays by the United States for oil, including transportation costs, increased from approximately $4.5 billion in

1972 to $8 billion in 1973. These outlays probably will considerably more than double in 1974, with the result that the U.S. current account, instead of showing a strong surplus, is likely to be in equilibrium or in moderate deficit. Nonetheless, once the necessary transitional adjustments have been completed, higher oil prices should not prove to be a special burden for the U.S. balance of payments. This conclusion follows from the fact that of all the oil-importing countries, the United States has the greatest possibilities on both the energy supply and demand side for reacting to higher oil prices in ways that will reduce oil imports.

The main underlying factor in this conclusion can be illustrated by using the 1974-85 projections discussed earlier in this chapter. At a long-term price of $5 for Saudi Arabian oil (equivalent to a delivered price of $6.25 on the U.S. East Coast), the non-Communist world market was projected to grow (in barrels per day) from 32 million in 1973 to 39 million in 1980 and 43 million in 1985. Of these totals, U.S. imports might amount to approximately 8 million barrels per day in both 1980 and 1985, the leveling off occurring because at this price there would be both a reduction in the rate of growth of energy consumption and a steady increase in domestic production of primary energy sources.

These projections would mean that by 1980 the cost of U.S. oil imports would amount to $18 billion (in 1973 dollars), or an increase of $10 billion over 1973. At the same time, however, the U.S. current account would be improved by offsetting transactions, principally an increase in U.S. sales of goods and services to the oil-exporting countries. The growth in the markets of those countries obviously would be large. Based on the data in Tables 14-1 and 14-3, by 1980 the oil-exporting countries in the aggregate would be spending about three-fourths of their oil revenues on imported goods and services, and by 1985 the proportion would rise to about 90 percent. Their purchases would constitute a market of $50 billion in 1980 and almost $65 billion in 1985, the predominant portion of which would be supplied by the industrialized countries.

Taking these projections into account, the impact of higher oil prices on U.S. current international transactions can be roughly estimated, using the following operating assumptions:

— U.S. exports of goods and services are calculated on the basis of U.S. shares in the markets of oil-exporting countries in 1970. The relatively small U.S. trade surplus of $3 billion in that year might be said to reflect a reasonable approximation of its equilibrium trade position and, hence, of its future competitive position in the markets of those countries.* In addition, the United States is assumed to supply, directly or indirectly, one-fourth of the goods and services purchased with the bilateral military and economic aid provided by the financial surplus oil-exporting countries.

— Transportation costs are estimated on the assumption that one-half of U.S. oil imports would come from Western Hemisphere sources, including Canada, at an average cost of 30 cents a barrel, and one-half from the Persian Gulf and North and West Africa at an average cost of $1.25 a barrel. Transportation earnings of U.S. companies, either as a return on capital or for other services, are assumed to be 10 percent of total non-Communist oil-transportation revenues. Japan (because of its predominance in tanker construction), and Western Europe (because of receipts from tanker construction and operation) would earn the bulk of oil-transportation revenues.

— Earnings of U.S. oil companies are calculated on the basis of a fixed margin of 35 cents a barrel assumed earlier in this chapter to cover profits from the production and marketing of crude oil. The U.S. companies are projected to account for 50 percent of the total and to repatriate the funds they earn.

On these assumptions, three-fourths of the impact of higher oil prices on the U.S. current account would be offset by oil-related foreign exchange earnings by 1980 and all of it by 1985. (See Table 14-6.) Much the same conclusion holds for capital transactions related to oil. The U.S. oil companies have been investing approximately $2 billion a year abroad, including reinvested earnings. The amount of these investments might decline in the future, but in almost any circumstances their effect on the balance of payments is likely, at the least, to be offset by investments made in the United States by oil-exporting countries. In other words, if the equilibrium export price of oil should be about $5 a barrel for Saudi Arabian oil, the resulting increase in U.S. oil-import costs, compared with 1973, should affect the U.S. balance of payments position very little, one way or the other.

These results are influenced heavily by assumptions relating to the U.S. international competitive position and to future prices of oil. As to competitive position, market shares based on the 1970 experience would appear to be a minimum expectation. In that

*Based on incomplete data, U.S. market shares in these countries in 1973, when the United States once again had a small trade surplus, seem to be fairly close to those in 1970.

Table 14-6. Estimated Impact of Oil on the U.S. Current
Account (in billions of 1973 U.S. dollars)

	1973	*1980*	*1985*
Expenditures:			
Oil Imports (net)	7.0	16.0[a]	16.0[a]
Transportation Costs	1.1	2.2	2.2
Total	8.1	18.2	18.2
Receipts:			
Exports of goods and services to oil-producing countries	4.4	12.4	14.0
Earnings from oil transportation	.8	1.0	1.2
Earnings of U.S. companies from production and sale of crude oil	3.5[b]	2.6	2.8
Total	8.7	16.0	18.0
Balance	+0.6	−2.2	−0.2

Sources: Author's estimates based on the assumptions stated in the text.
a. Based on a weighted average price of $5.50 a barrel for all imported oil, taking price differentials into account, and U.S. imports of 8 mbd.
b. Assumes a profit margin in 1973 of 50 cents a barrel and that U.S. oil companies in that year marketed 60 percent of total non-Communist crude production.

year, the U.S. trade surplus was in large measure the result of a recession in the United States and full employment in Western Europe and Japan. Consequently, equilibrium exchange rates in a situation in which all industrialized countries were at full employment may put the United States in a somewhat stronger competitive position in the markets of the oil-exporting countries. In addition, the oil-producing countries' increasing purchases of military equipment and their growing expenditures on domestic oil-refining capacity would suggest a higher market share for the United States in the future.

Alternative price assumptions give rise to somewhat different situations. Substantially higher prices would mean larger deficits in the short term, but probably would strengthen the U.S. balance-of-payments position in the medium and long term. For example, at an export price of $6.25 a barrel at the Persian Gulf (equivalent to $7.50 delivered to the U.S. East Coast), the analysis in Chapter 13 indicates that U.S. oil imports by 1980 would be between 5 and 6 million barrels per day and by 1985 about 3 million barrels a day. (At even higher prices, for example, the price of $9 a barrel existing in the first half of 1974, the United States probably would have achieved self-sufficiency in oil by 1985.) U.S. sales in the oil-exporting countries, however, probably would be reduced only marginally,* and this would result in a substantial surplus in U.S. current transactions attributable to oil.

The effect of oil prices substantially lower than $5 a barrel on the U.S. current account seems less predictable. The volume of U.S. oil imports would rise, but probably by less than the reduction in price, principally because the drive toward greater energy self-sufficiency in the aftermath of the oil embargo is no longer entirely reversible. Even at considerably lower oil-import prices, some moderation would probably occur in the growth of energy consumption in the United States, and an increase in U.S. domestic production of primary energy sources might to some extent be supported by subsidies or protectionist measures. Thus, lower oil-import prices would still mean reduced U.S. oil-import costs. Similar reactions might take place in other importing countries, although to a lesser extent. Consequently, the oil-exporting countries could receive lower oil revenues than they would if the price was around $5 a barrel, and this would result in a decline in their purchases abroad, including their purchases from the United States. On the whole, it is probable that U.S. oil-import costs would fall more than would U.S. sales to oil-exporting countries, so that in this case as well the U.S. current account would tend to be strengthened.

IMPLICATIONS FOR DEVELOPMENT ASSISTANCE

Higher oil prices will have wide-ranging economic consequences for the developing countries and hence for development assistance requirements. To examine these consequences, it is essential to distinguish between the oil exporters and the oil importers and between different groups within each category. Looking to the period through 1985, how would the oil-trade projections in this chapter affect each of these groups?

The oil-exporting developing countries, of course, will be in a vastly improved economic position. The financial-surplus LDCs, which are few in number and which contain approximately 15

---

*At prices substantially above $5 a barrel, the long-term price elasticity of world demand for oil imports is probably considerably greater than unity (e.g., a 10 percent increase in price would be accompanied by a decrease in imports of more than 10 percent), which would mean a decline in the oil revenues of the oil-exporting countries. Oil-producing country purchases abroad, however, would decline proportionately less. To maintain higher prices, the financial-surplus rather than the low-income countries would have to reduce exports the most. Consequently, the decline in total oil revenues would be reflected more in reduced financial surpluses than in reduced imports.

million people, will be developing countries only in the sense that they will continue to be non-industrial. Their per capita incomes will be high, their economies will be growing rapidly, and their annual capital exports will be almost equal to the net flow of official and private resources from all OECD industrialized countries to all developing countries in 1972. How and where these resources are invested could substantially affect the prospects of other developing countries in 1972.

For the other major oil-exporting LDCs, which contain approximately 250 million people, oil revenues will be an enormous source of additional real resources for internal investment. At the projected price of $5 a barrel, oil exports would *add* an average of $22 billion a year to their foreign exchange receipts, an amount equal to perhaps one-third of their present combined GNP. In addition, these oil receipts would greatly improve their international credit position and hence their ability to borrow abroad on commercial terms. The comparative significance of the oil bonanza will vary greatly among these countries, but in each it will be a major stimulant to economic expansion. These countries as a group now receive approximately $750 million in concessional aid (official development assistance). In the new oil situation, most of these resources presumably would have higher priority uses in other countries.

The smaller net oil-exporting countries, including Ecuador, Colombia, Egypt, and Trinidad-Tobago, contain approximately 100 million people. Their additional oil revenues, while comparatively modest, might rise to an average of $3 billion a year, or roughly 10 percent of their present GNP. The situation among countries in this group also varies widely, but for some at least, additions to oil-export revenues above the amounts received in 1973 are already or will become substantial. These countries received a total of $250 million in concessional aid in 1972.

The oil-importing developing countries, on the other hand, will be heavily burdened by higher oil prices—directly through the increased cost (in real resources and scarce foreign exchange) for oil imports and for other imports whose prices will rise because of the cost of oil, and indirectly through the dampening of world income and hence of export prospects in the industrialized countries. These adverse economic consequences in combination could be very serious, the amount of damage depending on (a) the speed and effectiveness with which the industrialized countries adjust to higher oil prices, (b) price trends for their export commodities, and (c) the availability of financing to mitigate the impact of oil costs. These countries contain approximately 1.5 billion people and have a combined GNP of perhaps $500 billion, or a little more than $300 per capita.

For the purposes of this chapter, the calculations concerning these countries are confined to oil-import costs and how they might be offset. Compared with 1973, the increase in the cost of oil imports for these countries might amount to approximately $8 billion in 1974, then decline during the remainder of the transitional period, and finally level off at perhaps $4 billion a year as they gradually reduce their rate of growth in energy consumption and increase domestic production of primary energy sources. In comparison with this potential new drain on resources, these countries received approximately $9 billion in concessional aid in 1973, an amount that has been virtually stagnant in real terms for almost a decade. How might they manage this new burden?*

For those developing countries with comparatively strong economies, recycling could be the major means of adjustment, as in the industrial countries. Special measures will be needed however to ensure that an adequate portion of oil-surplus funds flow directly or indirectly to these countries. This group, including countries like Brazil, Korea, Taiwan, Mexico, Argentina, Morocco, and the Ivory Coast, accounts for approximately three-fourths of the oil imports of developing countries. Increased oil costs will be a substantial burden for most of them. But generally speaking, they are in a position to contain the damage because their economies are either growing rapidly or prices for their major export commodities have improved, or their dependence on oil imports is small, or because of a combination of these reasons. To offset higher oil costs this group of countries, in the aggregate, might need perhaps $3 to $4 billion a year in additional financing over the transitional period, a large portion of which could be on commercial or intermediate credit terms. Over a longer period, they probably have the capacity to adjust to higher oil prices through increased exports.

For the economically weak developing countries, however, the adjustment problem will be far more difficult. This group, including India and the other countries in South Asia, the least developed countries in tropical Africa, the Philippines, and Kenya, contains approximately 1 billion people. Typically, these countries have per capita incomes below $200, slow growing economies, poor export prospects, severe resource constraints, and a thin margin for mobilizing domestic resources for investment. Higher oil prices will mean additional resource and foreign exchange requirements of

*The problems faced by the oil-importing less developed countries are examined in some detail in Chapter 10.

perhaps $2 billion a year during the transitional period and $1 billion thereafter. India will account for approximately one-fourth of the total for all countries in this group. In general, these countries do not have access to international capital markets and in any event would have great difficulty in servicing additional debt on commercial terms—either in the next few years or over the more distant future. Consequently, they will need additional intergovernmental financing on concessional terms if they are to avoid severe economic setbacks.

In reviewing the consequences of higher oil prices, three general considerations distinguish the problems of the developing countries from those of the industrial countries. First, the developing countries are not likely to be recipients of significant capital inflows from the oil-exporting countries. Consequently, without some form of assistance, they will have to bear most of the impact of higher oil-import costs through an immediate and probably disproportionately large reduction in income, consumption, and employment. Second, the increase in real resource costs for oil imports will be more difficult for them to sustain simply because their incomes are low. And third, a large number can do very little on their own to ease the adjustment process.

Measures that might be taken by the United States in cooperation with other nations to relieve the problems faced by the oil-importing developing countries are discussed in Chapter 19. As a group, these measures can be viewed as a special means of recycling a portion of the surplus oil revenues to the developing countries and hence as an integral part of the process through which the world economy makes the financial adjustment to higher oil prices, both in the transitional period and over the longer term. Both the industrial and the oil-exporting countries would have to share responsibility. If this process is to grow in politically acceptable and internationally negotiable ways, the IMF, the World Bank, and the other international financial institutions would have to exercise the central role. This in turn would require that the oil-exporting countries assume much greater financial responsibilities in these institutions and receive commensurate influence in the determination of their policies. Neither condition seems unmanageable and both would be necessary to enable the developing countries to make a reasonably orderly adjustment to higher oil prices.

CONCLUSIONS: HOW MANAGEABLE IS MANAGEABLE?

It should be stressed at the outset that the arguments advanced in this chapter do not hinge on the specific export price of oil used as the basis for the financial projections—that is, a long-term export price of $5 a barrel for Saudi Arabian light oil (in 1973 dollars). Higher prices would lead to reductions in export volume and probably to reduced total revenues and surpluses. At somewhat lower prices, the change in financial surpluses is less predictable. In either case the variation is probably not wide enough to alter in any significant way the financial projections in this chapter or their implications. These projections cast doubt on both the apocalyptic and the complacent assessments of the financial problems related to higher oil prices. The apocalyptic view is based on the proposition that the very large oil revenues and financial surpluses accruing to oil-exporting countries in 1974 will continue unchanged or might even be larger in the future. The inevitable result is a financial doomsday machine: surpluses eventually become so large that the oil exporters own a large portion of the world's productive assets and at the same time no longer have the incentive to continue selling enough oil to keep the world economy functioning, thus putting the value of their assets in jeopardy. At the other extreme is the view that financial flows arising from oil should not be considered too serious or too difficult a problem because solutions in the form of offsetting capital transactions take place automatically. Financial surpluses create their own investments, so to speak. Consequently, the only real need is to stay calm as the capital and oil markets work things out.

The analysis in this chapter suggests that, while the second approach is more nearly right, neither provides a practical guide to policy.

The doomsday approach is wrong because it fails to recognize that the international economic system is flexible—for oil as for anything else. Higher oil prices and supply uncertainties caused by embargoes bring about reactions which in turn reduce oil exports and prices. This is not to say that the world economy will go back to where it was before the oil crisis. Certainly, the position of oil producers over the future will be immeasurably enhanced because sooner or later they will receive from the rest of the world a much larger flow of resources in exchange for their oil exports. The adjustment process, however, will substantially modify the situation from what it is today, and events since the embargo clearly show that such a process is under way. Furthermore, the world economy is growing each year by large absolute amounts; what seem like large capital flows today will not seem so large in the future. In short, the problems arising from surplus oil revenues will diminish rather than grow over time.

While the stay-calm school has logic strongly on its side, the danger of relying on that approach stems from the strong possibility that this logic will not be perceived by the actors involved—individuals, industrial and banking executives, and government officials. Instead, they might see the surplus oil revenues, and their related financial flows and trade deficits, as overwhelmingly destabilizing forces and react by seeking to build national storm shelters and fending for themselves. In effect, there can be no assurance that oil's sudden and massive entry on the international economic scene will automatically generate optimal policy responses.

Thus, while the quantitative impact of higher oil prices— large as it is—seems manageable enough, the strains on the international system are likely to be very severe. Higher oil prices have the same effect on consumers as a new excise tax, and sooner or later this tax will have to be paid by foregoing consumption of other things. Large trade deficits—no matter how inevitable they are recognized to be—can lead a life of their own in creating uncertainty in financial markets and causing restrictionist policy reactions. And the impact on poor countries, while requiring comparatively small sums to offset, will come at a time when the industrialized countries will be struggling with their own problems and in no mood to take on additional responsibilities.

In sum, the unconventional character of the problem calls for unconventional policy norms, and its widespread international repercussions call for cooperative international measures on a hitherto unmatched scale. All this is within the present capabilities of the international economic system and its institutions. Whether the financial impact of higher oil prices proves to be manageable, however, will depend as much on how the problem is perceived as on its substance.

NOTES FOR CHAPTER 14

1. A list of military equipment now on order or being considered is shown in Dale R. Tahtinen, *Arms in the Persian Gulf* (Washington, D.C.: American Enterprise Institute for Public Policy Research, 1974). The cost of this equipment (including the order for 50 additional F-14 aircraft reported in the *Wall Street Journal* on June 12, 1974) has been estimated on the basis of U.S. procurement costs, as shown in various Department of Defense publications.

2. These policy forecasts and the resulting import requirements are based on an analysis contained in Jerome F. Fried, "Oil Income and Economic Development in the LDCs," July 1973 revised, a background paper prepared for The Brookings Institution.

3. This conclusion is strongly argued in Donald Wells, "An Estimate of Saudi Arabian Savings of Foreign Exchange by 1980," an unpublished manuscript prepared for Resources for the Future. Wells stresses that King Faisal's regime has been moving purposefully in the direction of internal social and economic development. In his view, "the record suggests that King Faisal has directed efforts toward introducing the population to the changing circumstances in a manner most consistent with Saudi traditions and religion rather than attempting to perpetuate the status quo."

4. In comparison, Wells, ibid., projects an increase of 25 percent a year in government expenditures to 1980. His analysis, however, assumes much higher oil revenues than are projected in this chapter.

5. Another indication of accelerated development is a recent report that the Saudi Arabian Planning Ministry is drawing up a five-year development plan (1975-80) that calls for the expenditure of $60 billion. The current five-year spending program totals $10 billion. *Wall Street Journal*, April 12, 1974, p. 1.

6. In March 1974, Kuwait increased the capitalization of its Fund for Arab Economic Development from $600 million to $3 billion. With borrowing authority, the Fund has a total lending potential of $10 billion. All developing countries are now eligible to borrow. Abu Dhabi announced in May 1974 that it increased the capitalization of its Fund for Arab Economic Development from $125 million to $500 million. Saudi Arabia announced that it would establish the Saudi Arabian Development Fund, which would be open to all developing countries, but the size of its capitalization has not yet been made public.

7. The requirements are outlined with exceptional clarity by Robert Solomon of the Federal Reserve Board in an address before the International Accounting Conference, March 13, 1974, on "The Oil Price Impact on the International Monetary System."

8. *New York Times*, May 31, 1974, p. 41.

9. Leonard A. Lupo, "U.S. Direct Investment Abroad in 1972," *Survey of Current Business*, September 1973, p. 23. This figure is substantially understated as measured in 1973 dollars because, for the most part, it is not adjusted for the devaluation of the dollar since 1971 and for inflation. In addition, U.S. government owned assets abroad in 1972 amounted to $36 billion.

10. Statement before the Annual Bankers Association International Monetary Conference, Paris, June 6, 1973.

11. Statement by Saudi Arabia's Minister of Petroleum and Mineral Resources, Ahmed Zaki Yamani, in an address given in London, May 8, 1974, entitled, "Producer-Consumer Relationships in the Oil Industry: A New Era." (Reprinted in *Middle East Economic Survey*, May 17, 1974.)

Chapter Fifteen

Oil and National Security

Oil is so vital to the functioning of the economies of the industrialized nations that interruption of any large part of the oil supplies of the United States and its major allies and trading partners must be regarded as a threat to the national security. This chapter will explore several possible contingencies affecting the oil imports of the United States, Western Europe, and Japan. The policy relevance of a politically motivated interruption of oil imports is obvious in the aftermath of the Arab embargo of 1973-74. Various wartime contingencies seem more remote from the perspectives of mid-1974, but they also must be examined as part of the long-term context in which U.S. energy policy must be made.

SUPPLY INTERRUPTIONS IN PEACETIME

An oil supply interruption could of course be used to exact higher prices from the United States and other oil-importing nations, but that contingency is part of a larger economic problem discussed elsewhere in this study. In the present national security context, two kinds of contingencies require particular attention: sabotage and an embargo.

Sabotage and Other Physical Disruptions

Even when the United States itself is not at war, U.S. oil supplies could be disrupted by physical damage under a number of conceivable circumstances. One possibility would be the sabotage of oil facilities in Iran in conjunction with some sort of domestic

Note: This chapter is based in large part on research and analysis conducted by Barry M. Blechman and Arnold M. Kuzmak.

turmoil. Another possibility would be damage during a war among the states at the head of the Persian Gulf—Iran, Kuwait, and Iraq. But the danger of physical disruption is most credible in the event of another Arab-Israeli war.

In such a conflict, there would be three relatively independent actors: Israel, the Arab governments, and Palestinian irregular forces. Of these, only the latter would have any incentive to damage oil-production facilities. Israel would not want to take actions that would make U.S. support more difficult, and the Arab governments could accomplish the same thing at less cost through administrative fiat. The Palestinians, on the other hand, could see an attack on oil facilities as a means of exerting pressure on the Arab governments for greater support for their cause, or as a means of compelling the United States to limit its support for Israel.

It is unlikely that such an attack could disrupt oil supplies sufficiently to create significant problems for the United States. The oil facilities are relatively isolated and distant from the primary focus of the Arab-Israeli conflict. There are few single targets whose destruction would have grossly disproportional effects. The only exceptions are the pipelines leading to the Mediterranean, but these are becoming less and less important to the oil-transport system as tanker shipments from the gulf increase. Thus, while a determined Palestinian effort would undoubtedly do considerable damage in dollar terms, its impact on U.S. oil supplies would not likely be major.

Embargoes and Other Political Manipulations of Oil Supplies

In theory, any oil-exporting country or group of oil-exporting countries could try to use oil as a political weapon. In practice, however, no oil-exporting country acting alone could expect to gain much political advantage by suspending shipments to the United States or other destinations, and it would risk losing its share of the international oil market to other suppliers. Moreover, with one exception, no conceivable group of oil-exporting countries appears to possess both the market power and the common political goals required for effective joint action.

The one exception is of course the Arab oil-exporting countries, which, despite their differences on many issues, find common ground in their opposition to Israel. In 1973-74, their reduction of total oil output and differential cuts in exports to various destinations gained international support for their goals with respect to Israel. The Arabs might find reason to resort to the oil weapon again. If they do, they need not use it in exactly the same way as they did in 1973-74. There are in fact a number of different ways the Arab countries could interrupt the flow of oil, each with very different implications for the United States, Western Europe, Japan, and the world oil-supply picture generally.

The Arab countries could, for example, decide to halt all shipments to the United States, but otherwise to follow their economic self-interests. In particular, they could increase shipments of oil to their other customers and thereby recoup the revenues lost by not shipping to the United States. By doing so, however, they would greatly reduce the effectiveness of the cutoff of the United States. Importing countries other than the United States could increase their imports of Arab oil and thus free non-Arab sources, including Iran, to meet U.S. needs. Consuming nations could also swap oil supplies, or the international oil companies could perform the same function, as they apparently did to some extent in 1973-74. Obviously, the Arab states would be aware of these mechanisms. They would view an interruption of this kind simply as a symbolic step—a means of indicating displeasure to the United States and, at the same time, helping to still domestic criticism and counter militant Arab pressures for more consequential (and costly) measures. It was, in fact, precisely this kind of embargo that some Arab producers directed primarily against the United States and Great Britain following the 1967 Arab-Israeli war.

Alternatively, the Arab countries could stop shipments to the United States and continue shipments to Western Europe and Japan, but refuse to increase shipments to the latter countries. Thus, they would not attempt to pressure or punish Western Europe and Japan, but neither would they be permitting the sort of adjustments described above that would vitiate the effectiveness of the cutoff. If the Arabs had followed this course in the fall of 1973, about two million barrels of oil per day would have been withheld, or roughly four percent of total non-Communist oil consumption.

In this situation, it is quite reasonable to assume that, with some time delay, additional capacity in non-Arab countries would be brought into production (possibly at higher prices) and would make up a large part, if not all, of the U.S. deficiency. To fill the gap while non-Arab production was being increased, the U.S. oil industry could operate for a time from its stocks. Drawing down commercial and pipeline stocks in the United States would, however, lead to inefficiencies, temporarily higher costs, and possible shortages in some areas for particular products, even before stocks fell below minimum working levels.

To the extent that existing excess capacity would not be adequate to satisfy the shortfall in supplies, two further consequences would result. First, energy prices in the United States would rise, which would lead to reduced consumption. This effect could be supplemented by specific government policies designed to conserve energy and to save oil in particular. These measures could include the relatively modest steps taken in the winter of 1973-74 (an oil allocation system, reduced highway speed limits, relaxation of restrictions on the use of coal, and the like) and might go as far as the rationing of petroleum products. Second, high U.S. prices would draw supplies away from other countries, resulting in higher prices and hence reduced demand in those countries. The international oil companies, which supply many countries, would presumably spread their available supplies in the way that would maximize profits. Thus, the shortfall in supply in the United States would tend to be spread throughout the oil-importing countries.*

Still another possibility would be for the participating Arab states to cut off oil exports to Western Europe and Japan, as well as to the United States. In 1973, this action would have affected over 15 million barrels of oil per day, or about a third of the total oil supply of the non-Communist nations. A cutoff of this magnitude would be too large to be handled by the sort of automatic mechanisms discussed above. Stringent government rationing would therefore be necessary in the United States and elsewhere. The impact on daily life would probably be as great as occurred in World War II. The impact on industry would be much greater, since industry generally would probably not be given the same degree of preferred treatment as was accorded war-related industries in World War II.

On the other hand, this sort of cutoff would also have the most severe impact on the revenues and, consequently, on the economic and military development of the producing nations. Foregoing the bulk of their oil revenues is not a decision those governments would take lightly. While growing currency reserves provide a substantial cushion, particularly for Saudi Arabia, the expectations of lost income would provide considerable incentive against such a wide oil supply cutoff. Also, the Arab nations are part of the world economy, and they could not hope to escape severe damage to both their domestic economies and the value of their rapidly increasing foreign investments if a shortage of oil plunged the

*If an OECD oil-sharing agreement was in effect at the time, it would provide an intergovernmental mechanism for facilitating the dispersion process.

United States, Western Europe, and Japan into a severe depression. Moreover, all but the more militant Arab governments would want to avoid the rupture of political relations with the major industrialized nations that might result from so destructive a use of the oil weapon.

The above review of three possible uses of the Arab oil weapon helps to explain what the Arabs actually did in 1973-74. A simple embargo of the United States alone would have had little effect. Even an embargo of the United States combined with a freezing of export levels to other major customers would have had only a small and temporary impact. On the other hand, a total embargo of all of the non-Communist industrialized nations would have been too costly economically and too dangerous politically for the Arabs themselves.

In 1973-74, the Arabs in effect adopted a strategy falling between the second and third alternatives discussed above. By reducing their total output by about 25 percent (susbstantially more than the share that had been going to the United States), they made certain that the adjustment mechanisms described above would not quickly reduce the direct impact of their total embargo on exports to the United States. At the same time, they calculated that by creating a worldwide oil shortage, they could induce Western Europe and Japan to press the United States to change its policy in the Arab-Israeli dispute. The Arabs took care, however, not to cut oil supplies so deeply as to precipitate an immediate economic breakdown in the industrialized nations. They also sought to keep those nations from reacting in a unified manner by supplying them with varying amounts of oil depending on whether they were classified as friendly, hostile, or neutral. And finally, acting in concert with the non-Arab oil-exporting countries, they pushed oil prices to new, very high levels, thereby more than compensating for the effect of supply cutbacks on their revenues.

It is too early for final judgments on the success of the Arab use of the oil weapon in 1973-74. They did achieve some change in the publicly proclaimed policies of Japan and several European countries toward the Arab-Israeli dispute. The Arabs also may believe that their embargo and production restrictions spurred the United States to work harder for a Middle Eastern settlement, although it can equally well be argued that U.S. diplomacy was driven more by a desire to defuse a dangerous threat to world peace than by fear of an oil shortage.

The most important consequences of the Arab oil embargo and production cubacks of 1973-74 may not be fully apparent for some time. The huge price increases that the production cutbacks made possible will undoubtedly stimulate both the development of alternative energy sources and efforts to economize on the use of energy in all oil-importing nations, including the United States. To the extent that high prices cause the United States to become more self-sufficient, the political utility of the oil weapon will be reduced.

History's ultimate judgment on the success or failure of the Arab venture in economic warfare in 1973-74 has, however, little to do with Arab willingness or unwillingness to use the oil weapon again. Several considerations argue in favor of such action in the event of major, new hostilities with Israel, or possibly if progress toward a satisfactory peace settlement appears to have ceased.

The unified action of most of the Arab oil-exporting states had a symbolic importance second only to the improved battlefield performance of Arab armies. Also, the political standing of the conservative Arab states in the Arab world—and the leadership claims of Saudi Arabia in particular—depends critically upon their willingness to use their oil to achieve Arab political goals. Perhaps most important, the oil weapon is widely believed by the Arabs to have been effective in 1973-74, and they have no other comparable weapon short of a renewal of hostilities against Israel.

If the Arab oil weapon is used again, the tactics pursued in 1973-74 may well be taken as the starting point. Refinements may be adopted to cut down leakages to embargoed destinations. The major dilemma confronting the Arabs next time—if there is a next time—could be deciding how hard to squeeze Western Europe and Japan in order to increase pressure on the United States.

If the United States in the meantime had become more dependent on Arab oil, keeping Western Europe and Japan from sharing their normal supplies with the United States would be sufficient. If the United States had moved considerably closer to self-sufficiency, the Arabs could get at the United States only by depressing the economies of its allies and major trading partners. In either situation, the existence of an effective OECD-wide system for conserving and sharing oil supplies in an emergency would complicate Arab calculations.

THE INTERRUPTION OF OIL SUPPLIES IN WARTIME

Some of the national concern over increased dependence on imported oil stems from the U.S. experience in World War II. Both civilian and military requirements for oil increased greatly, and for a time German submarines had considerable success in interdicting the movement of oil from the Caribbean and the U.S. Gulf Coast to the eastern seaboard.

The oil supply problems of World War II throw little light, however, on the problems that may arise in a future war. A new look must be taken at three questions: What in rough terms would be the increased demand for oil in wartime? What options are available to an enemy seeking to interdict the oil supplies of the United States and its allies? What might be the consequences of oil-supply interruptions in a war?

Wartime Demand for Oil

For very different reasons, limited wars and a major nuclear war lie largely outside the present analysis. Limited wars add relatively little to total U.S. oil requirements. The need to produce more of certain refined products (particularly jet fuels) and the need to deliver large quantities of fuel to the war zone can be met fairly easily by adjusting refining schedules and shifting transport patterns. In the event of a major nuclear war, the problem of oil supply would be submerged in the general disaster. Recent studies have concluded that the petroleum refining and transport capacity surviving a nuclear strike on the United States would be sufficient for the reduced population and shattered economy surviving the engagement.[1]

The wartime contingency of greatest interest in the present connection is a major conventional war of considerable duration. The specific case chosen for analysis is a conflict of global proportions, but centering on Europe, between NATO and the Warsaw Pact beginning in the mid-1980s. It is only in such a major protracted East-West conflict that the interdiction of oil shipments could become important as a means of waging war.

In the event of a protracted war in Europe, the increase in military demand for petroleum would not be as sharp or as severe as was the case during World War II. The United States maintains much larger standing peacetime military forces now than it did in the 1930s. Consequently, while total demand by military forces would probably exceed that of World War II (military forces are much more dependent upon mechanized vehicles and aircraft than in the past), the incremental demand would not be as severe. Nonetheless, military and war-related industrial incremental demands would be sizable, at least in absolute terms, as is shown below.

In mid-1974, there were approximately 200,000 U.S. Army troops in Europe, comprising 4-1/3 divisions plus special combat units and support elements. In the event of war in Europe, the United States plans to send at least 9 additional active Army and

Marine divisions, and various reserve units to the combat theater, bringing total strength up to perhaps 750,000 men.

For planning purposes, Army field manuals specify that about 7 gallons (one-sixth of a barrel) of petroleum products are required per day for each man in the combat theater.[2] Using this planning factor, the augmented U.S. ground forces in Europe could be expected to consume 125,000 barrels of petroleum per day, an increase of about 100,000 barrels per day over current consumption.

In mid-1974, the United States also maintained 21 Air Force fighter attack squadrons and 2 Navy carrier air wings in the European theater. In wartime, this force would be augmented by 4 Air Force squadrons based in the United States but designated for use in Europe, 16 squadrons from the strategic reserve, 39 squadrons from the Air National Guard and Air Force Reserve, 4 carrier air wings, and 1 Marine Corps air wing, bringing total strength up to about 2,100 aircraft. On the basis of the U.S. experience in Vietnam, the average aircraft in a combat zone consumes about 60,000 barrels of fuel per year. Total wartime consumption of U.S. tactical air forces in Europe may therefore be estimated at roughly 350,000 barrels per day, twice what the same number of planes would consume in peacetime.

Naval forces also consume more petroleum during war than in peace; ships spend less time in port and are more active when at sea. In fiscal year 1964, U.S. military forces consumed 65,000 barrels of residual fuels, Navy special fuel oil, and Navy standard distillate for every active-duty ship in the Navy. In fiscal year 1968, the height of the Vietnam war, this figure rose to 93,000 barrels. The difference between the overall figures for 1964 and the war years, however, underestimates incremental demand in wartime because it averages consumption of ships not in the combat theater with those operating on a wartime footing. Thus, a 75 percent increase to about 114,000 barrels per ship annually would not appear unreasonable.

In the event of a major war in Europe, we may assume that the entire active-duty Navy would be on a wartime footing. Assuming a force of 550 ships, it would consume about 63 million barrels per year or roughly 170,000 barrels of petroleum each day, of which about 70,000 barrels could be considered the incremental demand of the war. Moreover, some naval forces presently in reserve would be mobilized. If this force consisted of 50 ships—primarily destroyers and other escorts—another 16,000 barrels per day would be added to the incremental demand.

It is unclear the extent to which, if at all, a protracted war in Europe would impose incremental demands upon the oil consumed by U.S. industries.* The increase attributable to war should not, however, be large. In part this occurs because, once the economy is functioning near its full-employment rate, increased demand for military goods will result in compensating decreases in civilian production. Incremental demand is also somewhat limited because the wartime environment permits the government to impose rationing on civilian consumption, as was the case in World War II. Hence, assuming a 1 percent increase in demand for war-related industrial production would seem to be a generous estimate. This increase would be more than offset by reductions in civilian consumption imposed by likely government conservation measures.

Finally, the states of Western Europe would, of course, have sizable military forces engaged in any protracted war in Europe, and it would be in the U.S. interest to help supply their petroleum needs. It is not clear that this requirement would impose incremental demands on production, however, insofar as the economies and the societies of the nations of Europe are likely to be severely disrupted in any contingency such as that considered here and, consequently, total petroleum consumption is likely to decrease. During World War II, for example, German civilian consumption of gasoline was only one-eighth of what it had been before the war, and civilian diesel fuel supplies were reduced by two-thirds. The United States would, though, be confronted with the problem of helping to assure continued deliveries of even reduced demand in the face of attempts at interdiction.

Table 15-1 summarizes these calculations for incremental demand and for total deliveries to Europe. Generally, growth in demand is quite modest. Of course, should the war continue for a very long period of time, say more than one year, additional military forces could be conscripted, trained, and deployed to the combat theater and thus further increase demand. Moreover, war-related industrial needs are likely to continue to rise somewhat so long as the war continued.

The modest size of the incremental demand does not mean that the United States would not face difficulties in assuring the supply of petroleum in wartime contingencies. There might be some difficulties in providing sufficient amounts of the specific kinds of products needed by the military in the early stages of the conflict.

---

*During World War II, oil consumption in the United States increased some 22 percent,[3] a rise of only 4.5 percent per year. This figure, which includes the growth attributable to sharply increased military consumption, may be compared with the average annual increase of 4.1 percent in U.S. consumption between 1961 and 1971.

Table 15-1.   Incremental U.S. Petroleum Demands and Deliveries
in the Event of a Major War in Europe in 1985 (thousand barrels per
day)

	Incremental Demand	Required Deliveries to Europe
U.S. Military Forces		
Ground combat forces	100	125
Tactical air forces	175	350
Naval forces	86	—
SUBTOTAL	361	475
Other NATO Military Forces	0	1,080[a]
SUBTOTAL	361	1,555
NATO Civilian Demand	0	9,000 – 12,000[b]
TOTAL	361	10,555 – 13,555

[a]Assumes 2.5 million men in ground forces, 2,500 tactical aircraft, and 800
naval ships. These forces represent roughly a 10 percent decrease from the
present force levels (including reserves) of Belgium, Britain, Denmark, France,
West Germany, Italy, the Netherlands, and Norway.

[b]Assumes that in 1985, the states listed in note *a* consume roughly 20 million
barrels of petroleum per day, one-fourth of which is assumed to come from
domestic sources (mainly the North Sea). The degree to which this consumption
would be reduced—whether by physical destruction of industrial facilities or by
government-imposed rationing—is unclear. The lower figure in the table
represents a total cutoff of German consumption (such as would occur as far as
the United States was concerned if West Germany was overrun by Soviet troops)
and a 20 percent reduction in the import requirements of the other states. The
higher figure assumes only an overall 20 percent drop in import needs. This
could come about through a combination of increased domestic production
(perhaps with synthetics as Germany did in World War II), rationing, and
war-related damage.

More important, assuring delivery of even peacetime supply levels in
the United States could pose severe problems if the United States
continues to depend on imports for a substantial portion of its oil
requirements. Most of the incremental demand traceable to the war,
however, would impact on shipments to Europe. Perhaps as much as
11 to 14 million barrels of oil per day would have to be shipped
there to meet essential military and civilian needs.

Interdiction Scenarios

There are a number of ways that, in a wartime environment, an adversary with sizable and sophisticated military forces
could attempt to interdict the movement of petroleum between
producing areas and the United States and Western Europe. The
relative attractiveness of each of these options and the degree to
which an opponent might expect to achieve some success in each

would depend upon a wide variety of factors, including the performance characteristics of weapons, the amount of force committed to the engagement and the skill with which it was employed, and the countermeasures taken by the West. This section will outline the tactics available to an opponent and the counter-actions likely to be taken by the West, the pertinent forces that each side could bring to bear, and certain critical determinants of the outcome of the engagement.

For purposes of this analysis, the following assumptions have been made:

— The war between NATO and the Warsaw Pact has been going on for some time, and neither side is in immediate danger of defeat.
— No nuclear weapons are being used, and tacit rules of engagement preclude even conventional air strikes on the territories of the United States and the USSR.
— Major oil exporters have maintained shipments to the United States and Western Europe.

In this situation, the Soviet Union would have a number of options in attempting to disrupt the West's oil supply system.

One possibility would be to attempt to deny the oil fields themselves to the West. This could be done either by destroying the fields in air raids or by physically seizing the oil-producing areas. To the extent that either tactic would be a practical option for the Soviet Union, the fields along the Persian Gulf would seem, because of their proximity to the USSR and their great output, to be the most likely targets.

There are difficulties, however, in the air-strike option. The oil fields themselves would not seem to be particularly vulnerable to air raids with conventional munitions: they are comprised of many individual targets, each of which is relatively "hard" and individually of low intrinsic value. A somewhat more fruitful approach might be to attack the tanker-loading terminals, insofar as only a few such facilities (such as Kharg Island, Ras Tanura, Bahrain, and Abadan) will provide a large portion of the Persian Gulf's output in the 1980s. These targets would be relatively easy to repair, however, and would have to be attacked repeatedly.

An additional problem derives from deficiencies in Soviet capabilities. Most oil fields in the gulf area (even those in Iran) are beyond the combat range of Soviet fighter aircraft based in the USSR. Thus, the primary vehicles for such strikes—the 700 medium-range bombers assigned to the USSR's long-range air force,* less those used in other missions or held in reserve to deter China from initiating hostile action, would have to carry out the bombing without fighter escorts. In such a situation, they would be very vulnerable to interceptor aircraft. Even if no U.S. forces were deployed to the region, the Iranian air force, which presently is composed of 8 tactical fighter squadrons but reportedly will be expanded in the next few years, would be likely to inflict heavy losses on the attacking bombers.

The second means of denying the oil fields to the West—occupying them—would seem to be a somewhat better choice. The USSR probably would not attempt either an amphibious or an airborne invasion. The most likely course would be an invasion of Iran overland from the Caucasus region and then, assuming such a foray was successful and Iranian bases became available for Soviet use, expanding control to the other producing regions on the Persian Gulf one by one.

In peacetime, the USSR maintains 21 divisions in the Caucasus and West Turkestan regions. These troops alone, with suitable air support, should be capable of occupying Iran—which maintains 7-plus divisions at present—within several weeks. If the United States sent troops and airpower to the region, the outcome would be less clear, depending on the size of forces the United States and the Soviet Union felt they could commit to what, after all, would be a less important arena than Europe, the performance of the Iranian armed forces, and other factors. Conceivably, the USSR might gain the support of Iraq in such a contingency because of its deep-seated ethnic and political rivalry with Iran.

A third possibility would be for the Soviet Union to attempt to cripple the West's refining capability. This tactic was used with considerable success by the allies against Germany during World War II. That success is unlikely to be repeated in Soviet attacks against NATO-country refineries, however. For one thing, NATO's refining capacity is not as concentrated as was Germany's in World War II. Whereas there were fewer than 100 refinery targets at that time, there are already more than 700 contributing to Western oil supplies and that number is likely to have grown by the mid-1980s. Moreover, more than 40 percent of the West's refining capacity is

*It is assumed here that mid-1980 force levels will be the same as those now evident. This assumption overstates the number of forces likely to be available, as force levels in both the West and the Soviet Union have tended to decrease as more modern (and costly) systems have entered into service. In the present case, for example, the new Soviet bomber, *Backfire*, is likely to replace the *Badgers* and *Blinders* now in the force at less than a one-to-one ratio.

located in the Western Hemisphere, beyond the range of all but a few Soviet aircraft.[4]

Another set of options available to the USSR would involve the interdiction of oil tankers, both those moving crude oil to storage areas and refineries and those transporting finished products to consumers. In the early years of World War II, Germany adopted this tactic against the movement of oil from the Caribbean and the Gulf of Mexico to the northeast United States. Within four months of the beginning of this interdiction campaign (February 1942), tanker shipments were only 18 percent of what they were in December 1941. Shipments continued to decline for another year, reaching a low point of 6 percent of the December 1941 rate in May of 1943. Tanker shipments did not return to December 1941 levels until late 1944.[5]

The navy that brought about this marked drop in tanker movements began the war, in 1939, with only 57 submarines. The force was expanded rapidly, however, and reached a peak of 400 boats for a brief period in 1942—the heart of the interdiction campaign.[6]

In the mid-1980s, assuming that submarine construction rates continue as has been evident for the past five years and that the Soviet navy continues to retire its submarines when they are no more than 20 years old, the Soviet Union could have a force of between 180 and 200 attack submarines, almost all of which would be nuclear powered. These boats would probably have over 30 torpedoes each; additionally, about 70 of them would probably also be equipped with cruise missiles (perhaps 8 launchers each). Most submarines are also capable of delivering sea mines through their torpedo tubes. Finally, the USSR could make use of medium-range strike aircraft carrying air-to-surface missiles or bombs. There were 300 such aircraft assigned to the Soviet naval air arm in 1972. Many of these aircraft could also be used to carry mines.

The actual number of submarines and aircraft committed to an oil-interdiction campaign would depend, again, on the USSR's perception of other requirements for these forces, its assessment of the chances of successfully interdicting oil flows, and its reading of the strategic consequences of such a campaign should it be successful. It is conceivable that the USSR would have only limited resources available for an interdiction campaign; perhaps fewer than 50 submarines. Such a situation would obtain if, as many observers believe, the primary mission of the Soviet submarine and naval air forces was to attack western "naval groupings," such as aircraft carrier task forces.

Of course, if the war continued for a long time, or if it was preceded by a long period of tension during which the USSR considered it prudent to accelerate its submarine-construction rate, larger numbers of submarines could be available for the interdiction mission. Given the much greater technical complexity of modern-day submarines, however, it is very unlikely that the force could be expanded at a rate approximating that of the German undersea fleet in the early years of World War II.

If the USSR decided to attempt to interdict oil flows, it could choose among several potential choke points on the oil tanker routes. Three possibilities come to mind. One is the off-loading facilities in Europe. The average size of tankers has increased sharply during the past decade and will rise even more sharply during the remainder of the seventies. The reason for this increase is the introduction of very large crude oil carriers (VLCCs), ships ranging between 175,000 and 600,000 deadweight tons, as the main means of transporting petroleum between primary producing and consuming areas. These giant vessels require special deep-water facilities where they can disgorge their cargoes into pipelines or storage tanks, or transship them for final delivery by smaller ships. There are about 10 such facilities in operation in Europe, and about 10 more are planned. These facilities, each with several VLCCs tied up to them at any one time, would be tempting targets for Soviet bombers and would provide a reference point that submarines might use for locating incoming tanker shipments.[7]

Another possibility, and probably the most vulnerable point in the oil-flow network, is the entrance to the Persian Gulf—the Strait of Hormuz. This waterway is only 25 nautical miles wide (outside the 20-fathom line), and the channel used by deep-draft oil tankers is about 2 nautical miles wide. Soviet submarines could deploy to the Indian Ocean and use the entrance to the gulf as a reference for locating tankers, which they could then attack with torpedoes. More profitably, they could attempt to close the strait with sea mines. Assuming that Soviet sea mines have a lethal sweepwidth of 200 feet relative to a typical VLCC, say of 250,000 deadweight tons, and further assuming that Soviet naval doctrine calls for mine fields with a 25 percent kill probability per passage, only 200 mines would be required to close the 25-mile strait. Another 50 mines might be added to this requirement to hedge against technical failures, misplaced drops, and other errors. If the mine fields were laid by submarine, only 6 to 8 submarine loads would be required.* The USSR could also employ air-dropped mines

*Present-day Soviet nuclear submarines can carry up to 64 mines apiece,[8] but some of this capacity would likely be used for torpedoes for self-defense or targets of opportunity.[8]

to close the gulf. Medium-range *Badger* aircraft can probably carry about 20 mines apiece (if they are not equipped with air-to-surface missiles).* Thus, only 12 plane loads would be required.

Clearly, if it was unopposed, the Soviet Union would have more than ample resources to close the Persian Gulf. The situation is not quite that simple, however. Unless they were deployed before the initiation of hostilities, Soviet submarines would have to pass through barriers established by the NATO navies between Soviet naval bases and the submarines' expected operating areas, such as the various straits between the Pacific and Indian Oceans or the area between Greenland, Iceland, and the United Kingdom. Moreover, NATO and Iranian antisubmarine forces could patrol in the Arabian Sea near the mouth of the gulf. If aircraft were used, they would face problems similar to those discussed for aircraft attempting to bomb the oil fields or loading terminals at the head of the gulf. In fact, aircraft on mine-laying missions would face greater difficulties in that they would have to penetrate farther south. On the other hand, they might avoid a large portion of Western defensive interceptors by overflying India, or perhaps Afghanistan and Pakistan, before heading for the Strait of Hormuz. This tactic could be partially countered by the use of an aircraft carrier in the Arabian Sea, but carrier-based air defenses would be relatively inefficient if the attacking aircraft remained over land as much as possible.

Obviously, some Soviet submarines and aircraft would be destroyed, and others would succeed in depositing their mines. Since the number of successful missions required is quite low (6 to 8 for submarines, 12 for aircraft), it should be expected that the USSR would be relatively successful in this mission; and although mine countermeasure forces (minesweeper ships and specially equipped helicopters) could be deployed to the area, their prospects for clearing the mine field would be slight. Modern sophisticated mines are extremely difficult to sweep, and the U.S. Navy has been reducing its mine countermeasure forces (perhaps because of their limited effectiveness). In addition, of course, it would be necessary to hold back a large portion of these forces for mine-clearing operations in the United States and Western Europe.

The final tactic available to the Soviets would be to attack tankers on the high seas, somewhere between their origin and destination. As noted previously, during a war in the mid-1980s, roughly 11 to 14 million barrels (1.4 to 1.9 million metric tons) of oil per day would have to be delivered to Western Europe. Assuming that tankers averaged 250,000 deadweight tons in 1985, then 8 to 10 deliveries would have to be made each day. If the average trip from

---

Badgers without air-to-surface missiles are reported to carry 9 tons of ordnance internally.[10]

producing to consuming area took 30 days, then at any one time there would likely be some 500 to 600 potential targets on the high seas. If the United States was not self-sufficient, additional tankers would of course be at sea carrying oil to U.S. ports.

In World War II, ships sailed in convoys for protection and each convoy was escorted by destroyers. There might be some difficulty adapting this tactic to VLCCs, however, in that the great size of these vessels makes them difficult to maneuver and, therefore, risky to sail close to one another. More likely, the defensive tactic adopted would be to attempt to keep the general area of the oil routes clear of submarines. This would be attempted by attacking the Soviet submarines force as it passed through the barriers mentioned earlier and by searching the oil routes with antisubmarine aircraft, based at sea or on land. In some areas, these aircraft could be assisted in locating submarines by fixed underwater detection systems. However, a force of 50 to 100 submarines could be expected to sink a considerable number of tankers before being destroyed by Western forces.

The Consequences of Oil-Supply Interruptions in Wartime

The most potent threat posed by the USSR with regard to the movement of oil in a major NATO-Warsaw Pact conflict would clearly be to close the entrance to the Persian Gulf with mines. The United States and Western Europe could require 17 million barrels per day of imported oil in the assumed contingency in the mid-1980s.* Even if the NATO powers could obtain all of the oil exported by areas outside the Persian Gulf states (and outside the Far East, whose exports would presumably go to Japan) they would be about 3 million barrels short. The actual deficit would, however, be much larger, since the requirements of other oil importing countries could not be ignored. A shortage of over 10 million barrels per day, or roughly one-third of total U.S. and West European requirements, would be entirely possible.

There is no question that oil shortages played a significant role in weakening the German war effort in World War II. Consumption of petroleum products by the *Luftwaffe* and the *Wehrmacht* had to be cut sharply beginning in May and June of 1944 and even deeper cuts were made later in the year and in the early part of 1945. These reductions in consumption limited the mobility, flexibility, and overall effectiveness of the German armed forces.

*Includes 11 to 14 million barrels per day for Western Europe, plus 4 million barrels per day for the United States.

Limitations on oil supplies cut more deeply into the German war economy, as the military generally received first priority for available stocks; thus oil restrictions severely weakened Germany's ability to provide equipment and supplies to its armed forces.[11]

According to the U.S. Strategic Bombing Survey, Japan's dependence on imported oil after depletion of stocks existing in 1941 "proved to be a fatal weakness." Although Japan managed to increase imports until the end of 1943, it never did match consumption levels. Consequently stocks were drawn down and, by the fall of 1944, sharp cutbacks in consumption by military forces had to be made. By April 1945, supplies of naval fuel oil were so low that only one of the five battleships remaining in the Japanese fleet was able to participate in the Battle of Okinawa.[12]

The interruption of German and Japanese oil supplies toward the latter part of the war far exceeded, in percentage terms, the portion of U.S. and NATO supplies that would be affected by a blockage of the Persian Gulf. In January 1945, German production and imports of petroleum products were less than one-fourth the monthly average in 1943.[13] Similarly, in the spring of 1945, Japanese production and imports of all petroleum products totaled less than one-fourth the prewar rate.[14] A cutoff of imports from the Persian Gulf would, under the above assumptions, still leave two-thirds of U.S. and West European demand unaffected. Moreover, despite the sharp increase in the dependence of modern-day military forces on petroleum—due to the introduction of helicopters and mechanized personnel carriers, and greater reliance on armor, aircraft, and other vehicles—total military demand is small relative to civilian demand. Consequently, a one-third cutoff is unlikely to affect the consumption of the military itself. It would, however, bite strongly into civilian consumption. The degree to which such a reduction would be debilitating to the war effort would depend upon the scope and effectiveness of compensatory measures.

NOTES FOR CHAPTER 15

1. Richard A. Schmidt, *Support of Energy Program Planning* (Calif.: Stanford Research Institute, 1972), p. XIV-1; U.S. Cabinet Task Force on Oil Import Control, *The Oil Import Question, a Report on the Relationship of Oil Imports to National Security* (Washington, D.C.: Government Printing Office, 1970), p. 35.

2. U.S. Department of the Army, Headquarters, *Staff Officers' Field Manual—Organizational, Technical and Logistical Data—Unclassified Data*, Field Manual FM-101-10-1 (Washington, D.C., 1969), p. 5-70; para. 5-26c.

3. American Petroleum Institute, *Petroleum Facts and Figures* (New York, 1959), p. 213.

4. "World-wide Oil at a Glance," *Oil and Gas Journal,* December 25, 1972, pp. 82-83.

5. John W. Frey and H. Chandler Ide, eds., *A History of the Petroleum Administration for War: 1941-1945* (Washington D.C.: Government Printing Office, 1946), pp. 87-88.

6. F. H. Hinsley, *Hitler's Strategy* (London: Cambridge University Press, 1951).

7. See, CDR Bernard Frankel, USN, "Offshore Tanker Terminals: Study in Depth," *U.S. Naval Institute Proceedings,* Vol. 99 (March 1973), pp. 56-64; see also several studies reproduced in *Deep Water Port Policy Issues,* Hearings before the Senate Committee on Interior and Insular Affairs, 92 Cong. 2nd. sess. (1972).

8. See LCDR Robert D. Wells, USN, "The Soviet Submarine Force," *US Naval Institute Proceedings,* Vol. 97 (August 1971), pp. 63-79.

9. Byron E. Calame, "Deadly & Sophisticated, Navy's Mines Can Do Everything but Smell," *Wall Street Journal,* May 10, 1972, p. 1.

10. See John W. R. Taylor and Gordon Swanborough, *Military Aircraft of the World* (New York: Charles Scribner's Sons, 1971), p. 1.

11. U.S. Strategic Bombing Survey, *The Effects of Strategic Bombing on the German War Economy* (Washington, D.C., 1945), pp. 76-81.

12. U.S. Strategic Bombing Survey, *The Effects of Strategic Bombing on Japan's War Economy* (Washington, D.C., 1946), pp. 36-47.

13. U.S. Strategic Bombing Survey, *Oil Division Final Report* (Washington, D.C., 1947), Figure 16.

14. U.S. Strategic Bombing Survey, *The Effects of Strategic Bombing on Japan's War Economy,* p. 34.

Part Six

Nuclear Energy

Chapter Sixteen

The Growth of the International Nuclear Power Industry

Nuclear power is not yet a major source of energy, but its rapid expansion is a virtual certainty in the remainder of the 1970s and in the 1980s. By 1985, nuclear power could provide close to one-seventh of the non-Communist world's energy requirements. This revolutionary change in the world's sources of energy is in part the result of government policies, but increasingly the initiative lies with a major new nuclear industry, international in scope and rapidly expanding in scale. The future structure of this new industry and of the emerging international market for reactors and nuclear fuels will have an important bearing on the adequacy, reliability, and price of nuclear energy supplies and on the effectiveness of international efforts to safeguard nuclear facilities and materials.

As the governments of the major industrialized countries approach these problems, they find themselves in a somewhat curious position. Their close relationship with the nuclear industry and their desire for both increased domestic energy supplies and larger export markets cause them to promote the growth of commercial nuclear power. At the same time, their broader security interest in avoiding further proliferation of nuclear weapons causes them to impose restraints on the industry in an effort to channel its growth in less dangerous directions.

The problem of safeguarding commercial nuclear energy will be explored in the next chapter. This chapter will concentrate on the economic and technical aspects of the impending rapid growth of nuclear energy.

Note: This chapter is based in large part on research and analysis conducted by Jerome H. Kahan.

TRENDS IN COMMERCIAL NUCLEAR POWER

Under existing technology, civil nuclear energy is usable almost exclusively as a means of generating electricity. In 1973, nuclear energy provided less than 2 percent of electric generating capacity in the world, but its projected growth both in absolute terms and in its relative position as a primary fuel for electricity production is striking (see Table 16-1).* The U.S. Atomic Energy Commission (AEC) forecasts that, by 1985, nuclear power will account for some 25 percent of total electric generating capacity in the principal non-Communist countries. The projected increases in nuclear power production may not be attained on schedule, due primarily to the numerous and complex technical difficulties that have plagued the industry and that continue to cause delays in plant construction and operation—not to mention possible delays stemming from environmental concerns. On the other hand, the oil embargo and steep increases in the prices of hydrocarbons have provided strong additional inducements to accelerate the development of nuclear power.

The desire for greater energy security in the nations of Western Europe and in Japan, which—unlike the United States—must import the bulk of their fuel, is one of the primary reasons why those countries are finding nuclear power facilities increasingly attractive. Equally important, nuclear generating costs are comparing favorably with fossil-fuel costs in highly industrialized nations, particularly where economies-of-scale exist.** Moreover, the economic viability of nuclear power generators will be enhanced if fossil-fuel costs remain high.***

Thus far, the use of nuclear power has not reduced the demand for oil and other hydrocarbons very much. In 1970, the hydrocarbon savings attributable to nuclear power were well under 1.0 percent of total oil consumption in both the United States and

*Nuclear energy accounts for a larger percentage of electricity production than it does of installed generating capacity. This reflects the greater operating efficiency of nuclear plants as compared with fossil-fuel plants.

**At present, 500 to 600 megawatts is considered the minimum economic size in many situations, and the trend is toward larger reactors in the industrialized nations. The investment cost in 1973 U.S. dollars of a typical 1,000-megawatt nuclear reactor is estimated at about $500 million in the United States and $400 million in Western Europe and Japan. (Differences in construction time and labor productivity account for the lower foreign costs.) The electricity generated by a 500-megawatt reactor costs perhaps 15 percent more per kilowatt-hour than that generated by a 1,000-megawatt reactor.

***In contrast, the cost of nuclear generated electric power is relatively insensitive to the price of uranium ore. Even a doubling of the cost of uranium would increase the cost of delivered electric power by only 5 percent.[1]

Table 16-1. Projected Nuclear Power Capacity in Principal Non-Communist Countries

Country	1980			1985		
	MW (e) Total Electric Capacity	MW (e) Nuclear Power Capacity	Nuclear Power as % of Total Electric Capacity	MW (e) Total Electric Capacity	MW (e) Nuclear Power Capacity	Nuclear Power as % of Total Electric Capacity
Austria	12,000	700	5.8	14,800	3,200	21.6
Benelux	28,900	2,800	10.0	39,100	11,000	28.1
France	61,600	8,400	13.6	78,200	19,000	24.3
West Germany	80,600	20,000	24.8	105,300	44,000	41.8
Italy	53,400	3,200	5.9	70,300	14,000	19.9
Sweden	24,900	6,400	25.7	30,900	12,800	41.4
Spain	33,700	7,400	22.0	46,500	13,200	28.4
Switzerland	17,000	3,700	21.8	21,500	7,600	35.3
Japan	182,200	27,000	14.8	272,800	74,000	27.1
United Kingdom	100,000	16,300	16.3	126,500	36,000	28.5
U.S.A.	680,000	102,000	15.0	845,000	260,000	30.8
Others[a]	529,600	21,300	4.0	680,100	75,200	11.1
Total Non-Communist Countries	1,803,900	216,000	12.0	2,331,000	570,000	25.0

Note: These projections were made in February 1974 after the Arab crude oil production cutbacks and selective embargo of lage 1973 and after the vast crude oil price increases of the winter of 1973-74. Although the AEC assumed that there would be a major push for expansion of nuclear power to reduce dependence on oil, the 1974 projections for 1980 are lower than the previous forecasts made in 1972. This is because nuclear power plants have not gone into operation in the United States or abroad as rapidly as the AEC had anticipated. The 1974 forecast for total non-Communist nuclear power capacity in 1985 is also lower than the 1972 forecast. The 1974 forecast for Japan's capacity in 1985 is higher than the 1972 forecast.

[a]Includes Argentina, Australia, Brazil, Canada, India, Mexico, Pakistan, and others.

Japan and only 1.6 percent of Western Europe's oil consumption. This situation will change markedly in future years, however. By 1985, on the basis of the latest AEC projections, nuclear power may be saving hydrocarbons—principally oil—equivalent to over one-half of projected total oil requirements in the United States and about one-quarter of such requirements in both Western Europe and Japan.

Not surprisingly, the governments of the major West European states have made major commitments to nuclear energy. In its "Second Illustrative Nuclear Program," issued in December 1972, the Commission of the European Communities called for greater reliance on nuclear power by 1985. Pointing to the fact that nuclear energy could reduce dependence on petroleum imports, the Commission concluded that meeting this goal would require accelerating the nuclear plant order rate, greater capital commitments, and coordinated planning.

In Japan, the long-range atomic energy program of 1972 highlighted the supply and environmental problems associated with petroleum as an energy source and stated the government's decision to increase the use of atomic energy.[2] Table 16-1 shows that Japan's nuclear power projections are perhaps even more striking than those of Western Europe; the 1985 projection for Japan is well ahead of that for every major West European nation in terms of absolute nuclear energy capacity and equals almost 35 percent of Europe's total nuclear energy output projected for that year.

By Western standards, total nuclear power estimates for the Communist countries in 1985 seem modest, barely equaling Japan's 1985 projection.[3] The Soviet Union's nuclear power program has been slow in expanding, largely because the country is rich in conventional energy sources and the large amounts of capital required for nuclear power plants were needed elsewhere. Solid information on possible Chinese plans to launch a civil nuclear energy program is lacking.

Elsewhere in the world, the distribution of nuclear energy capacities is uneven. Canada has a nuclear program surpassing that of a number of West European nations, and India is mounting a major nuclear program that is well beyond that of other developing nations and is comparable to many programs in Western Europe. Brazil, Mexico, Pakistan, Argentina, Australia, and Taiwan also have significant nuclear power programs or plans. Other nations that have already undertaken research reactor programs are coming to view nuclear power projects with increasing favor, partly as a means of obtaining greater energy security, and partly as a symbol of political stature and independence. India's explosion of a nuclear device in

May 1974 dramatically demonstrated how a civil nuclear power program can be used to advance toward a nuclear weapons capability.

THE INTERNATIONAL REACTOR INDUSTRY

Predominance of Light Water Reactors

Based on the projections cited above, cumulative capital expenditures in the non-Communist world for nuclear power plants and equipment could exceed $250 billion by 1985, with approximately 50 percent of that sum expended in the United States, 30 percent in Western Europe, 12 percent in Japan, and the remainder in various other countries.[4] The United States has dominated the international reactor market. Canada and a number of countries in Western Europe have developed an indigenous reactor capability, but Canada, with its CANDU heavy water reactor (HWR),* is the only country besides the United States that has had any appreciable success with exports. The West European manufacturers have essentially been able to sell only to their respective domestic markets.

As shown in Table 16-2, over 75 percent of the power reactors currently operational, under construction, or on order in the non-Communist world are light water reactors (LWRs). These reactors use ordinary water as both moderator and coolant and slightly enriched uranium as fuel. Looking ahead, it appears that LWRs will continue to dominate the nuclear market in the industrialized nations for many years despite competition from advanced reactor types and from HWRs. West Germany, Sweden, France, Italy, and Japan in recent years have switched from natural uranium reactors to light water reactor systems. Light water reactors will probably also be acquired by most developing nations for the majority of their nuclear energy projects for the simple reason that LWRs hold the dominant position in both the United States and in the emerging reactor supply centers in Western Europe and Japan. Nevertheless, concern over dependence on foreign sources of enriched uranium has already led some developing nations, particularly those with domestic supplies of natural uranium, to procure heavy water reactors. The Indian government, for example, is committed to Canadian-type HWRs. India has small but sufficient

*HWRs use natural uranium as fuel and heavy water as moderator and coolant. These reactors are more capital intensive than reactors using enriched uranium but they have lower fuel costs. In a molecule of heavy water, one or both of the two hydrogen atoms is a heavy isotope of hydrogen known as deuterium.

Table 16-2. Distribution of World Power Reactors by Type and Size (100 MWe or more)

Country	Operating			Under Construction			On Order			Sub-Total			Proposed			Total		
	Units	LWRs	Ave. Size (MWe)	Units	LWRs	Ave. Size (MWe)	Units	LWRs	Ave. Size (MWe)	Units	LWRs	Ave. Size (MWe)	Units	LWRs	Ave. Size (MWe)	Units	LWRs	Ave. Size (MWe)
U.K.	22	0	192.5	8	0	636.2	2	0	660.0	32	0	332.7	8	0	1051.2	40	0	476.4
France	7	1	364.4	1	1	850.0	3	3	900.0	11	5	554.6	0	0	—	11	5	554.6
W. Germany	6	5	344.5	6	6	1186.3	5	5	984.8	17	16	829.9	15	15	908.4	32	31	866.7
Italy	3	2	202.3	1	1	783.0	0	0	—	4	3	347.5	1	1	750.0	5	4	428.0
Benelux	1	1	450.0	3	3	810.0	0	0	—	4	4	720.0	3	3	550.0	7	7	647.1
Sub-Total	39	9	254.1	19	11	856.3	10	8	894.4	68	28	516.5	27	19	905.0	95	47	627.0
Switzerland	3	3	335.3	1	1	850.0	1	1	850.0	5	5	541.2	1	1	850.0	6	6	592.7
Spain	3	3	357.6	6	6	902.0	0	0	—	9	9	720.6	3	3	1110.0	12	12	817.9
Sweden	2	2	600.0	3	3	686.6	4	4	820.0	9	9	726.7	5	5	740.0	14	14	731.4
Other W. Europe	0	0	—	1	1	440.0	1	1	720.0	2	2	570.0	5	4	537.5	7	6	548.3
Sub-Total	8	8	409.8	11	11	796.5	6	6	808.3	25	25	674.9	14	13	754.8	39	38	703.8
U.S.	22	21	607.2	54	53	901.2	68	62	1030.4	144	136	917.3	18	16	1088.6	162	152	936.3
Japan	7	6	426.9	8	8	731.9	6	6	939.5	21	20	689.5	25	25	961.6	46	45	837.4
Canada	6	0	415.0	0	0	—	4	0	750.0	10	0	549.0	0	0	—	10	0	549.0
Latin America	0	0	—	1	1	318.0	2	2	713.0	3	3	581.3	6	6	466.7	9	9	504.9
Asia and Other	5	2	261.0	3	3	—	4	4	591.0	14	9	445.8	9	7	458.2	23	13	450.6
TOTAL	87	46	383.1	98	87	841.2	100	88	962.7	285	221	744.0	99	86	863.3	384	307	775.1

Sources: U.S. Atomic Energy Commission, *Nuclear Power 1973-2000* (Washington, D.C.: Government Printing Office, 1972); and additional data provided by the Commission's Division of International Programs, May 1973.
Note: Table does not include estimates for Communist countries.

uranium resources and has essentially achieved independence in all stages of the fuel cycle, including a heavy water plant. Argentina has also acquired CANDU heavy water reactors and, although not so advanced as India, is moving toward fuel-cycle self-sufficiency.

By 1970, Japan and the nations of Western Europe (except the United Kingdom, which is relying on advanced gas-cooled reactors produced domestically) had not only decided to move forward with nuclear power based on LWRs, but had begun to acquire greater independent capabilities to design and build nuclear facilities. The U.S. government, under the auspices of the "Atoms for Peace" program launched in the mid-1950s, actively encouraged the sale abroad of U.S.-built reactors by providing extensive technical assistance, attractive financing through the Export-Import Bank, and long-term supplies of enriched uranium fuel at stable prices. During this period, foreign firms entered into licensing arrangements with U.S. firms in order to acquire the capability to produce reactors. Many of the early U.S. licensing arrangements, notably those with West German and Japanese firms, have already expired or will expire in the 1980s, and foreign firms are now or soon will be completely independent and able to bid against U.S. manufacturers in the export market. Thus far, the Soviet Union has confined its reactor exports to Eastern Europe and has not attempted to compete with Western suppliers in non-Communist countries.

Diversification of Sources of Supply

Current trends toward a greater number of centers of nuclear supply can be seen by examining how nuclear power plants are being constructed abroad today (Appendix Table B-1). In the past, two U.S. manufacturers (General Electric and Westinghouse) were responsible for the majority of light water reactors constructed in non-Communist nations. In early reactor projects, major plant equipment was produced in the United States, exported abroad, and assembled at the site by a U.S. reactor manufacturing team. In more recent reactor projects in Western Europe and Japan, however, there has been a steady movement toward increasing reliance on locally fabricated components and material. Indeed, most ongoing foreign reactor plants can no longer be characterized as "U.S. built," although many are still managed by U.S. reactor firms and from 75 to 80 percent of the components might be imported from the United States.

Under licensing arrangements, U.S. reactor firms generally provide technical advice and a small number of components, but from 50 to 100 percent of the reactor construction is indigenous. Thus, the U.S. firm may receive only royalty payments and sell a limited amount of components. For many overseas reactor projects, one foreign firm has primary managerial responsibility, and components and equipment are purchased from a variety of non-U.S. sources. In other projects, U.S. and foreign firms combine forces to develop improved reactor designs and establish joint manufacturing and marketing companies. Some foreign governments encourage their utilities to buy reactors from domestic licensees of U.S. companies rather than directly from U.S. firms. Japan, for example, strictly limits the number of units of a particular type that a given utility may import and insists that subsequent units be produced domestically under license.[5]

The first and most dramatic step toward foreign independence in the light water reactor field came in 1969 with the formation of the West German Kraftwerk Union (KWU) organization, which is owned jointly by Siemens and AEG (Allgemeine Electricitats Gesellschaft)—a company that operates without importing U.S. equipment and with no apparent dependence on the original General Electric or Westinghouse licensing arrangements.[6] The KWU organization has acquired a powerful technical capability for building complete light water reactor systems and has succeeded in obtaining contracts for all West German utility orders since 1969 and for two reactors in other European countries. The KWU is also attempting to capture a share of the non-European reactor market and has bid against U.S. companies for such projects.

The KWU organization is the only independent non-U.S. manufacturer of LWRs in the non-Communist world. Italian, French, Swedish, and possibly Spanish manufacturing firms will also certainly achieve a substantial degree of independence before the end of the 1970s, however, and can be expected to produce nuclear power systems both for domestic and export markets. In Japan the trend is equally clear. Toshiba and Hitachi, still under license to U.S. firms, are prime contractors for a number of Japanese power plants under construction or on order, and essentially all of the equipment for recent Japanese nuclear plants is being fabricated domestically.[7]

Going one step further, European reactor suppliers are coordinating industrial efforts across national boundaries in order to gain greater efficiencies and resources and to compete more effectively against the United States for nuclear projects at home and in the world market. A notable example of such coordination is the agreement among West Germany's KWU, the United Kingdom's TNPG (The Nuclear Power Group), and small industrial firms in Belgium and Italy. These organizations agreed to pool their resources "for projects in countries where they have common interests," and they have already begun to compete for the non-European market.[8]

The growth of foreign reactor markets, as well as competition from foreign firms, has led U.S. companies to establish subsidiaries or affiliates abroad that often operate virtually autonomously and compete with each other and with U.S.-based companies for foreign reactor projects. For example, a GE affiliate in Spain is contributing to Spanish nuclear power plant projects, and Westinghouse has established a successful Brussels-based subsidiary, which has been awarded a number of key contracts for reactor systems in Spain and elsewhere.[9]

Some U.S. reactor and nuclear equipment companies also have developed ownership interests in foreign firms without any formal subsidiary ties, and through this technique are participating in the foreign markets.[10] With these kinds of interlocking company connections, for example, a U.S.-based nuclear power company can enter into a broad technology and marketing licensing agreement with a foreign company in which it may already have an ownership interest.

Less dramatic but nevertheless clear trends toward independence can be seen in some developing nations. In India, for example, a heavy water reactor is being built with minimal reliance on imported equipment or technical assistance from the Canadian firm that licenses the Indian manufacturer.[11] Argentina could attempt to follow suit as it gains experience with its second heavy water reactor. Most developing nations, however, will have to rely on foreign sources to satisfy their reactor requirements for at least the next fifteen years.

Out of the maze of industrial connections in the nuclear power field, one conclusion emerges forcefully: by 1985, Western Europe and Japan will be able to manufacture essentially all of the components required for their nuclear power facilities. The industrial capabilities of these nations are reflected in the fact that the United States now imports some reactor pressure vessels from Western Europe and Japan.

On a worldwide basis, the major centers of supply will remain in the industrialized areas. The precise pattern of industrial organization that will emerge is, however, uncertain. If companies in Western Europe and Japan develop along independent and nationalistic lines, there conceivably could be dozens of firms in the non-Communist industrialized nations competing with each other for local reactor contracts and for a share of export sales to other areas.* On the other hand, if, as appears to be more likely, the KWU-TNPG precedent of coordination eventually becomes the norm and if Japanese firms combine forces, by the mid-1980s the international market could be dominated by perhaps eight giant organizations in the United States, Western Europe, and Japan. The heavy capital requirements and the extremely high level of technological sophistication that is needed (and which implies continuation of extensive research and development programs) point to the eventual development of an international nuclear industry structure not unlike the present U.S. domestic market—i.e., dominance by a few giant firms, with a number of smaller companies manufacturing some components.

The advanced reactor programs under way abroad (discussed below), particularly those for the fast breeder reactor and the high temperature gas reactor, are exhibiting the same mixed picture of fragmentation and consolidation as is evident in the light water reactor field. Again, the high development and capital costs and the level of technology required point to the ultimate emergence of a few large firms that will dominate the world market.

Whether an international reactor industry composed of a few large firms is preferable to one composed of many smaller firms is not clear. On the one hand, consolidation could lead to a cartel that would keep prices and profits above competitive levels. On the other hand, larger firms would be more likely to have the staff and financial resources needed to advance nuclear technology, including the development of efficient small reactors for some of the less developed countries.

From the standpoint of security, it might be easier to administer safeguards in an industry consisting of only a few large firms. Also, the more intensive competition that might occur between a large number of smaller firms could create temptations to cut corners on safeguards requirements. At the same time, it must be recognized that the concentration of economic power in a few firms might make it easier for those firms to resist safeguards that they regarded as inconvenient or costly.

---

*A few East European nations have acquired or are planning to acquire nuclear power plants from the USSR, including Czechoslovakia, East Germany, and Bulgaria; these nations might attain substantial independent manufacturing capabilities if the Soviet Union permits the Communist equivalent of licensing arrangements to be instituted.

ADVANCED NUCLEAR SYSTEMS

Although light water reactors are spreading internationally and will remain the dominant source of nuclear power through the year 2000, the United States, Western Europe, Japan, and the USSR are pursuing advanced reactor projects in an effort to reduce the cost of nuclear power, to conserve uranium resources, and to build still greater national self-sufficiency in nuclear power generation. The high temperature gas-cooled reactor (HTGR) is expected to become commercially available worldwide by 1980 and could provide approximately 10 percent of total nuclear power generation in the industrialized nations in the 1990s. Breeder reactors, if successfully developed, could provide a growing share of nuclear power generation in the 1990s and beyond.[12]

High Temperature Gas Reactors

In the United States, HTGR technology is at an advanced stage. With some limited AEC support, one U.S. company, General Atomic Corporation (formerly the Gulf General Atomic Corp.), has successfully engineered and refined HTGR systems and demonstrated their commercial viability. A small experimental HTGR is already operational, a larger demonstration plant is nearing completion, and a number of domestic utilities have placed orders for commercial HTGR systems. Many experts now consider this reactor type extremely attractive in terms of electric generating costs.

The HTGR is a gas-cooled, graphite-moderated reactor that uses a fuel mixture of highly enriched uranium (a minimum of 90 percent ^{235}U) and thorium, the only natural mineral resource besides uranium from which nuclear fuels can be made, rather than the slightly enriched uranium (2 to 5 percent ^{235}U) used in light water reactors. Fuel for the HTGRs is provided by the AEC.

Operating on a thorium cycle, HTGRs produce ^{233}U, a uranium isotope that can be recovered and used as a reactor fuel, much as plutonium is produced and recovered as by a by-product in light water reactors. The advantages of this system over the light water system are its greater energy conversion and thermal efficiency, its lower fuel costs, its use of the new resource thorium, and its avoidance of the serious thermal pollution problem associated with the water cooling systems of LWRs.

Foreign HTGR efforts are behind those in the United States but are advancing.[13] Development programs are under way in several West European countries and should lead to a number of operational prototype plants in the next few years. European experts feel that HTGRs are particularly suited for their countries because they are small enough to operate efficiently in the many separate electrical networks that exist in Western Europe. The KWU organization in West Germany is among the organizations that have recently set up companies to develop HTGRs. In addition, several cooperative programs have been established in Western Europe. Major utility organizations from West Germany, Italy, and the United Kingdom, for example, have organized a company (Euro-HKG) to study the HTGR's potential, and the powerful KWU-TNPG alliance is engaged in joint activities in this field. Japan has also launched an indigenous HTGR program, and a Japanese utility company has announced its intention of becoming associated with the Euro-HKG group.

Possibly overshadowing the indigenous HTGR efforts in Western Europe, however, is the flurry of licensing and joint-ownership arrangements that have recently been instituted between French and West German firms and General Atomic. Through such arrangements, the first commercial HTGR in West Germany will be built. Licensing agreements are also likely to be made for fuel fabrication for HTGRs. As in the early years of light water reactors, these industrial associations with U.S. companies and the anticipated U.S. sales of HTGRs abroad reflect the present U.S. advantage in reactor technology. The market pattern of light water reactors is likely to be repeated again as foreign industries gain technological independence in the HTGR field through a combination of their own efforts and connections with U.S. companies.

Breeder Reactors

Whatever the future of HTGRs or improved conventional reactor designs, the breeder reactor could well have the most dramatic and far-reaching consequences for the international nuclear industry during the decades ahead. In his Energy Message of June 1971, President Nixon stated that "our best hope for meeting the nation's growing demand for economical clean energy lies with the fast breeder reactor."[14] As a result of President Nixon's decision on priorities, increased funds were made available for the construction of a 350 to 500 megawatt U.S. demonstration breeder plant by 1979. The first fully commercial U.S. breeder reactor is officially expected to be in operation in 1986, although many observers regard this date as very optimistic.

The U.S. government's effort is devoted to one specific type of breeder—the liquid metal-cooled fast breeder reactor (LMFBR).[15] This effort, expected to cost well over $4 billion over the next ten years, is being undertaken through cooperative arrangements among the AEC, utility companies, and reactor manufacturers. Fuel for the LMFBR will consist of a mixture of plutonium and uranium and a blanket of ^{238}U (which is placed around the reactor core); liquid sodium is used as a coolant.* As in light water reactors, breeders form plutonium from uranium as a by-product of the fission process, which produces heat for electric power. Unlike light water reactors or other so-called converter reactors, however, breeders produce more nuclear fuel, in the form of plutonium, than they burn. Initially, breeders will use the plutonium, now being stockpiled, that is produced by existing light water reactors.** This will increase the value of the plutonium and forge an economic link between breeders and converters. About three metric tons of plutonium are required as initial fuel for a 1,000-megawatt breeder. Eventually, light water reactors will probably also use recycled plutonium as fuel.

Research and development in the breeder field is characterized by intense, highly nationalistic commercial competition. In addition to the major LMFBR program in the United States, there are six large demonstration projects in Western Europe, including both national and consortia efforts. Although Soviet progress in this field has been uneven, the USSR announced in July 1973 that a 350-megawatt breeder reactor was operating successfully.[16]

Breeder reactors are expected to be quite efficient, extracting anywhere from 50 to 80 percent of the energy content of natural uranium, and therefore significantly conserving uranium resources. Not only can this feature eventually reduce requirements for natural uranium and help avert possible future world shortages, but it also makes electric power generating costs with breeders even more insensitive to the price of uranium than is true of other reactor types. Because of their lack of large indigenous uranium supplies, West European nations and Japan find breeders particularly attractive.

Breeder reactors have their critics, however. Many experts question the economic justification for rapid movement into breeders.[17] To begin with, they do not agree that the breeder reactors must be developed in order to avoid a serious shortage of uranium in the next few decades. They point out that, as is discussed below,

---

*Liquid sodium is an extremely difficult material to work with—it ignites on contact with air and water and reacts with many metals.

**Plutonium is not now used as a fuel for LWRs because of the safety problems associated with handling this highly toxic material and because uranium fabrication is cheaper, but recycling of plutonium in fuel elements is expected to become commercially feasible before the end of the decade.

reasonably assured world uranium resources are more than sufficient to fuel light water reactors until well past the year 2000. They also dispute the claim that the LWRs are very wasteful of uranium, since most of the uranium in used fuel rods can be recovered and enriched again for reuse. Furthermore, it is likely that LWRs will be able to produce electricity more cheaply than the breeders, in large part because of the breeders' high capital costs. The breeders will be at a special disadvantage in Western Europe and Japan, since they are expected to have very large installed capacities, which are ill-suited to the smaller electrical grid systems of those areas.

More importantly, critics of the breeder express serious concern over the high toxicity of plutonium and the security problems associated with the large quantities of plutonium that will be generated as a result of widespread commercialization of this reactor type. Estimates indicate that as much as 80,000 kilograms of plutonium could be generated annually by breeders by the end of the century.[18] Plutonium is not only one of the most toxic substances in the world, raising potential safety hazards during storage, reactor operations, fuel shipments, and waste disposal, but it can be manufactured into nuclear weapons after reprocessing. Thus, the problems associated with effectively safeguarding the use of nuclear materials would be increased.

THE WORLDWIDE NUCLEAR FUEL CYCLE

In addition to the emerging international market for nuclear reactors, a world market is also developing for nuclear fuel supplies. A potentially important move to institutionalize this market was made in January 1974 when the World Nuclear Fuel Market (WNFM) was established in Atlanta, Georgia. Initial participants in the WNFM are expected to include 125 utilities and industrial organizations from the United States, Canada, Europe, Japan, Australia, and South Africa. The WNFM is designed to provide participating members with "prompt, ready access to all sectors of the nuclear fuel market." Trading in nuclear materials (uranium oxide, enriched uranium, and plutonium) for future delivery is anticipated eventually.[19]

Although all nuclear fuel was at one time manufactured in U.S. government-owned plants, there is a strong trend toward turning this activity over to U.S. private industry. At the same time, foreign firms have in recent years begun to compete with the United States, and a lively international trade in nuclear fuel supplies and services has developed. Because nuclear fuel is derived from the same fissionable materials that are used in making bombs, there are obvious international security risks associated with the worldwide flow of nuclear fuel.

The manufacture of nuclear fuel is complex. Uranium must undergo several processes before it can be used in a reactor. The key stages of the uranium fuel cycle are uranium mining and milling, fabrication of fuel elements for use in nuclear reactors, and reprocessing of spent fuel, which involves recovery of plutonium produced by nuclear power plants. For the vast majority of reactors currently in use (LWRs), two intermediate steps—conversion and enrichment of uranium—must precede fuel fabrication.*

Uranium Resources and Trade

Uranium ore is the starting point of the nuclear fuel cycle. Uranium ore is found naturally in the form of uranium oxide (U_3O_8), usually mixed in with other material. After mining, the ore is milled to increase the U_3O_8 concentration. The end product, called "yellowcake" or uranium ore concentrate, has a U_3O_8 content of between 70 and 90 percent and is the form in which uranium moves from mills.

An OECD estimate of uranium ore reserves in non-Communist countries indicates that proved reserves recoverable at a price of under $10 per pound presently total slightly more than 1 million short tons of U_3O_8.[20] The OECD further estimates that additional resources almost equal to proved reserves will be found in unexplored extensions of known deposits or in undiscovered deposits in known uranium districts.** No estimate has been made of the world's ultimately recoverable uranium resources, since most parts of the world have not yet been extensively explored for uranium.

Estimates of total cumulative uranium requirements in the non-Communist world, 1973-2000, range from 2.1 to 2.7 million short tons of U_3O_8.[21] To support power requirements into the 1990s and beyond, new uranium reserves will have to be proved, but potential reserves at the $10 price appear to be substantial. Despite the generally low level of uranium exploration and prospecting in the late 1960s and early 1970s, estimates of uranium reserves steadily increased. Total world reserves will undoubtedly be greater than present estimates indicate.

*Fuel costs for a typical light water reactor can be apportioned as follows: mining and milling, 30 percent; enrichment, 40 percent; fuel fabrication, 15 percent; and reprocessing and "miscellaneous," 15 percent.
**Proved reserves are defined by the OECD as "uranium which occurs in known ore deposits of such grade, quantity and configuration that it can, within the given price range, be profitably recovered with currently proven mining and processing technology."[22]

Earlier optimism about the speed at which nuclear power would develop and thus stimulate demand for uranium ore proved unjustified. During most of the past decade, the price of U_3O_8 in the world market fluctuated between \$5 and \$6 a pound—substantially below the \$8 range that prevailed in the early 1960s, when demand was vigorous. In 1973, however, the market turned around and by early 1974 the price of U_3O_8 was above \$10 a pound.

The United States is self-sufficient in uranium and is likely to remain so for many years. The U.S. price, protected by a prohibition against the importation of U_3O_8 for domestic use, has averaged between 10 and 20 percent above the world price. Because of the higher domestic price, the United States exports very little natural uranium and does not normally supply the uranium for the fuel fabrication and enrichment services it performs for foreign customers. A gradual lifting of the embargo is expected, however, in the near future.

The foreign supply-and-demand picture is considerably more complicated than the domestic U.S. situation. Western Europe and Japan must seek external sources of uranium supply for their expanding reactor programs. Western Europe's indigenous reserves, largely located in France, can serve less than 20 percent of the region's projected needs for the 1980s; Japan's uranium reserves are negligible. Substantial reserves exist, however, in Canada, South Africa, and Australia, and lesser reserves are found in Gabon, Niger, and the Central African Republic. Among the most significant trade flows, as shown in Appendix Table A-11, are South African and Canadian exports to both the United Kingdom and Japan and French shipments to West Germany and Japan.* A few nations with modest nuclear programs, such as India and Argentina, can satisfy their own needs.

Because of the geographical and political diversity of suppliers, the danger of a uranium cutoff to Western Europe or Japan is probably minimal. Nevertheless, the fact that the principal users of uranium ore (except the United States) are heavily dependent on foreign suppliers could cause concern in consuming countries over the long-term reliability of their uranium supplies and could threaten their goal of greater energy security.

Recognizing their need for secure uranium supplies, foreign industrialized nations are engaging in what one expert has

*In the early 1970s, Canadian mining companies concluded extensive, long-term contracts for supplying West German utilities; these transactions are not yet included in official trade statistics and thus do not appear in Appendix Table A-11.

called "uranium politics," by developing diverse sources, entering into long-term supply contracts, and involving their own companies in overseas uranium exploration and production ventures.[23] French firms have made extensive investments in uranium mines in Gabon and Niger (former French colonies). Through its overseas Uranium Resource Development Company, Japan is prospecting abroad, both independently and in concert with foreign firms and governments, while West German firms have interests in Australian and Canadian exploration firms.

Uranium trade also takes place in forms other than yellowcake, and uranium in some form might pass through several countries between initial mining and use in a reactor. Natural uranium purchased from South Africa by a given country, for example, might first be sent to the United Kingdom for conversion into gas, then on to the United States for enrichment and fuel fabrication, and then finally to the purchasing country.

Conversion, Fabrication, and Reprocessing

After extraction and milling, the U_3O_8 destined for light water reactors first undergoes conversion into uranium hexafluoride gas (UF_6) and then enrichment.* (Enrichment will be discussed separately in the following section.)

Conversion, fabrication of fuel elements, and reprocessing of spent fuel elements are carried out by various private and government-owned firms in the United States and several other countries. Most of the foreign firms perform their services for domestic customers and compete with U.S. firms for business abroad.

Presently, the United States, Canada, the United Kingdom, and France perform all the non-Communist world's conversion services, and the total capacity of their facilities should be sufficient to satisfy overall non-Communist world demand through the 1980s. While the United States provides conversion services for foreign customers, an increasing amount of the uranium arriving in this country for enrichment has already been converted into UF_6. Significantly, in what appears to be something of a loophole in the prohibition on imports of uranium U_3O_8 for domestic consumption, U.S. utilities in recent years have imported roughly one-fourth of their uranium for domestic commercial energy purposes in the form of UF_6 manufactured from foreign uranium, principally Canadian.[24]

---

*U_3O_8 for use in natural uranium reactors, such as the Canadian heavy water reactor, is not converted into UF_6, but is fabricated directly into fuel elements.

Looking ahead, the spread of conversion facilities abroad appears certain. Conversion is a relatively simple industrial process, and neither the United States nor the other major industrialized countries have a technical advantage impelling firms in other countries to rely on licensing agreements. For the near term, however, many nations may remain content to rely on external sources as long as the supplying nations offer reliable and economical services.

The first fuel-fabrication capabilities were developed by the United States, the United Kingdom, and France for their military reactor programs, but commercial fabrication plants in these nations —and Canada—now serve civilian nuclear power needs at home and abroad.* The initial fuel loading of a reactor and possibly fuel rods for one reloading are usually supplied directly by ·the reactor manufacturer. As the major exporters of reactors, therefore, U.S. reactor companies have sold fabricated fuel rods for light water reactors to numerous nations. British and Canadian firms also fabricate natural fuel rods for the reactors they export. There is no technical requirement that fuel rods be provided by the reactor manufacturer, however, and several independent U.S. fuel companies are operating successfully in this field.

In most instances, fabrication plants have been established under licensing arrangements with the fuel organizations of major reactor suppliers, such as General Electric and Westinghouse. By the mid-1980s, the industrialized nations of Western Europe and Japan will almost certainly have enough fuel fabrication plants to supply their domestic needs and contribute to the export market. Six fabrication plants are in operation in the continental European Community states, for example. Several other nations, including Brazil, Argentina, and India, already have either pilot plants or small commercial facilities in operation.

The U.S. companies will continue to offer fuel fabrication services for foreign clients, but the extent of this effort already shows signs of diminishing as independent facilities become available and increasingly compete with U.S. firms in the world market. In 1971, for example, only 30 percent of the uranium enriched in the United States for foreign clients was fabricated into fuel rods; 70

*Fabrication of fuel for nuclear reactors involves a set of mechanical, chemical, and physical processes by which uranium—enriched or natural, depending upon the type of reactor—is manufactured into fuel rods. A fabrication plant capable of supporting a 20,000-megawatt reactor program might cost $25 to 30 million to construct. Economies-of-scale are not as demanding for fabrication as they are for chemical reprocessing or for gaseous-diffusion facilities for uranium enrichment.

percent was shipped as UF_6 and fabricated into fuel elements in foreign facilities.[25] By the early 1980s, it is expected that virtually all fuel fabrication for foreign light water reactors will be done abroad.

Chemical reprocessing is more difficult to master than fuel fabrication. The primary purpose of reprocessing is to recover the plutonium produced by nuclear power plants and the unused uranium. Fuel elements are highly radioactive when taken out of a reactor and must be stored for over three months before reprocessing can begin. Remote-control equipment is used in the various procedures. The fuel elements must be chemically dissolved to separate the plutonium and residual uranium, and it might be as much as one year before plutonium extracted from fuel rods becomes usable. Currently, the recovered plutonium is stockpiled, but it is eventually expected to be used as fuel for light water reactors and, more importantly, breeder reactors.

The problem of disposal of nuclear wastes occurs in this phase of the fuel cycle. No adequate permanent disposal system has been developed for the radioactive waste products that remain after reprocessing is completed. This was not a problem for Western Europe and Japan when all their reprocessing was done in the United States, but it will become a serious and costly problem as new reprocessing plants are built abroad.

With an eye on growing future requirements, the British, West German, and French governments have formed United Reprocessers—a trinational commercial venture whose objective is to ensure proper use of French and British reprocessing plants, to construct a third large facility in West Germany to help meet European demands in the 1980s, and to acquire non-European reprocessing business. In Japan, a sizable reprocessing plant, to supplement a small plant already in operation, is currently under construction with French assistance. As for the less developed countries, India has a small plant, and a few countries, e.g., Argentina, have small pilot plants.

Enrichment

Before it can be used as fuel for light water reactors, U_3O_8 must be enriched to a ^{235}U content of 2 to 5 percent from the approximately 0.7 percent found naturally in uranium.* In the gaseous diffusion technique—the only method currently in commercial use—uranium is enriched by diffusing UF_6 through thousands of

---

*Over 99 percent of natural uranium consists of the non-fissionable isotope ^{238}U.

porous barriers in order to separate out some of the heavier ^{238}U molecules and leave a higher concentration of ^{235}U.

This process is far more difficult technically than the processes associated with other stages of the nuclear fuel cycle. Moreover, an enrichment plant using the gaseous diffusion technique must be extremely large. A diffusion plant with an annual capacity of 5,000 tons, the minimum judged to be commercially viable, could service over 40 large light water reactors and would cost well over $1 billion to construct. Another potentially important enrichment technology—the gas centrifuge method, which is currently under development both in the United States and abroad—will require much smaller plant facilities and a far lower capital investment.

Of greater importance than technical complexities and costs, from the foreign policy standpoint, however, is the fact that access to enrichment technology has serious security implications, because uranium used in nuclear weapons must be enriched (much more highly enriched than is necessary for LWR fuel, but the technology is essentially the same). Thus, the U.S. government, in line with its strong support of non-proliferation policies, has not been willing to disseminate this particular technology abroad or, until recently, to turn it over to the U.S. private sector.[26]

Since the emergence of the commercial nuclear power industry, three gaseous diffusion plants owned and operated by the Atomic Energy Commission have met nearly all of the non-Communist world's enriched uranium requirements.* France and the United Kingdom built small enrichment plants in connection with their weapons programs, but these plants cannot supply domestic commercial needs and expansion apparently is not economically feasible.

When foreign utilities first became interested in nuclear power, the United States, through its Atoms for Peace program, entered into a number of bilateral agreements for sales of nuclear fuel. Contracts made pursuant to these agreements obliged the United States to cover the enrichment requirements of the recipient's nuclear power projects for an operating lifetime of 30 years.** Charges for enrichment services were attractively low. Among the many justifications given for this enriched uranium supply policy was

---

*In 1971, the U.S. government launched a policy of attempting to induce private industry to construct enrichment plants. The response from the private sector as of mid-1974 was not encouraging, and the AEC has indicated that it will provide additional facilities if a private venture does not materialize.[27]

**The agreements also required that the recipient country agree to safeguard its nuclear facilities and refrain from transferring nuclear materials to third countries without U.S. approval.

the fact that foreign reliance on U.S. enrichment services would help to prevent proliferation of nuclear weapons by ensuring that safeguards were placed on U.S.-supplied nuclear facilities and by reducing incentives for construction of enriched uranium facilities abroad. Foreign sales and services of enriched uranium also helped the government to recover the cost of its three large enrichment plants and stimulated foreign interest in U.S. light water reactors.

World demand for enriched uranium could exceed the capacity of the three U.S. plants in the late 1970s or early 1980s. Exactly when additional U.S. capacity will be needed depends on a variety of factors, including the speed of development of the domestic nuclear power industry, the extent to which foreign countries will continue to depend on the United States for enrichment services, and the amount of plutonium recycled as nuclear fuel. On the basis of its current projections of requirements for enriched uranium in the United States and abroad through the 1980s and 1990s (see Appendix Table B-2), the AEC estimates that, at a minimum, an additional facility with an approximate capacity of 8,500 tons per year will have to be built and fully operational by 1984 and that another plant of equivalent size should be in operation one year later. Because of the long lead times involved, construction of the first of these facilities, which will use the gaseous diffusion technology and cost over $2 billion, must be begun by 1975, if this schedule is to be met.

Strong interest in Western Europe and in Japan in indigenous enrichment facilities has been kindled by a desire for nuclear independence and by the AEC's rather precipitous price increases in 1971 and its requirement for advance commitment to long-term contracts for enrichment services. The price increases in 1971 and subsequently, which greatly annoyed foreign customers, were considered necessary to pave the way for putting the industry on a commercially viable basis and to elicit interest from the private sector in building a new facility.

The major effort in Western Europe to develop indigenous enrichment capabilities has been directed at perfecting the gas centrifuge method of isotope separation. This process involves the use of thousands of small high-speed centrifuges operating in tandem to remove some of the ^{238}U from the UF_6, leaving a higher concentration of ^{235}U.

In early 1970, the governments of the United Kingdom, West Germany, and the Netherlands signed a Tripartite Centrifuge Agreement establishing two commercial organizations composed of companies from each nation. The prime contractor, CENTEC, will develop and maintain the centrifuges and overall centrifuge facilities. The more well-known Uranium Enrichment Company, URENCO, will own and operate the plants and assume sales and marketing responsibilities. These organizations, in turn, have formed a centrifuge study association, whose membership includes organizations from other countries interested in making supply contracts with the tripartite group, in entering into joint ownership arrangements with URENCO for enrichment plants, or in acquiring licenses to construct centrifuge plants. Thus far, it appears that this unique commercial venture into centrifuge technology is being pursued successfully. Manufacturing facilities in the three countries are producing centrifuge units, and three small demonstration plants are presently under construction in Britain and the Netherlands and should be in operation in the mid-1970s. Although West Germany is playing a major role in centrifuge manufacturing, it appears that most of URENCO's initial capabilities will be obtained by adding to the demonstration facilities at Almelo in the Netherlands and Capenhurst in the United Kingdom.[28]

The production problems associated with centrifuge technology are formidable. The small pilot facilities (90-ton capacity) in Western Eruope could involve a total of 10,000 centrifuge units, and CENTEC might have to produce approximately 100,000 machines annually during the late 1970s and as many as 500,000 annually during the early 1980s to support the tripartite program. The cost of operating centrifuge plants is still uncertain; they require less initial investment and use less electricity per unit of output than do gaseous diffusion plants, but the reliability of the centrifuges themselves has not been fully tested. It appears likely, however, that such centrifuge facilities will in time become competitive with gaseous diffusion plants.[29]

Assuming that the tripartite program continues to go forward successfully, the three sponsoring nations could in theory absorb all of URENCO's output through the mid-1980s.[30] In actual fact, URENCO will not be able to take over the entire nuclear fuel market of the United Kingdom, West Germany, and the Netherlands. URENCO has therefore already developed its contract terms and has begun negotiations with consumers in other countries, including at least one utility company in the United States.

Despite the progress of the tripartite centrifuge program and the fact that centrifuge facilities—which use relatively little electricity and water—seem to be economically well suited for European conditions, the possibility of utilizing gaseous diffusion technology is still alive in West European circles. The French government has generated some European support for the construction by the French-sponsored Eurodif of a large gaseous diffusion plant in Western Europe based on French technology.

Japan is also actively seeking to assure reliable long-term supplies of enriched uranium for its burgeoning nuclear power program.[31] Although it is not happy with the new U.S. supply policy, Japan has nonetheless committed itself to major purchases of U.S. enriched uranium services and will probably continue to remain the major U.S. foreign client through at least the mid-1980s. At the same time, it is trying to diversify its future enriched uranium sources. Centrifuge development is being seriously undertaken with the goal of establishing a 40-ton pilot plant by 1980 and a commercial facility with an annual capacity of 1500 tons by 1985. Even if they become commercially operational on schedule, these initial centrifuge facilities will not begin to meet Japan's requirements, and Japan can be expected to pursue various joint venture arrangements with foreign companies, most likely in the United States.

The Soviet Union could become a significant source of uranium enrichment services for non-Communist nations. The Soviets have a substantial capacity to produce enriched uranium in the large gaseous diffusion facilities originally built to support their military programs. At a Vienna conference in 1969, a Soviet spokesman announced that the USSR was prepared to enter the enriched uranium supply field, and in 1971, France signed a short-term agreement with the Soviets for the enrichment of 80 tons of uranium. More recently the Soviets have concluded fuel supply contracts with utilities in West Germany and Italy and have expressed an interest in selling enriched uranium to Japan.[32]

In response to the impending emergence of enrichment facilities in Western Europe, the AEC in 1971 announced its willingness to explore with other nations the possibility of providing U.S. technology for a multilateral gaseous diffusion project abroad. Foreign interest was not strong, for a number of reasons, including the AEC's technical security and financial requirements and its insistence that a specific proposal would have to be submitted by a group of nations before more detailed data could be made available. Nevertheless, interest in this proposal could become keen if the URENCO project takes longer than expected, or if the AEC becomes more flexible in its approach. The AEC has also indicated that it would consider foreign participation in a new plant located in the United States. Japan appears to be the most likely potential participant in such a venture.

Finally, in response to the apparent success of centrifuge efforts abroad, the U.S. government has begun to place much higher priority on domestic centrifuge efforts. For many years, U.S.

centrifuge activities have been tightly classified for national security reasons, but the AEC is known to be pursuing a large development program in this field, and a large demonstration plant under construction at Oak Ridge, Tenn., is expected to begin operations in 1976.

Both the centrifuge and the gaseous diffusion technologies are included in the AEC's industrial access program. This will permit U.S. industry to evaluate the commercial potential of centrifuge technology for domestic production of enriched uranium and could help the United States maintain its competitive position in the international enriched uranium market.[33]

U.S. NUCLEAR EXPORTS

Despite U.S. dominance in the commercial nuclear field during the 1950s and 1960s, U.S. nuclear-related exports were rather limited because world demand was modest. From 1959 through 1970, U.S. exports of nuclear power equipment, goods, and services totaled approximately $1.5 billion (see Appendix Table B-3). Between 1971 and 1973, the value of U.S. reactor exports increased by about 40 percent; and in fiscal year 1973 the total value of nuclear exports was almost as large as the cumulative value of nuclear exports for the entire 1959-70 period. While U.S. nuclear-related exports seem to be expanding, this trend must be examined in a wider context before conclusions can be drawn.

Enriched Uranium Exports

As the supplier of enriched uranium for the non-Communist world, the United States, by the end of 1972, had entered into a total of 70 enrichment supply contracts involving over 81,000 tons of separative work units for users in Western Europe and Japan.* (Japan is presently the largest single customer of the United States—with total commitments greater than those of all the European Community nations combined; West Germany also purchases substantial amounts.) Sales of U.S. enriched uranium services (in 1973 dollars) may average $70 million annually until 1975, rise to over $100 million by 1976, exceed $300 million in 1980, and reach $600 million by 1985. While these annual enrichment sales are not dramatic, they represent substantial sums over the long run, since enrichment contracts cover periods up to 30 years. As of early 1973,

*Separative work unit (SWU) is an index that measures the effort expended in a plant to produce a given amount of a certain ^{235}U concentration from an initial quantity of natural uranium feed.

for example, the value of all U.S. enrichment contracts when fully executed amounted to more than $3.0 billion, and by 1980 this figure could rise to almost $30 billion.[34]

The timing, size, and pattern of future non-U.S. enrichment sources cannot be predicted with high confidence at this time, but it now appears that foreign suppliers will be able to capture only a small share of the market in this decade. The AEC now estimates that the United States will provide 80 percent of foreign enriched uranium requirements in non-Communist countries in 1980, and about 60 percent in 1985.[35] Appendix Table B-3 displays the past pattern and likely future course of U.S. enrichment exports.

Reactor Exports

The total value of U.S. nuclear-related exports will be influenced most strongly by plant and equipment sales abroad. The growth of competitive industries in Western Europe and Japan will inevitably decrease the flow of reactor equipment from the United States to industrialized nations.[36] Although complete reactor systems may not be exported to these areas, U.S. steam supply equipment, turbine generators, and other power plant components will not be totally shut out of the overall market of the industrialized nations.

The U.S. nuclear industry is beginning to pay increasing attention to potential markets in those developing nations that possess little industrial potential for domestic manufacture of reactor systems or related fuel-cycle activities.[37] However, U.S. reactor companies will receive vigorous competition from foreign firms for both power plants and nuclear fuel. One important step U.S. firms could take to enhance their competitive position would be to standardize their nuclear power equipment. Until now, each reactor has been of a unique design. Standardization of equipment design should lower costs, and, with utility companies moving away from the practice of buying a complete reactor package, should facilitate sales of individual reactor components by U.S. firms.

While no foreign company has to date won a light water reactor competition against U.S. firms in any area outside Western Europe and Japan, it is only a matter of time before this U.S. export monopoly will be broken. If foreign firms produce efficient small-size reactors before U.S. companies do, the U.S. competitive position in less developed countries could be further weakened. Moreover, some developing nations might find natural uranium reactors from Canada more desirable than light water reactors, and U.S. companies have no capacity at present to produce these reactor types for the commercial market. Finally, there is always the possibility that the USSR will reverse its policy of providing reactors only to East European countries and compete against Western reactor suppliers in the Third World.

The AEC predicts that the United States will capture between zero and 15 percent of LWR and related nuclear equipment sales to supplier nations (Japan, West Germany, France, Italy, and Sweden—the United Kingdom does not presently manufacture light water reactors) through most of the 1970s.[38] This estimate of U.S. sales to supplier nations appears reasonable. However, the AEC's "most likely case" assumption that U.S. firms will supply 100 percent of the remaining world market appears too optimistic. Moreover, in projecting market shares, it would be useful to separate the less developed countries from the European nations in the "non-supplier" category on grounds that, in time, European importers are more likely to buy from European suppliers than from the United States. (Note, for example, that the Benelux countries, which are not supplier countries, are in the Common Market; also, Switzerland seems a natural market of West Germany or France.) On the other hand, the less developed countries appear to offer better prospects as far as U.S. firms are concerned.

Based on a narrower range of estimates than the more optimistic AEC forecasts concerning the share of various foreign markets that might be captured by U.S. firms, Appendix Table B-3 indicates a range of approximately $1.4 to $1.9 billion for light water reactors and related sales for 1980. By 1985, U.S. market shares are expected to decline dramatically. Despite the vast increase in the size of the total foreign market, U.S. reactor exports—mostly to the less developed countries—will probably only be in the range of $1.4 to $2.2 billion.

In sum, combining enriched uranium and reactor estimates for both industrialized and developing areas, the value of U.S. nuclear-related exports in 1980 could be in the range of $1.7 to $2.2 billion. By 1985, the figure would range between $2.0 and $2.8 billion. This would amount to a relatively small proportion of U.S. exports for those years. While nuclear exports clearly are not going to make or break the U.S. balance of payments, the degree of success of U.S. firms in the world market may be one indication of the ability of the United States to compete in world markets for goods and services involving a high degree of capital intensiveness and technological sophistication.

*   *   *   *   *   *

The rapid economic and technological changes described above will both create new problems for U.S. foreign policy and reduce the ability of the United States to solve them unaided. The problems, which in one way or another involve reconciling partially conflicting economic and security interests, will be explored in some detail in the next chapter. The principal conclusion to be drawn from this chapter is that the United States is losing its near monopoly position in two key areas of nuclear industry—the production and export of light water reactors and the provision of uranium enrichment services—but that it will probably retain a strong position in both areas for many years.

Foreign reactor manufacturers, originally launched through licensing arrangements with U.S. companies, are becoming independent as their licenses expire and are starting to compete with their former parent companies for export sales. As a result of this competition, the U.S. share of the reactor market will probably drop dramatically over the next fifteen years. An international reactor market is emerging and is likely to be dominated by some half-dozen large U.S., Japanese, and West European companies.

European and Japanese electric utility companies will have to continue to rely principally on the U.S. Atomic Energy Commission for enriched uranium for another decade before substantial alternatives become available. It appears certain, however, that centrifuge enrichment facilities will be built by the British-Dutch-German combine. At least one gaseous diffusion plant, using either French or U.S. technology, may also be built in Europe. Additional enrichment plants will almost certainly be built in the United States, but whether by private industry or by the AEC, and whether foreign corporations will participate, is uncertain.

As a consequence of these and other possible developments, the United States will lose its ability to control the future shape of the international nuclear energy industry. But it will remain in a position to exert a strong and even leading influence over events.

NOTES FOR CHAPTER 16

1. For a discussion of the economics of nuclear power plants and comparative costs of nuclear and fossil-fuel electrical generation see, Resources for the Future, *Energy Research Needs* (Washington, D.C.: Government Printing Office, 1971); U.S. Senate, Committee on Interior and Insular Affairs, *Energy Research Policy Alternatives*, Hearings, 92nd Cong., 2nd sess. (July 1972); NUS Corporation, *Scope of the International Nuclear Industry* (Rockville, Md., 1972); and Manson Benedit, "Electric Power From Nuclear Fission," *Technology Review*, October-November 1971.

2. Japan Atomic Energy Commission, "Outline of Long-Range Program on Development and Utilization of Atomic Energy" (Tokyo, 1972).

3. Information provided by U.S. Atomic Energy Commission.

4. NUS Corporation, *Scope of the International Nuclear Industry.*

5. *Nuclear News*, September 1972, p. 39.

6. The formation of the Kraftwerk Union is discussed in *Nuclear Industry*, November-December 1968, pp. 90-91.

7. *Nuclear News*, May 1972, p. 51.

8. *Nuclear Industry*, July 1971, pp. 57-58.

9. *Nuclear Industry*, February 1971, p. 62.

10. *Nuclear Industry*, October 1972, p. 39.

11. *Nuclear Industry*, September 1972, p. 13. See also Stockholm International Peace Research Institute (SIPRI), *The Near-Nuclear Countries and the NPT* (New York: Humanities Press, 1972), pp. 16-24.

12. U.S. Atomic Energy Commission, *Nuclear Power: 1973-2000*, p. 5.

13. For a discussion of HTGRs see, U.S. Atomic Energy Commission, *Major Activities in the Atomic Energy Programs, January-December 1971* (Washington, D.C.: Government Printing Office, 1972); and U.S. Congress, Joint Committee on Atomic Energy, *AEC Authorizing Legislation, Fiscal Year 1973*, Hearings, 92nd Cong., 2nd sess. (1972), p. 1236; and Resources for the Future, *Energy Research Needs*, Chapter VI. *Nuclear Industry*, October 1972, and *Nucleonics Week*, January 18, 1973, discuss General Atomic's foreign connections.

14. U.S. Atomic Energy Commission, *Major Activities, 1971*, p. 103.

15. For general information on breeder reactors, the U.S. program, and the pros and cons of rapid breeder developments, see U.S. Atomic Energy Commission, *Background Information on Development of LMFBR* (Washington, D.C., 1972); "The Fast Breeder Reactor: Signs of a Critical Reaction," *Science*, April 28, 1973; and Mason Willrich, *The Global Politics of Nuclear Energy* (New York: Praeger Publishers, 1971), pp. 107-28.

16. *New York Times*, July 17, 1973. Foreign breeder programs are described by Milton Shaw (AEC), Statement of March 14, 1973, before U.S. Congress, Joint Committee on Atomic Energy, *AEC Authorizing Legislation Fiscal Year 1974*, 93rd Cong., 2nd sess. (1973). Further evaluation of foreign breeder efforts can be found in U.S. Senate, *Energy Research Policy Alternatives*, pp. 700-704.

17. See Thomas B. Cochran, *The Liquid Metal Fast Breeder Reactor: An Economic and Environmental Critique* (Baltimore: The Johns Hopkins University Press, 1974), and Resources for the Future, *Energy Research Needs.*

18. U.S. United Nations Association, *Safeguarding the Atom, A Soviet-American Exchange* (New York, 1972). The SIPRI study (see note no. 11) estimates that 130,000 kilograms of plutonium could be available in the world by 1980.

19. Information provided by Nuclear Assurance Corp., Atlanta, Georgia, which took the lead in organizing the WNFM.

20. Organisation for Economic Co-operation and Development, *Uranium Resources, Production and Demand*, joint report by the European

Nuclear Energy Agency and the International Atomic Energy Agency (Paris, 1973), p. 9.

21. U.S. Atomic Energy Commission, Forecasting Branch, Office of Planning and Analysis.

22. Organisation for Economic Co-operation and Development, *Uranium Resources, Production and Demand*, p. 10.

23. Ryukichi Imai, *Nuclear Safeguards*, Adelphi Paper No. 86 (London: International Institute for Strategic Studies, 1972), pp. 26-28.

24. U.S. Atomic Energy Commission, *The Nuclear Industry: 1971*, WASH 1174-71 (Washington, D.C.: Government Printing Office, 1971), pp. 24-26.

25. Ibid., pp. 46-47.

26. The AEC's industrial access program is described in *Major Activities in the Atomic Energy Programs, January-December 1972* (Washington, D.C.: Government Printing Office, 1972), and discussed in *Nuclear Industry*, July 1971, pp. 5-10, and January 1973, pp. 28-31.

27. Press briefing of December 8, 1972, by then AEC Chairman James Schlesinger.

28. *Nuclear News*, January 1973, p. 47; and *Nuclear Engineering International*, April 1973, p. 275.

29. For estimates of comparative costs of centrifuge and gaseous diffusion plants, see *Nucleonics Week*, August 6, 1973.

30. *Nucleonics Week*, March 8, 1973, p. 2.

31. Ibid.

32. See *Nuclear Industry*, April 1971, pp. 39-40 for the Soviet-French arrangement; the *Washington Post*, April 10, 1972, and *Nuclear Engineering International*, May and August 1972, for the Soviet-West German deal; and the *Washington Post*, June 22, 1973, p. 31, for reports of a possible Soviet-Japanese deal.

33. For a summary of the AEC's recent report on centrifuges (released by the Joint Committee on Atomic Energy of the U.S. Congress), see *Nuclear Engineering International*, July 1972, pp. 521-28.

34. U.S. Atomic Energy Commission, *Major Activities, 1971*, put this figure at $1.8 billion. Latest estimate based on information provided by the Commission is over $3.0 billion.

35. The AEC's projections for the U.S. share of the future world enriched uranium market are given in *Nuclear Power: 1973-2000*, pp. 11-13 and 21.

36. U.S. Congress, Joint Committee on Atomic Energy, *Nuclear Energy and Related Problems*, Hearings, 92nd Cong., lst sess. (December 1971), p. 105.

37. Statement of Myron Kratzer, assistant general manager for International Activities (AEC), *Hearings* before the U.S. House of Representatives, Subcommittee on International Cooperation in Science and Space, 92nd Cong., 1st sess. (1971), p. 335.

38. Based on conversations and correspondence with the staff of the Atomic Energy Commission, Division of International Programs.

Chapter Seventeen

International Security Problems Associated with the Growth of Nuclear Power

Since 1945, the world has grappled with the problem of how to harness atomic energy for peaceful purposes and at the same time reduce the security dangers inherent in nuclear technology. Although the number of nations with nuclear weapons has remained constant at five for almost a decade, the spread of civil nuclear capabilities has given several more nations the technical and industrial potential to build atomic weapons relatively easily and rapidly. This fact was dramatically brought home to the world in May 1974 when India exploded its first nuclear device. The Government of India, in announcing this event, declared that it has no intention of developing nuclear weapons. India has, however, consistently refused to sign the Non-Proliferation Treaty, and the explosion of a nuclear device brings India within a short step of full membership in the nuclear club.

Whether other countries will follow India's example, or move directly to acquire a nuclear weapons capability, is not yet clear. An ongoing nuclear weapons program would have to be based on a supply of plutonium derived from an autonomous nuclear fuel cycle or a supply of weapons-grade uranium produced by a national enrichment facility. Besides India which is a special case, Argentina and Canada are the only non-nuclear-weapons nations known to have the technological capability for developing autonomous fuel cycles, and only the Netherlands (in collaboration with the United Kingdom

Note: This chapter is based in large part on research and analysis conducted by Jerome H. Kahan.

and West Germany) appears to be on the verge of acquiring uranium enrichment facilities.* The number of countries with the technological and industrial capability to produce nuclear weapons will undoubtedly grow. The key commercial nuclear capabilities of the countries most advanced in this field are summarized in Table 17-1.

The further proliferation of independent nuclear weapons capabilities would pose a clear danger to international stability. Once a nation produces nuclear weapons—or achieves the capability to produce them fairly quickly—its adversaries and the entire international community become suspicious of its intentions, and serious political tensions can arise. It is even conceivable that the fear of a nation's "going nuclear" could invite military countermeasures by its opponents.[1]

Gaining the technical prerequisites for manufacturing nuclear weapons, of course, does not necessarily imply that a nation intends to embark on a nuclear military program and may not increase the chances that such a decision would be made in a given situation. Decisions to initiate nuclear weapons programs stem from a complex set of considerations, including military and political calculations that vary widely from nation to nation and that go beyond the technical difficulties or costs of producing bombs. Moreover, acquiring a nuclear weapons force involves delivery systems, as well as bombs, and in many circumstances that would be the most expensive and difficult hurdle to overcome.

Deterring further nuclear weapons proliferation depends in large part upon a host of factors totally unrelated to commercial nuclear programs and policies, such as settlements of regional disputes, security guarantees, and progress in controlling the arms race. An international safeguards system, such as the existing one instituted under the auspices of the International Atomic Energy Agency (IAEA), can make it more difficult for a nation to use commercial nuclear facilities to support a nuclear weapons program. But it cannot by itself prevent a government that is determined to acquire nuclear weapons from doing so.

---

*Countries with only reactors and the means of extracting plutonium from used fuel rods could obtain material needed for a minimum weapons capability, and the basic principles of how nuclear weapons operate have been published in the scientific literature. The necessary engineering design, fabrication, and testing would, however, require skilled manpower and other resources beyond the means of many countries that have, or may acquire, reactors.

Table 17–1. Commercial Nuclear Capabilities of Selected Non-Nuclear-Weapons States, January 1974

| | | Nuclear Reactors | | | | Nuclear Fuel Facilities | | |
| | | Operating/Under Construction | | Ordered/Proposed | | | | |
Type	*Type*	*Units*	*Total MWe Capacity*	*Units*	*Total MWe Capacity*	*Enrichment*	*Fabrication*	*Reprocessing*
Argentina	Natural Ur.	1	318	3	1,800		1 pilot plant	1 pilot plant
Australia	Enriched Ur.			1	500			
Belgium	Enriched Ur.	5	1,533				2 commercial facilities	1 small commercial facility (Eurochemic plant)
Brazil	Enriched Ur.			2	1,252		1 small commercial facility	
Canada	Enriched Ur.	1	13				2 small commercial facilities	Commercial facility planned
	Natural Ur.	8	2,535	12	9,000			
West Germany	Enriched Ur.	12	9,147	20	18,550	Commercial centrifuge plant likely (URENCO)	4 commercial facilities	1 small commercial facility; large plant by early 80s
India	Natural Ur.	2	150					
	Enriched Ur.	2	380				2 small commercial facilities	1 small commercial facility; 2 others under construction
Israel	Natural Ur.	4	613	3	640			
	Natural Ur.	1	8.7					
Italy	Enriched Ur.	6	1,205	1	750		2 commercial facilities	Commercial facility under construction
	Natural Ur.	1	210					
Japan	Enriched Ur.	15	8,707	31	29,676	1 small pilot centrifuge facility by 1980; commercial plant by 1985	5 commercial facilities	1 commercial facility under construction

Table 17–1.   Commercial Nuclear Capabilities of Selected Non-Nuclear-Weapons States, January 1974 (continued)

| | | Nuclear Reactors | | | | Nuclear Fuel Facilities | | |
| | | Operating/Under Construction | | Ordered/Proposed | | | | |
	Type	Units	Total MWe Capacity	Units	Total MWe Capacity	Enrichment	Fabrication	Reprocessing
Mexico	Enriched Ur.			2	1,200			
Netherlands	Enriched Ur.	3	952	2	1,200	1 small pilot centrifuge plant; expansion under URENCO likely	1 small commercial facility	
Pakistan	Enriched Ur. Natural Ur.	1	125	1	200			
South Africa	Enriched Ur.	1		1	500	1 small pilot facility (ion-exchange technique)		
South Korea	Enriched Ur.	1	564	3	1,764		Tentative plans to build	Tentative plans to build
Spain	Enriched Ur. Natural Ur.	8 1	6,005 480	5	5,730		1 commercial facility planned	1 pilot plant
Sweden	Enriched Ur.	6	3,277	11	8,580		1 commercial facility	Commercial facility planned
Switzerland	Enriched Ur.	4	1,856	6	5,150			
Taiwan	Enriched Ur. Natural Ur.	2 1	1,208 13	2	1,600			

Note: Includes all power reactors and research reactors of 8MWe (25MW+) or more.

POTENTIAL DIVERSION OF NUCLEAR
MATERIALS FOR WEAPONS PURPOSES

Nuclear weapons can be fabricated using either plutonium or highly enriched uranium, or possibly a combination of both.[2] Between 5 and 10 kilograms of plutonium containing at least 90 percent ^{239}PU and less than 10 percent ^{240}PU are normally used in manufacturing a fission weapon with a yield of approximately 20 kilotons. For a weapon using enriched uranium, between 20 and 30 kilograms of highly enriched uranium are required—that is, uranium with a ^{235}U content of 90 percent in contrast with the 2 to 5 percent normally used to fuel light water reactors. Bombs fabricated from highly enriched uranium are based on a simpler concept than plutonium weapons, and enriched uranium has the additional advantage of being easier and safer to handle than the highly toxic, chemically unstable plutonium.

Plutonium
A relatively small light water reactor (100 megawatts) typically generates 20 to 30 kilograms of plutonium annually—enough to produce 3 or 4 weapons—and a larger, more efficient 1000-megawatt reactor will generate 200 to 300 kilograms of plutonium each year. Natural uranium reactors produce approximately twice as much plutonium per megawatt of output as do light water reactors, and for this reason, in addition to bypassing the need for enriched uranium fuel, these systems are considered to be more suitable for military programs. Moreover, fuel rods can be replaced in natural uranium reactors without shutting down the reactor, thus permitting easier and less visible operations to produce weapons-grade plutonium than do light water reactors. However, the expected availability of diverse enriched uranium supplies within the next decade, the widespread construction of light water reactors throughout the world, and the anticipated introduction of plutonium as fuel for light water reactors suggest that these standard commercial reactors may pose at least as great a proliferation danger in the future as natural uranium reactors. In addition, breeder reactors, if they become commercially important, will produce large quantities of mixed-grade plutonium, portions of which may be relatively uncontaminated with ^{240}PU,* and will increase the problem of applying effective international safeguards against plutonium diversion.

*Plutonium-240 is not a fissionable material and therefore is useless, or worse. The percentage of ^{240}PU in a plutonium mixture indicates the extent of contamination.

With the use of plutonium as part of the fuel mix for light water reactors and breeders, plutonium will be found in fuel fabrication plants as well as reprocessing facilities, and increased quantities of plutonium in various forms will be in transit within a given nation and between nations. Plutonium cannot be used for weapons purposes until spent fuel is reprocessed—a procedure which takes at least one year and which is now performed in only a few countries. Many nations can soon be expected to obtain their own reprocessing plants, however, and even if foreign reprocessing services are used, plutonium is generally returned to its originating country. Even small "pilot" fabrication and reprocessing plants, which might not be commercially sound, could provide sufficient plutonium to support weapons programs.

Apart from an overt program, nations could follow two routes to plutonium-based weapons—plutonium diversion from commercial facilities or clandestine construction of separate facilities for military programs. Diversion of plutonium from a reprocessing plant appears most likely since the material could be used directly for weapons purposes. Diversion of irradiated fuel rods from commercial nuclear reactors would be technically feasible, but this might cause noticeable changes in normal operations and would adversely affect the functioning of the power plant, particularly if fuel rods were removed more frequently than usual in an attempt to generate higher weapons-grade plutonium. In addition, reprocessing would still be required before the diverted plutonium could be used for a military project. International safeguards on a nation's commercial facilities could help deter and possibly detect such diversions, although small diversions might still easily occur.

Construction of clandestine facilities using commercially available technology provides another approach to a covert weapons program. In principle, a small reprocessing plant could be built in a remote area or within a large industrial complex and used to process fuel diverted from commercial reactors. A clandestine reactor could also be constructed, but fuel would have to be obtained through diversion from fabrication plants or construction of a clandestine fuel plant. For light water reactors, acquiring enriched uranium fuel could pose special problems, even if natural uranium could be freely obtained. If the fuel problem was solved, it would still be difficult to prevent detection of a clandestine reactor facility by the international community, particularly the intelligence systems of the United States, the USSR, and the United Kingdom.

Enriched Uranium

The security implications of enriched uranium use have generally not been viewed with the same seriousness as the problem of plutonium use since the slightly enriched uranium widely available as fuel for light water reactors is itself useless for weapons fabrication without further enrichment. While enrichment technology was still not available to non-nuclear-weapons states, this route to weaponry was not considered likely. It is now virtually certain, however, that enrichment facilities, particularly centrifuge plants, will eventually be located in non-nuclear-weapons states. If high temperature gas reactors take hold, moreover, there will be a need for highly enriched uranium fuel for commercial purposes; fuel fabrication facilities could then become potential sources of diversion, along with enrichment plants.

Gaseous diffusion enrichment facilities currently being planned could be designed to produce only slightly enriched uranium; this would probably be done for economic as well as safeguards reasons in any diffusion plants built abroad on an open, multilateral basis. Such facilities could be modified by adding "upper stages" to produce weapons-grade uranium, but by the nature of the process neither covert installation of upper stages nor diversion of enriched uranium from an operating gaseous diffusion facility would offer viable evasion schemes.[3] Even less feasible would be the clandestine construction of an entire gaseous diffusion facility because of the enormous size and electric power requirements of such an installation.

Centrifuge enrichment facilities, on the other hand, could be constructed secretly without much difficulty. Because of their small size and low electric power requirements, such facilities could be built in remote sites or subdivided into separate sections and hidden within industrial complexes. A small 100-ton centrifuge facility, for example, could produce 500 kilograms of highly enriched uranium annually—enough material for approximately 20 bombs.

Centrifuges can operate more efficiently if slightly enriched uranium rather than natural uranium is provided as an input, since 80 percent of the separative work would already have been performed. Thus, slightly enriched uranium, widely available as the basic fuel for commercial light water reactors, will have growing security significance.

In short, although concerns over the security implications of plutonium have been dominant and should remain so, the likely spread of centrifuge technology will soon make the enriched uranium route, as well as the plutonium route, to weapons programs possible. Nations will have increased opportunities for diversion or clandestine operations at virtually any stage of the fuel cycle—aiming for enriched uranium before it is utilized in reactors or for plutonium as a reactor by-product. Apart from easier handling and bomb design, nations might actually prefer enriched uranium for a weapons option because the difficult and time-consuming reprocessing step required for plutonium could be bypassed.

Other Security Threats

As commercial nuclear power programs expand, subnational groups—terrorists, organized criminals, and others—might try to steal or hijack nuclear material. They might hope to further their objectives by threatening either a nuclear explosion or the release of radioactive material in populated areas. Some of the same means whereby governments might clandestinely produce or divert weapons-grade material might also be used by subnational groups.[4] If a nuclear theft or blackmail effort involved direct threats against a government by a revolutionary group based in another country, war might result. If a subnational group detonated a nuclear device at sea or underground, war might break out because of misunderstanding of the source of the explosion.

Other possibilities also deserve attention. A government party to the Non-Proliferation Treaty might divert nuclear material from a domestic commercial facility and then claim that the missing material had been stolen. Revolutionary groups or military cliques in small countries might hire criminals to commit nuclear thefts in the major nuclear countries. The appearance of an international black market for nuclear material is a real possibility during the next decade.[5] Finally, in the event theft occurs in the course of international transport or if nuclear thieves seek asylum abroad, problems of coordinating governmental policies, protective measures, and security and intelligence forces would arise.

THE INTERNATIONAL SAFEGUARDS SYSTEM

The concept of applying safeguards to ensure that nuclear material exported for peaceful applications will not be used for weapons purposes had its origins in the U.S. Atoms for Peace Program of the mid-1950s. Consistent with the provisions of the Atomic Energy Act of 1954, safeguards were included in bilateral "Agreements for Cooperation" to assist the United States in verifying that nuclear materials, equipment, and facilities exported by the AEC or private industry would be used for peaceful purposes. These bilateral arrangements called for the United States to assist recipient countries in instituting national safeguards systems and permitted the United

States to examine records, review designs of facilities, and conduct on-site inspections to obtain assurance of peaceful uses.[6]

In 1962, a decision was made to transfer to the International Atomic Energy Agency the responsibility for safeguarding items provided under U.S. bilateral agreements, and the first "trilateral" agreement went into effect with Japan one year later.* This policy of seeking to utilize safeguards administered by the IAEA—and to encourage all nations to rely on international safeguards—was motivated by a number of considerations, including the greater effectiveness and credibility of international systems and the benefits to be derived from developing uniform standards and avoiding the discrimination and degradation of standards that could result from competing bilateral arrangements. At the end of 1972, twenty-five out of thirty U.S. bilateral agreements outside Euratom had been superseded by trilateral agreements, and all British and Canadian bilateral agreements (the only others in existence) with their customers had also been superseded by trilateral arrangements with the IAEA.[7]

The development of international safeguards was closely tied to the evolution of the IAEA—an agency established in 1957 with the goal of furthering the peaceful applications of nuclear energy and materials throughout the world. The agency plays a significant role as a cooperative forum for information exchange on all facets of nuclear energy, as a source of assistance in the nuclear field to the developing nations, and most importantly, as the principal authority for the establishment and application of international safeguards. The IAEA's membership has now grown to over 100 nations; twenty-five states are represented on the Board of Governors. With the technical and financial support of the United States, the USSR, and other nations, the IAEA instituted an initial international safeguards system in 1961, which was revised in 1965 and strengthened in 1968.

Safeguards and the NPT

With the negotiation of the Non-Proliferation Treaty in 1968 and the entry into force of that agreement in March 1970, the IAEA's safeguards responsibilities were significantly increased. Within eighteen months of ratification of the NPT, all non-nuclear-

*The West European countries that had formed the European Atomic Energy Community (Euratom) in 1957 were by and large reluctant to accept international safeguards. The United States entered into a U.S.-Euratom agreement, which absorbed the earlier bilateral agreeements between individual Euratom member states and the United States.

weapons parties are required under Article III of the treaty to negotiate a safeguards agreement with IAEA for the purpose of verifying that they are observing their obligation to refrain from manufacturing or acquiring nuclear weapons or other nuclear explosive devices; Article III also commits all parties (both nuclear weapons states and non-nuclear-weapons states) to refrain from exporting fissionable material or nuclear equipment to non-nuclear-weapons countries unless IAEA safeguards are applied. The NPT does not contain any provisions requiring inspection for safeguards purposes of the five nuclear weapons states, nor does it subject their imports to safeguards. Article IV guarantees to all NPT members the right to the fullest possible exchange of equipment, material, and technological information for peaceful uses and obliges the technologically advanced countries to promote the development of nuclear energy in non-nuclear-weapons states.

Although the Non-Proliferation Treaty bans the acquisition by non-nuclear-weapons states of any nuclear explosive device, Article V of the NPT provides that such states should be able to utilize nuclear explosives for peaceful purposes (such as for extraction of underground natural gas) under the auspices of an appropriate international body. In practice, this provision suggests that the industrial use of nuclear explosives would have to be conducted under IAEA safeguards procedures using explosive devices provided by the nuclear weapons states.*

Soon after the NPT went into effect, a special Safeguards Committee, with representatives from fifty member states, was established to formulate a revised safeguards system consistent with the IAEA's responsibilities under the treaty. In designing safeguard procedures, the committee had to resolve a number of difficult problems, including the need to strike a balance between the effectiveness of safeguards and their acceptability to the nations involved. As a result of this effort, in May 1971 the agency issued a "model" safeguards agreement for conclusion between the IAEA and parties to the NPT.[8]

The IAEA safeguards system developed prior to the NPT called for relatively complete and effective procedures, but they applied only to specific materials and facilities provided under particular agreements or projects. Many non-nuclear-weapons states

---

*The NPT has not been interpreted as permitting non-nuclear-weapons states to acquire nuclear explosives even for peaceful purposes. India's effort to distinguish between peaceful and military nuclear explosive capabilities would, if accepted by adherents to the NPT, subvert a central feature of the treaty.

feared that the IAEA procedures under the NPT would be excessively intrusive and burdensome when applied to an entire nuclear industry. Accordingly, the NPT safeguards procedures depart from the original IAEA safeguards concept by placing primary responsibility on national—or in the case of Euratom, multinational—systems. The agency then verifies compliance by checking the records and reports of the national systems and by conducting independent measurements and inspections at strategic points in the nuclear fuel cycle.

By February 1974, thirty-two states that had either signed or ratified the treaty had concluded safeguards agreements with the IAEA and twenty-seven non-nuclear states were negotiating such agreements. (See Table 17-2.) Of prime significance was the safeguards agreement between the Euratom nations (represented by the Commission of the European Communities) and the IAEA, which was concluded in April 1973.[9] This lengthy and complex negotiation process—which required a change in Euratom's longstanding hostility toward externally imposed safeguards—had not only prevented West Germany, Italy, Belgium, and the Netherlands from ratifying the NPT, but had also led Japan and other nations to take the position that in order to avoid discriminatory treatment they would delay initiating safeguards discussions with the IAEA until Euratom negotiations were completed.[10]

Two stumbling blocks to NPT adherence have always been concern over the intrusive effects of safeguards on nuclear plant operations and fears that nations free of external safeguards would gain technological as well as commercial advantages. To demonstrate the feasibility of applying safeguards to commercial nuclear facilities and to help blunt charges of discrimination, the United States has accepted IAEA safeguards on a number of research facilities, an operating power reactor, and a reprocessing plant. Of more significance in persuading non-nuclear-weapons states to adhere to the NPT was the offer made by President Johnson in 1967 to "permit the IAEA to apply its safeguards to all nuclear activities in the United States . . . excluding only those with direct national security significance. . . ."[11] (The United Kingdom has made a similar offer; the USSR has rejected this policy as being irrelevant to nonproliferation.)

The Nixon administration reiterated the 1967 U.S. proproposal but practical problems must be solved before the offer can be acted upon.[12] Perhaps the most crucial difficulty is the fact that all-inclusive applications of IAEA safeguards to U.S. commercial nuclear facilities "would mean spending a grossly disproportionate

Table 17-2. Status of Non-Proliferation Treaty and Safeguards Agreements, July 1974

	Non-Proliferation Treaty[a]			NPT Safeguards Agreement		Other Safeguards		Existence of Unsafeguarded Nuclear Material/Facilities[c]
	Signed and Ratified	Signed	Comments	Concluded	Negotiating	IAEA[b]	Euratom	
Argentina	x		Unlikely to sign[d]			x (3)		
Australia		x	—	x		x (2)		
Belgium		x	Expected to ratify[e]				x	
Brazil			Unlikely to sign[d]	x		x (3)		
Canada	x	x		x		x		
Egypt		x	Not likely to ratify[f]					
Fed. Rep. Germany		x	Expected to ratify[e]			x (2)	x	Large research reactor
India			Very unlikely to sign[g]					Large research reactor
Israel			Very unlikely to sign[g]			x		
Italy		x	Expected to ratify[e]	x			x	
Japan		x	Likely to ratify		x	x (5)		
Mexico	x	x	—	x		x		
Netherlands		x	Expected to ratify[e]	x			x	
Pakistan			Unlikely to sign			x (2)		
South Korea		x	Likely to ratify[h]			x		
South Africa		x	Uncertain					Possible pilot enrichment plant
Spain			Uncertain			x		
Sweden	x	x	—		x	x		
Switzerland		x	Likely to ratify[i]	x		x		
Taiwan	x	x	—	x		x		
Sub-total	5	13		8	2	14 (17)	4	
Others	75	84	Includes most of Eastern Europe[j]	24	25	19 (20)	1	
TOTAL	80	97		32	27	33 (37)	5	

Sources: International Atomic Energy Agency, *Annual Report* (Vienna, 1972); Mason Willrich, ed., *International Safeguards and Nuclear Industry* (Baltimore: Johns Hopkins University Press, 1973), Appendix D; and Stockholm International Peace Research Institute, *The Near-Nuclear Nations and the NPT* (New York: Praeger Publishers, Inc., 1972).

a. Non-nuclear weapons states only; the United States, the United Kingdom, and the USSR are also parties.

b. IAEA trilaterals, project agreements, or unilateral submission. These are "limited" safeguards, not legally applied to *all* nuclear facilities/material as in the case of NPT safeguards. Number of agreements shown in parentheses.

c. Nuclear material/facilities without external safeguards (International Atomic Energy Agency, Euratom, or bilateral). Excludes both research reactors that have no real military potential and minor quantities of material. Includes fuel facilities and stocks of enriched uranium or plutonium, but excludes natural uranium.

d. Strong commercial/political opposition.

e. As a consequence of IAEA–Euratom Agreement.

f. Tied to Israel's position.

g. Strong security opposition.

h. Should follow ratification by European Community and Japan.

i. Delay stems from legalistic, not substantive concerns.

j. Except Albania.

share of the agency's money and effort for a purpose that has nothing to do with the NPT's objectives."[13] The IAEA recognizes that comprehensive inspection of U.S. facilities could indeed cripple the agency's ability to apply safeguards in non-nuclear-weapons states, but some nations have at times put pressure on the United States to accept total safeguards on commercial facilities before they would agree to drawing up a safeguards agreement with the IAEA.

In seeking a practical compromise, the AEC, in consultation with U.S. industry, considered a number of alternatives, including such schemes as safeguarding those U.S. facilities involved in putting a specific supply of nuclear material through a complete fuel cycle. Another possible approach would be to concentrate IAEA's safeguards upon a rotating set of plants that utilize advanced technology or are engaged in international exports.[14] Successful negotiations of a safeguards agreement between the United States and the IAEA would put pressure on other nations to negotiate IAEA agreements of their own, or at least keep other nations from using the absence of such an agreement as an excuse for not ratifying the NPT.

In sum, it seems clear that the international safeguards system tied to the NPT can provide a framework for managing the continued expansion of world nuclear power with reduced risk. Within a few years, many national nuclear power programs will be wholly within the international system, and a major portion of the international export business will take place within a safeguarded environment. By early 1974, three nuclear weapons states—the United States, the United Kingdom, and the USSR—were parties to the Non-Proliferation Treaty, as were over eighty non-nuclear-weapons states. France and China have refused to join in the treaty. As shown in Table 17-2, among the non-nuclear-weapons states expected to have ratified the treaty and concluded an agreement with the IAEA within five years are the European Community nations, Sweden, Switzerland, Canada, Australia, Japan, Mexico, and all European Communist nations, except Albania. This group encompasses many major nuclear energy producers, suppliers, and users, as well as nations with modest commercial nuclear resources who nonetheless pose potential proliferation problems.

A number of crucial non-nuclear-weapons states (India, Pakistan, Israel, Spain, Portugal, Egypt, South Africa, Brazil, and Argentina) have thus far refused to join the NPT, and it is uncertain when or whether these countries will participate. In most of these countries, however, IAEA safeguards exist on most (but not all) peaceful nuclear facilities as the result of trilateral agreements covering all U.S.-origin equipment or material, similar trilaterals on materials supplied by Canada or the United Kingdom, or safeguards derived from IAEA project agreements. In addition, the Latin American Nuclear Free Zone treaty obligates parties to accept IAEA safeguards. (Argentina and Brazil have signed but have not yet ratified this treaty.) In fact, at the end of 1973, all operating commercial reactors and nuclear fuel facilities in non-nuclear-weapons states were under some form of external safeguards. Since treaty parties have an obligation not to export material or equipment without IAEA safeguards, the coverage of international safeguards will gradually expand with the growth of the nuclear energy industry in both NPT and non-NPT countries.

Megawatt-range research reactors, which are thus far free from external safeguards, currently exist in only two countries—India and Israel. The importance of this gap in the international safeguards system was highlighted by the recent Indian nuclear explosion, which was apparently made possible by diversion of plutonium from the unsafeguarded research reactor.[15]

The international safeguards system has other imperfections. Important nuclear-supply nations—notably France as a source of power plants and South Africa as a source of natural uranium—remain outside the system and could therefore legally transfer unsafeguarded equipment or material to non-nuclear-weapons states that are not parties to the NPT. With technological assistance from industrialized nations as permitted under the NPT, furthermore, additional nations could establish autonomous commercial nuclear energy capabilities and develop separate fuel cycles that would fall outside any external safeguards. By the mid-1980s, unsafeguarded reactors could appear in several countries that by then may be able to produce Canadian-type heavy water reactors. Unsafeguarded LWRs could appear in nations that refused to accede to the NPT if France decided to supply them. Fuel fabrication and reprocessing plants could, in a number of nations, remain free of external safeguards. If South Africa succeeds in its efforts to develop a new uranium enrichment process and refuses to join the NPT, an unsafeguarded source of LWR fuel could become available on the world market. Finally, India could in time become a supplier of nuclear reactors, technology, and even nuclear explosive devices.

Scope and Character of IAEA Safeguards

The IAEA's newly developed NPT safeguards system has the stated purpose of "timely detection of diversion of significant quantities of nuclear materials from peaceful nuclear activity to the manufacture of nuclear weapons or other nuclear explosive devices or for purposes unknown, and deterrence of such diversion by the risk of early detection."[16] This carefully constructed statement of objectives reflects the recognition that realistic verification systems cannot be expected to prevent violations with perfect reliability. By exposing possible diversions and giving early warning that a weapons program is being undertaken in violation of NPT obligations, however, the IAEA's safeguards system can raise the costs of violations and dissuade government leaders from undertaking such a course. The system can also help provide assurance to the international community that a nation is in fact adhering to its NPT obligations and not diverting material to a weapons program to develop nuclear weapons or explosive devices.[17]

The NPT safeguards agreement is based on three basic principles.[18] First, the NPT signatory state must establish a national safeguards system that meets IAEA standards and agree to provide the IAEA with plant design data, material inventories, and records and reports as specified in the agreement. In carrying out its verification activities, the IAEA is then expected to make "full use" of the state's accounting and control system, but it retains the right "to conduct independent measurements and observations" in accordance with agreed procedures. Within these limits, the actual extent and frequency of IAEA inspections are determined by the effectiveness of the national safeguards system in question—that is, the better the national system the less stringent the IAEA safeguards. The detailed application of agency safeguards for particular facilities is indicated in subsidiary agreements worked out between the IAEA and the state.

Second, the IAEA seeks to implement its safeguards without hampering technical nuclear developments in a nation, preventing the efficient economic application of nuclear energy, or unduly interfering with the operation of commercial facilities. Strict precautions, such as maintaining confidential files, are taken to protect industrial secrets. Prior notice of IAEA inspection is generally given, and national representatives are allowed to accompany agency officials.

Third, to reduce the degree of intrusiveness, the IAEA tries to concentrate on facilities where weapons-diversion risks are greatest, such as those utilizing or processing plutonium or highly enriched uranium, and conducts less-intensive inspections where natural or slightly enriched uranium is involved. In addition, under normal conditions, agency inspectors do not have unrestricted access to facility areas, but are supposed to focus their attention on certain "strategic points" where key measurements can be made to verify the amount and composition of safeguarded nuclear material present.

Principally due to the objections of South Africa and other natural uranium producers, NPT safeguards are not applied to unconcentrated uranium ore or mining operations. However, starting with yellowcake, safeguards must be applied to all facilities involved in other steps of the nuclear fuel cycle—conversion, enrichment, fuel fabrication, power generation, reprocessing, and storage. Safeguards are terminated when the IAEA has determined that the nuclear material has been either consumed or sufficiently diluted (e.g., depleted uranium) to pose no military or explosive threat.

Aside from small quantities of nuclear material for non-explosive military purposes and for medical, industrial, or research uses that are excluded from safeguards, advance notice and reporting requirements must be met when nuclear material is exported or imported. The IAEA has the right to verify the declared quantity and composition of such nuclear materials through direct inspection when the material is packed or unpacked. Movement of material within a country is physically safeguarded by national systems; IAEA verification is based upon shipping and receiving reports, which are checked against net inventory assessments from all facilities in a nation's safeguarded fuel cycle.

Fundamental to nuclear safeguards is the concept of material balance accountancy—an approach aimed at determining the flow and inventory of nuclear materials at critical locations or processing stages.[19] This is a standard industrial technique used by nuclear industries and is presumably part of any domestic safeguards system. Nevertheless, the NPT safeguards agreement to be negotiated between the IAEA and a particular state calls for the establishment of specific Material Balance Areas (MBAs) for each safeguarded facility and for the designation of strategic points at the periphery of those areas where nuclear material measurements can be made. By organizing and evaluating information provided by the states on the amount, composition, and location of material in individual facilities, and by examining inventory, operating, and transfer records, the IAEA can develop a picture of a nation's entire fuel cycle. This procedure enables the agency to assess the accuracy and credibility of national reports by looking for consistency in the overall nuclear program, and it increases the chances that forged or altered records will be spotted or diversions of materials detected.

The significance, methodology, and effectiveness of safe-guards differ throughout the various phases of the nuclear fuel cycle, and the frequency and intensity of inspections set by the IAEA take this into account. The NPT safeguards committee of the IAEA developed mathematical formulas to determine the maximum routine inspection effort—expressed in number of man-days per year—for various categories of commercial nuclear facilities. The maximum man-days of permitted inspection for each category is set on the basis of these formulas. Within a given category, however, the agency can concentrate its inspection effort on any facility about which suspicions have arisen.

Uranium enrichment facilities, which can produce highly enriched as well as reactor-grade material, have not been safeguarded because all of the existing enrichment plants are located in nuclear weapons states, whose nuclear facilities are not subject to safeguards under the NPT. However, when such plants are built in non-nuclear-weapons states, safeguards could be effectively applied on those using gaseous diffusion technology. Gaseous diffusion plants designed to produce slightly enriched uranium could not easily be modified to produce weapons-grade material without detection. Furthermore, the serial nature of diffusion plant operations makes input-output measurements an effective means of ensuring that diversion has not occurred or that highly enriched material has not been produced. On the other hand, as will be discusssed in more detail in Chapter 20, there are serious questions regarding the feasibility of imposing adequate safeguards on enrichment facilities using centrifuges.

Finally, the jurisdictional limits of the IAEA should be recognized. To begin with, because of intrinsic verification uncertainties and the possibility of legitimate explanations for missing material, the IAEA does not attempt to draw clear conclusions regarding compliance with the NPT at the first sign of a possible diversion. If diversion is suspected, the IAEA procedure is first to ask the nation in question to provide assurance of non-diversion; specific information may also be requested, and special inspections may be made as permitted in the agreements. If uncertainty cannot be resolved through such procedures—or if the state refuses to cooperate with the IAEA in resolving the uncertainties—then the IAEA Board of Governors can file a report of noncompliance with the United Nations and institute sanctions as specified in the IAEA statutes. The NPT itself contains no specific sanctions for noncompliance. Possible sanctions could include suspension or termination of nuclear material transfers by the IAEA and its members, but the agency is not empowered to recover nuclear material or act in any way as an international security force. Thus, the IAEA alone could not prevent a nation from proceeding to manufacture nuclear weapons with diverted material already in hand.

In addition, the agency is responsible only for safeguards on facilities declared to be conducting peaceful nuclear operations. The IAEA is not empowered to act as an intelligence organization for purposes of discovering clandestine weapons plants within the borders of a particular country, nor would it be feasible for the IAEA to undertake intelligence work if authorized to do so. Furthermore, the IAEA cannot assume police or custodial powers. This constraint not only precludes the agency from recovering diverted material, but places responsibility for the physical security of nuclear materials and protection against domestic theft with the national safeguards systems.

One glaring gap in the international safeguards system is in the area of nuclear information and technology exchanges. The NPT does not require safeguards of any kind in this vital area. In 1972, the United States led the way when the AEC issued new regulations requiring prior AEC approval for firms engaging directly or indirectly in foreign activities involving fuel reprocessing, uranium enrichment, and heavy water—areas that can be considered critical to a nuclear weapons program and for which the technology is not yet widely available outside the United States. The criteria that govern the AEC's consideration of export applications for unclassified nuclear information in these areas include such considerations as whether the recipient nation is a party to the NPT or will accept IAEA safeguards on the project in question, and the availability of comparable technical information from other sources. The new export regulations place the United States in a position of leadership in attempting to deal with the problem of potential weapons proliferation and represent a decision to give greater weight to national security considerations.

*   *   *   *   *

The new international safeguards system under the NPT is just getting under way. While many countries have negotiated agreements with the IAEA, a good many others have yet to do so. Two nuclear weapons states—France and China—have made plain their unwillingness to sign, as have India and a number of key countries with large nuclear power programs.

In addition to concern about the scope of the emerging safeguards system is the question of depth, or the degree of confidence provided by the system. Under the NPT, each party's national safeguards system is the foundation for the IAEA's system for that country. The stronger the national system the less stringent the inspection levels and other procedures actually conducted by the IAEA, although the agency will supplement national efforts to some degree in all cases. Just how effective and confidence inspiring are the various national safeguards systems? Available evidence suggests that the U.S. system—generally regarded as the toughest of any of the large industrialized nations—needs a good deal of strengthening.[20]

Other important questions concerning the future of nuclear safeguards also remain unanswered. Will member nations be willing to finance the IAEA adequately so that it can develop and implement the very complex inspection procedures, reporting systems, and technical devices needed to put teeth into the system? Will nuclear industries, which tend to look upon safeguards as a nuisance and as an avenue for theft of commercial secrets, be able to weaken the system before it is fully launched? Of much greater importance, will the system be able to keep up with the rapid increases in nuclear power facilities expected over the next ten to fifteen years, and to cope with the extremely difficult security implications of several emerging technologies? Policy options concerning these and related questions are discussed in Chapter 20.

NOTES FOR CHAPTER 17

1. George H. Quester, "Some Conceptual Problems in Nuclear Proliferation," *American Political Science Review,* June 1972, pp. 490-97.

2. This discussion draws heavily on the chapter by Victor Gilinsky, "Military Potential of Civil Nuclear Power," in *Civil Nuclear Power and International Security,* ed. Mason Willrich (New York: Praeger Publishers, Inc., 1970), pp. 14-41.

3. See remarks by Congressman Craig Hosmer (Joint Committee on Atomic Energy), in *Preventing Nuclear Theft: Guidelines for Industry and Government,* ed. Robert B. Leachman and Phillip Althoff (New York: Praeger Publishers, Inc., 1972), pp. 11-16.

4. This set of problems is explored in detail in another energy study sponsored by the Ford Foundation: Mason Willrich and Theodore B. Taylor, *Nuclear Theft: Risks and Safeguards* (Cambridge, Mass.: Ballinger Publishing Company, 1974). See also, James E. Lorett, "Who are the Enemy," in *Preventing Nuclear Theft,* ed. Leachman and Althoff, pp. 207-18; Theodore B. Taylor, "Diversion by Non-Governmental Organization," in *International Safeguards and Nuclear Industry,* ed. Mason Willrich (Baltimore: Johns Hopkins

University Press, 1973), pp. 176-200; and Ralph E. Lapp, "The Ultimate Blackmail," *New York Times*, Magazine Section, February 4, 1973.

    5. See remarks on the danger of a plutonium black market by Congressman Craig Hosmer, in *Preventing Nuclear Theft*, ed. Leachman and Althoff, pp. 11-16; and speech by the Congressman before the annual meeting of the Nuclear Materials Management Association, Gatlinburg, Tenn., March 25, 1970.

    6. For background on U.S. bilateral safeguards and information on the structure of typical Agreements for Cooperation, see U.S. Congress, Joint Committee on Atomic Energy, *International Agreements for Cooperation: 1967-1968*, Hearings before Subcommittee on Agreements for Cooperation, 90th Cong., (March 1967 and June 1968).

    7. The International Atomic Energy Agency's *Annual Report* (Vienna, 1972), p. 47, lists trilateral and other IAEA "Safeguards in Force" as of June 1972.

    8. The NPT Safeguards Model Agreement (INFCIRC/153) is reprinted in Willrich, ed., *International Safeguards and Nuclear Industry*, Appendix B. See also, Mason Willrich, *The Non-Proliferation Treaty: Framework for Control* (Charlottesville, Va.: Michie Company, 1969), pp. 67-98, for a detailed discussion of the various NPT restrictions.

    9. *Washington Post*, April 6, 1973, p. 2.

    10. Glenn T. Seaborg, *Man and Atom* (New York: E. P. Dutton & Co., Inc., 1971), pp. 320-21, et. seq.

    11. Ibid., p. 309.

    12. *Nuclear Industry*, May 1972, p. 43.

    13. Ibid.

    14. Leachman and Althoff, eds., *Preventing Nuclear Theft*, pp. 164-66.

    15. *New York Times*, August 3, 1974, p. 2.

    16. NPT Safeguards Agreement, para. 28, as reproduced in Willrich, ed., *International Safeguards and Nuclear Industry*, Appendix B.

    17. For a general treatment of the purposes and principles of international safeguards, see John G. Palfrey, "Assurance of International Safeguards," in *Nuclear Proliferation: Prospects for Control*, ed. Bennett Boskey and Mason Willrich (New York: Dunellen Publishing Co., Inc., 1970), pp. 81-86.

    18. The NPT Safeguards Agreement is described and analyzed in considerable detail by Paul C. Szasz, "International Atomic Energy Safeguards," in *International Safeguards and Nuclear Industry*, ed. Willrich, pp. 73-141.

    19. For discussions of MBAs and other technical aspects of safeguards see Ryukichi Imai, *Nuclear Safeguards*, Adelphi Paper No. 86 (London: International Institute for Strategic Studies, 1972); W. Gmelin, *et al*, "A Technical Basis for Safeguards," in *Peaceful Uses of Atomic Energy*, 15 vols., Proceedings of the Fourth International Conference, United Nations and the International Atomic Energy Agency (New York, 1972), Vol. 9, pp. 487-510; and Herbert Scoville, Jr., "Technical Capabilities of Safeguards," in *Nuclear Proliferation: Prospects for Control*, ed. Boskey and Willrich, pp. 52-63.

20. Edwin M. Kinderman, "National Safeguards," in *International Safeguards and Nuclear Industry,* ed. Willrich, pp. 142-50; Willrich and Taylor, *Nuclear Theft: Risks and Safeguards;* and U.S. General Accounting Office, *Improvements Needed in the Program for the Protection of Special Nuclear Material* (Washington, D.C., 1973).

Part Seven

U.S. Energy Policy

Chapter Eighteen

Major Uncertainties

In a few short years, energy has changed from a minor to a major concern of U.S. foreign policy. The simple explanation for this development is of course the recent shift of the United States from a position of near self-sufficiency in energy to one of substantial dependence on imported oil. More fundamentally, however, the emergence of energy as a major foreign policy problem must be explained in terms of the unexpected appearance of serious discontinuities and uncertainties in what had long been regarded as a smooth and relatively predictable historical process.

During the present century, oil has gradually replaced coal as the world's major source of energy. Until recently, it was generally assumed that oil would in turn gradually lose its primacy to nuclear power. In the interim, before nuclear power took over, it was further assumed that constantly increasing amounts of cheap oil would be readily forthcoming to meet the world's growing energy needs.

This comfortable view of how the world could smoothly shift from one source of energy to another was shattered by a series of events that culminated in the October 1973 war in the Middle East, the Arab use of oil as a political weapon, and the huge increase in oil prices unilaterally imposed by the oil-exporting countries at the end of 1973. Less dramatically, but of comparable importance, technical difficulties and public apprehensions have delayed the anticipated rapid development of nuclear power, and growing security problems cast a further shadow over its future.

It can no longer be assumed that problems of energy supply will work themselves out without significant governmental intervention. Deciding precisely what should be done, however, is made difficult by the uncertainties that cloud the future. And even if a clear view of the future could be had, the choices confronting U.S. foreign policy would still be hard ones.

THE FUTURE OF COAL AND NUCLEAR POWER

Since the reliability and cost of oil as an incremental source of energy are now uncertain, the feasibility of reversing or retarding the decline of coal and of speeding up the adoption of nuclear power becomes an important determinant of the nature and magnitude of U.S. energy problems internationally. An optimistic view concerning both coal and nuclear power would of course tend to validate the ambitious goals of Project Independence and in addition reinforce expectations of a decline in oil prices. A pessimistic view would regard both prolonged U.S. dependence on imported oil and high international oil prices as more likely.

The future of coal in both the United States and elsewhere depends upon a complex of largely unpredictable factors, including governmental policies on taxes, subsidies, and wages; public attitudes on environmental issues; and progress in the technologies of coal gasification and liquefaction and in the control of pollutants produced by coal-burning electric power and manufacturing plants.

Similar uncertainty exists concerning the rate with which nuclear power can be developed. Governments may or may not be able to relieve public fears concerning the safety of nuclear reactors or reduce the long lead time between initiation and completion of nuclear power projects. A single serious accident could set back nuclear power schedules by several years. Also, one theft of fissile material by a criminal or terrorist group could lead to a suspension or slowing down of new nuclear power projects until stricter safeguards could be established.

FUTURE OF THE INTERNATIONAL OIL MARKET

Uncertainties exist concerning both the demand for oil and the terms on which oil will be supplied in future years. Demand will depend not only on the rate of growth of the world economy, but also on the reactions to current high oil prices in the oil-importing countries. The basis for estimating the long-term price elasticity of demand for petroleum products probably does not exist. And the difficulty in predicting the rate at which alternative sources of energy will be developed has already been noted.

Forecasting the future supply of oil is no easier. One important question is whether the oil-exporting governments will be able to agree to restrict output in order to maintain high prices. Another is how much money will be invested in expanding oil-production capabilities, and by what interests—governments, major oil companies, or independents. No clear answer to either question is possible; both depend on future political and economic developments that cannot be confidently predicted.

An additional uncertainty concerns the future roles of the Soviet Union and China. If one of these countries should become either major importers or major exporters of energy (more likely the latter), future conditions in the international oil market could be significantly affected.

POLITICAL UNCERTAINTIES

The agreements negotiated in the spring of 1974 to separate first the Egyptian and Israeli forces and later the Syrian and Israeli forces have justifiably increased optimism concerning prospects for lasting peace in the Middle East. These military agreements, however, do not begin to deal with the fundamental issues between the Arab nations and Israel. Renewed hostilities are an ever-present possibility. And even if a new war does not break out, cessation of progress toward a settlement could cause the Arabs once more to use oil in an effort to achieve their political goals.

U.S. energy policy is also strongly conditioned by broader political events. If the United States enjoys good relations with other nations, a number of cooperative endeavors in the field of energy become more than theoretical policy alternatives. No one can be sure today, however, that conditions will be favorable some years from now for cooperation among the industrialized nations in—for example—safeguarding new nuclear technologies or sharing oil supplies in an emergency. Nor is it certain that the political climate will be right for negotiating an oil agreement, if and when the proper conditions arise in the international oil market.

BASIC ASSUMPTIONS

Although none of the uncertainties described above can be completely resolved on the basis of known—or knowable—facts, informed judgments can be made about all of them. Thus, in Chapter 13 the projections of future production and consumption of coal, nuclear power, and other alternatives to oil represent a compromise between the most optimistic and the most pessimistic possibilities. Similarly, the projections in the same chapter of the future demand for and supply of oil are consistent with known physical and political realities.

Certain assumptions are, however, essential to the making of judgments on international energy policy. This need not be a cause for concern, so long as the nature of the assumptions is kept clearly in mind along with an awareness of the possibility that events may prove some of them to have been wrong.

The following assumptions are basic to this study's analysis of U.S. energy policy alternatives in the period 1975-85:

— Oil will continue to provide a large part of the increase in total world energy requirements.
— High oil prices are not inevitable; declining oil prices are quite possible.
— A stable and lasting settlement of the Arab-Israeli dispute is far from assured, and the possibility of a new Arab resort to the oil weapon will continue to exist whether or not a new round of fighting breaks out.
— Relations between the United States and other nations, and especially relations with the other major industrialized nations, will not preclude efforts to find cooperative solutions to common energy-related problems.

Chapter Nineteen

International Oil Policy

The minimum objective of U.S. oil policy is of course to ensure adequate and reliable supplies of oil for the U.S. economy at reasonable cost and without serious damage to U.S. relations with other nations. In an interdependent world, however, this objective is not enough. The United States must also do what it can to keep oil from becoming a source of disruptive and dangerous international controversy and to create conditions in which all oil-importing countries can obtain the oil they need.

The United States and the other oil-importing countries share two basic problems: a substantial part of their oil supplies is vulnerable to politically motivated restrictions and military interdiction, and their market power is at present inferior to that of the oil-exporting countries. Several additional problems are in large part consequences of these two basic problems:

— Fear of recurrent oil shortages has intensified competition for available supplies and could damage relations among major oil-importing countries.

— The sudden, huge increase in the price of oil in late 1973 and early 1974 has created adjustment problems for all oil-importing countries and for the international monetary system.*

*These financial problems go beyond the field of oil policy and have been discussed in some detail in Chapter 14.

—   The high cost of oil also seriously threatens the welfare and growth prospects of some of the poorer nations.
—   The declining power of the international oil companies and the Arab use of oil as a political weapon raise the question of whether and in what ways the governments of the importing countries should deal directly with the governments of the exporting countries on questions of oil supply.
—   The adequacy of existing international institutions to deal with the serious new problems affecting the production and marketing of oil is at least open to question.

The United States has three broad alternatives in the field of oil policy. It can (1) seek self-sufficiency in oil, (2) it can accept continued partial dependence on imported oil and try to solve its international oil problems by unilateral actions, or (3) it can continue to import some of its oil requirements and seek solutions to associated problems through cooperation with other nations. Policies involving various combinations of these broad alternatives of course are possible. It seems best, however, to focus initially on each alternative in its unadulterated form and only later to consider possible mixed policies.

A POLICY OF SELF-SUFFICIENCY

Before the October 1973 war in the Middle East, the United States was importing approximately six million barrels of crude oil and refined products daily, which represented about one-third of its total consumption. Since consumption was rising and domestic production had leveled off, a continued increase in the dependence of the United States on imported oil seemed virtually certain. Published estimates varied widely, but one plausible and widely accepted estimate projected U.S. oil requirements at 26 million barrels per day in 1985, only about half of which would be produced domestically.[1]

The war and the associated Arab embargo and supply restrictions caused President Nixon in November 1973 to launch Project Independence with the goal of energy self-sufficiency by 1980.[2] Precisely how self-sufficiency is to be achieved had not been worked out by the administration by mid-1974. Preliminary approaches to the problem, however, called for a combination of measures to conserve energy and increase the production of various forms of energy. Some future energy requirements that under past projections would have been met by oil were to be satisfied by other fuels.[3]

Whether the ambitious objective of Project Independence is attainable on schedule—and at what cost—lies outside the scope of this study, which is concerned only with the international consequences of possible U.S. oil policies. But even if Project Independence is carried out successfully, U.S. oil-importing requirements would probably rise for a time before beginning their decline to zero. For a number of years, therefore, the United States would face much the same oil-related problems as it faces today. By the late 1970s, however, both the U.S. oil-import bill and direct U.S. vulnerability to politically motivated supply restrictions would begin to disappear.

As U.S. demand on international oil supplies was reduced, the oil-exporting countries would find it more difficult to keep prices high. If prices in fact fell, Western Europe, Japan, and the other oil-importing countries would of course benefit, but their vulnerability to oil embargoes and production cutbacks would not be eliminated. Nor would the United States be any less exposed to the indirect consequences of denial of oil to its major trading partners. An oil-induced recession in Western Europe and Japan would still have severe economic repercussions in the United States.

The withdrawal of U.S. demand from international oil markets would automatically eliminate the danger of controversy over oil supplies between the United States and other industrialized nations. This gain might be offset, however, by a decline in U.S. ability to influence the future of an important sector of the international economy.

Deciding on the merits of a policy of self-sufficiency therefore requires weighing a number of disparate factors, both domestic and international. If such a policy is adopted, it must be viewed as a long-term endeavor and not merely as a reaction to a passing emergency.

A UNILATERAL INTERNATIONAL OIL POLICY

The United States could decide that a high self-sufficiency policy is too expensive or involves too great a sacrifice of environmental goals. In that event, the key issue would be how the United States could best obtain increasing quantities of imported oil. Possible means of solving this and other oil-related problems through multilateral international cooperation will be examined in the next section of this chapter. The question to be addressed first is what could the United States do on its own to ensure that its needs for imported oil are met.

A unilateral international oil policy for the United States might have four principal components: (1) special oil supply arrangements with selected oil-exporting countries, (2) efforts to improve bargaining relationships with the exporting countries, (3) measures to prepare for interruptions in oil imports, and (4) efforts to defuse the Arab oil weapon. Some of the possible courses of action to be discussed under these headings could also fit into or supplement a policy that emphasized cooperation with other nations. They are taken up here because the United States could adopt them on its own initiative even in the absence of a broader framework of international cooperation.

Special Supply Arrangements with Selected Oil-Exporting Countries

The content of special oil supply arrangements could vary widely, but essentially they would require that the United States accord some kind of preferential treatment in return for the guaranteed supply of specified quantities of oil at agreed prices over a fairly long period of time. The preferences extended by the United States could include assured access to the U.S. market at favorable prices plus other measures, such as technical assistance, military aid, security commitments, and diplomatic concessions or support. Thus, in one way or another, special arrangements would probably involve paying more than commercial prices for oil.

An appraisal of the feasibility and desirability of special oil supply relationships cannot be made in the abstract but must focus on specific cases. On grounds of both security and tradition, the place to begin is this hemisphere. A special arrangement with Venezuela would be the main component of a Western Hemisphere oil policy. Special arrangements might also be possible with one or more of the smaller Latin American oil-exporting countries. As was brought out in Chapter 6, however, Canada would have little interest in a government-to-government energy arrangement with the United States, at least as long as oil shortages appear likely.

A special arrangement with Venezuela would be designed not only to ensure that most of that country's oil came to the United States, but also to facilitate large-scale production in the tar belt. An agreement serving both purposes appears to be feasible. Venezuela would promise to give the United States first claim on its oil exports at prices to be set in accordance with agreed principles and procedures. Certain assurances would also be required concerning the security of private U.S. investment in the tar belt and elsewhere, although perpetuation of the present system of concessions would not be essential (and probably would not be feasible given the dominant political sentiments in Venezuela toward foreign companies).

A major benefit accorded Venezuela under such an agreement would be assured access to the U.S. market at reduced or zero import fees. Venezuela might also seek preferential treatment for exports of manufactured goods, either for itself alone or for all Latin American nations.[4] The United States could not, however, grant such treatment for only the developing nations in Latin America without violating Article I of the General Agreement on Tariffs and Trade.

Through special arrangements, the United States might obtain three-fourths of Latin America's gross capacity to export oil, or about 3 million barrels a day in both 1980 and 1985. An additional 0.9 million barrels per day might be obtained from Canada in 1980 on a purely commercial basis, and that amount could increase to 1.2 million barrels per day in 1985.* The possible results of a Western Hemisphere policy might then be projected as follows:

	1973 actual	*1980*	*1985*
	(million barrels per day)		
Latin America	2.6	3.0	3.0
Canada	1.3	0.9	1.2
TOTAL	3.9	3.9	4.2

The above figures naturally give rise to two questions: Are special arrangements with Venezuela and possibly other Latin American oil exporters worthwhile if U.S. imports from that area can be increased by only 0.4 million barrels per day by the 1980s? Is there any way to get more oil from the Western Hemisphere?

The answer to the first question of course depends in part on what the United States has to pay for the projected special arrangements, which could be determined only through actual negotiations. In judging the value of such arrangements, however, it must be recognized that in their absence oil exports to the United States from Latin America might decline. Some kind of long-term

*These figures are the mid-points of official Canadian estimates of exports to the United States in 1980 and 1985, respectively.[5] If Canada carries out its plan, announced in September 1973, to ship Albertan oil to Montreal by pipeline, Canadian exports to the United States would be reduced by about 0.5 million barrels a day. Oil from Latin America, which would otherwise have been imported into Eastern Canada, could then be shifted to the United States, leaving U.S. imports from within the Western Hemisphere roughly the same.

commitment may be needed just to hold Latin American exports to areas other than the United States to their present level of about 1 million barrels per day. Possibly more important, new arrangements may be needed to encourage development of Venezuela's tar belt. In fact, without production from the tar belt, total Latin American output in 1985 would be 0.5 million barrels per day lower than has been projected above, and exports to the United States might also be somewhat reduced.*

The tar belt and the Canadian tar sands probably hold the answer to the second question, since they are the major known and unexploited oil resources in the Western Hemisphere outside the United States. Exploitation of both will require large amounts of capital, which might not be forthcoming in the absence of a special initiative and some financial support from the U.S. government. However, neither of the other two governments concerned could be counted on to respond favorably to such a unilateral U.S. approach. In any event, overcoming the technological obstacles to rapid exploitation of both the tar sands and the tar belt will take time as well as money. It is important to note that the production estimates underlying the export projections cited above already assume that 0.3 to 0.5 million barrels of oil per day will be extracted from the Canadian tar sands by 1980 and 0.5 million barrels per day from the Venezuelan tar belt by 1985. Stimulating greater output than this may well be beyond the unilateral capabilities of the U.S. government.

A Western Hemisphere policy, it must be concluded, would make only a marginal difference in the amount of oil that the United States could obtain from relatively secure, nearby areas. Thus, if as is assumed here, U.S. import requirements continue to rise, the United States must look increasingly to the Eastern Hemisphere as a source of oil.

If the United States sought special oil supply arrangements in the Eastern as well as the Western Hemisphere, the most likely partners would be Saudi Arabia, Iran, and the small Arab states bordering on the Persian Gulf. Special arrangements with Iraq or Libya are presumably out of the question, so long as they remain under anti-American governments. Whether Algeria, Indonesia, or Nigeria could be drawn into a special oil supply relationship with the United States is questionable. These countries prize their neutrality and probably believe that their independence is best protected by maintaining a multiplicity of international ties and avoiding special links with any single great power.

*Production from the tar belt is assumed to be negligible in 1980.

On the public record, negotiating a special oil supply arrangement with Saudi Arabia would appear to be quite feasible. The Saudi Arabian Minister of Petroleum and Mineral Resources, Sheik Ahmad Yamani, proposed just such an agreement in September 1972 at a conference in Washington, D.C.[6] The U.S. government neither accepted nor rejected Yamani's offer, and the offer appears not to have been withdrawn. It was, however, overtaken by the Arab-Israeli war in October 1973 and the Arab embargo of oil exports to the United States.

An improvement in U.S.-Saudi relations set in even before the embargo was lifted, principally as a result of U.S. efforts to bring about a settlement between Israel and the Arab states. In April 1974, the U.S. and Saudi Arabian governments jointly announced an agreement under which the United States would provide technical assistance in support of Saudi Arabia's economic development, as well as aircraft and other military equipment to the Saudis.[7] The United States emphasized that this agreement did not give it any preferential status with respect to Saudi Arabian oil and reiterated its often expressed hope for multilateral solutions to oil problems. This agreement was formalized and given further symbolic importance in June 1974 during the Washington visit of Prince Fahd Ibn Abdel Aziz, second deputy premier of Saudi Arabia. Prince Fahd and Secretary of State Kissinger signed a new agreement that established joint economic and military commissions to promote U.S.-Saudi cooperation. The agreement did not mention oil.[8] Nevertheless, it clearly revived the possibility of a special U.S.-Saudi oil supply arrangement.

Iran already enjoys a special political and military relationship with the United States. Diplomatic ties are close and cordial, the United States is the main source of weapons for the Iranian armed forces, and the qualified U.S. security commitment to Iran included in a 1959 bilateral agreement of cooperation remains in effect.[9] Iran would therefore have little to gain politically or militarily from a special oil supply arrangement with the United States. It might in fact hesitate to forge yet another link with the United States, fearing that to do so would polarize the international politics of the gulf area and make it an active arena of great power rivalry.

Iran would also appear to have little economic reason to favor a special oil supply relationship with the United States. It has encountered no difficulty in obtaining private investment and technical assistance from a variety of sources and presumably would not see a special oil supply arrangement with the United States as a means of giving a needed impetus to its economic development program.

A special oil supply arrangement with Iran would appear possible only as a sequel to the establishment of such an arrangement with Saudi Arabia. Iran would then have less fear of the international consequences and would, moreover, want to keep Saudi Arabia from gaining any undue commercial or political advantage with the United States. In fact, the United States would seriously damage its relations with the Shah if it sought a special arrangement with Saudi Arabia and did not make a similar proposal to Iran.

The smaller Persian Gulf states would not want to be the first to enter into special oil supply relationships with the United States, but at least some of them would tend to follow the lead of Saudi Arabia. Kuwait might be inhibited by fear of the reaction of its large resident Palestinian population. Abu Dhabi and the other members of the United Arab Emirates might, however, be willing to make arrangements with the United States similar to any arrangement worked out with Saudi Arabia. Even more than Saudi Arabia, these small states would see security advantages in closer ties with the United States.

There is no way to predict how much Eastern Hemisphere oil could be committed to the U.S. market through special supply arrangements. Depending largely on political developments, the answer could range from very little to the full difference between total oil-import requirements and the amounts obtainable in the Western Hemisphere. The durability of any arrangements that might be negotiated would be uncertain. Apart from political threats to their survival, such arrangements could easily be destroyed by disagreements over price.

Even if it is assumed somewhat optimistically that the United States could cover most of its oil-import requirements, over and above Western Hemisphere imports, through special supply arrangements, this "gain" must be weighed against substantial risks and costs. The dependence of the United States on oil from the Arab states of the Persian Gulf would probably be greater than would otherwise be the case, and U.S. vulnerability to supply restrictions and embargo would be correspondingly increased. Also, the United States would have acquired a greater stake in the survival of the governments entering into the special arrangements.

But more important than these risks would be the fact that the United States would have made a major departure from the principle of multilateralism in international economic relations and entered into a cut-throat competition with Western Europe and Japan to tie up the oil output of the Persian Gulf. This consideration would of course have much less force if many of the other major oil-importing nations had already embarked on a similar policy.

Efforts to Improve the U.S. Bargaining Position in the International Oil Market

In principle, the United States, acting alone, could try to improve its bargaining position in the international oil market by operating directly on supply and demand or by changing the institutional arrangements for determining prices. Both strategies will be considered here.

The most effective and readily available unilateral measures the United States might take to increase supply and reduce the demand for oil are domestic. The strong impact that a U.S. policy of seeking self-sufficiency would have on the international oil market was noted earlier in this chapter.

Internationally, the possible results of direct, unilateral U.S. actions to affect supply and demand are more speculative. Several possibilities deserve consideration. Specifically, the United States might:

— Provide technical assistance to developing countries that are interested in exploring promising offshore areas and that appear likely to pursue expansionist oil-production policies. One means of carrying out this policy would be to give special tax advantages to private U.S. oil companies willing to provide the needed technical assistance to selected developing countries.

— Facilitate the development of Soviet and Chinese oil (and gas) resources by helping to mobilize needed capital from public and private sources (possibly backed by government guarantees) and by providing technical assistance in exploring and exploiting promising offshore areas in the Arctic and off the coast of East Asia.

— Facilitate the more rapid adoption of nuclear energy in other countries (and thereby reduce the demand for oil) by expanding the uranium enrichment facilities of the Atomic Energy Commission and by helping private interests to build such facilities. Encouraging U.S. manufacturers to develop efficient small (200-300 Mwe) nuclear reactors would serve the same end by making nuclear power an attractive alternative for more LDCs.

It is impossible to quantify the effects of these and other possible unilateral U.S. efforts to operate directly on supply and demand. All that is certain is that results would not be achieved easily or quickly. It is therefore all the more important to examine possible indirect measures to influence the institutional arrangements under which oil is produced and marketed internationally.

Professor M. A. Adelman of the Massachusetts Institute of Technology argues that OPEC could not function as an effective cartel in the absence of its members' long-term agreements with the companies, which set a floor under oil prices. Adelman believes that the cartel would break up if the companies gave up what remains of their concessions and other privileged positions and became merely producers under contract and purchasers of crude oil from the cheapest sources.[10] The governments of the exporting countries may in fact bring about the change in the status of the oil companies that Adelman would prefer, but there appears to be little that the U.S. government could do, if it wanted to, to move events more rapidly in that direction. In any event, changing the role of the companies would at most remove an institutional obstacle that would tend to retard market reactions to a threatened condition of excess supply; it would not keep the oil-exporting countries from maintaining high prices by restricting production, if enough of them were able to agree on such a course of action.

A more drastic institutional change than any effort to modify the role of the oil companies would be for the U.S. government to establish an oil-import monopoly to buy crude oil and refined products from the cheapest available sources and resell its purchases to refiners and distributors in the United States at cost. Under some circumstances, the buying power wielded by the import monopoly could probably be used to force prices down, but that would probably not be easy. In order to use its power to maximum effect, the monopoly would have to go around the companies and seek to change the terms under which OPEC and its member governments permit the companies to produce and export oil. How effectively the monopoly could intervene in the OPEC-companies relationship is at best uncertain.

Measures to Prepare for Interruptions in Oil Imports*

The Arab embargo of oil shipments to the United States in the fall and winter of 1973-74 served as a convincing reminder of

*The analysis of stockpiling and other measures to prepare for interruptions in oil imports is based in large part on the work of Barry M. Blechman and Arnold M. Kuzmack.

the vulnerability of oil imports to sudden interruptions. A unilateral U.S. oil policy might therefore properly include measures specifically designed to meet this problem.

In theory, the United States could respond militarily to a threatened or actual interruption of oil imports. Such an action would be hard to justify legally or morally, and its political costs would be both high and long lasting.* Moreover, it is by no means certain that a resumption of oil shipments could be brought about quickly through military means. Initial successes might well be followed by prolonged guerrilla warfare featuring sabotage of oil-production and transport facilities. Nevertheless, the mere possibility that the United States could be driven to use military force to break a prolonged embargo could be a restraining influence on the oil-exporting countries.

A more practical way to deal with an embargo or other interruption of oil imports is to limit its effects through advance stockpiling, supplemented by energy conservation measures.** Table 19-1 presents rough calculations of the cost of a stockpile designed to replace different amounts of 1985 oil imports from Arab countries. The table indicates that—subject to the assumptions stated—for a cost of $12 billion (or $1.2 billion per year), the United States could create a stockpile that would enable it to replace Arab imports for about a year in 1985, even if those imports reached the improbably high level of 6 million barrels a day.

Storing crude oil in steel tanks is the simplest form of oil stockpiling and the form that can be fitted most easily into the existing supply system. Other stockpiling alternatives also merit consideration. There are many natural salt dome formations and abandoned coal mines in the United States that would provide much cheaper storage than steel tanks.[11] Since most of these underground facilities are unlikely to be near refineries, however, special pipelines would have to be built, which would narrow the cost advantage of natural storage over storage in steel tanks. Coal stockpiles and stockpiles of refined petroleum products are alternatives to stockpiles of crude oil. The former would be very cheap to maintain, and would be desirable principally near electric generating plants that

---

*These costs would be lower if the United States acted jointly with other major oil-importing countries. Somewhat paradoxically, however, while the need of those countries to break a prolonged embargo would in most cases be much greater than that of the United States, it is even less likely that they would be both able and willing to use military force for this purpose.

**An emergency stockpile would be in addition to stocks normally maintained for commercial purposes, which would provide an additional cushion in an emergency. In 1973, such stocks represented about 30 days of normal consumption.

Table 19-1.   Cost of Stockpiling Oil in Steel Tanks to Replace
Assumed 1985 U.S. Imports from Arab Countries (billions of
US dollars)

Days Covered	With no Conservation Measures	With 10% Reduction in Total Oil Consumption
90	$ 5.4	$ 3.0
180	10.7	6.0
360	21.5	12.0

Source: U.S. Cabinet Task Force on Oil Import Control, *The Oil Import Question, A Report on the Relationship of Oil Imports to National Security* (Washington, D.C.: Government Printing Office, 1970), Appendix J.

Note: These estimates are based on a very simple formula: Total cost is equal to the cost per barrel of oil storage, including the cost of the oil, times the number of barrels per day that must be stored, times the number of days for which protection is desired. The cost of any reserve transportation system that might have to be established is not included. Total U.S. oil consumption in 1985 is assumed to be 26 million b/d, of which half will be imported—essentially a "worst case" from present perspectives. Imports from Arab countries are assumed to be 6 million b/d. Two assumptions are made concerning oil conservation (demand reduction) measures: none, and a 10 percent reduction. Oil is assumed to cost $6 per barrel, and the cost of storage for ten years is set at $4 per barrel ($2.50 for tanks, $0.25 for land, and $1.25 for operating costs). Estimates of storage costs are drawn from the U.S. Cabinet Task Force on Oil Import Control.

All figures are in constant 1973 dollars. The $6 per barrel cost used for the oil to be stockpiled rests on the assumption that by 1985 oil prices will be significantly lower than was the case in mid-1974.

could switch quickly from oil to coal. For many oil users, however, including almost all of the transportation sector, a rapid change of fuels is not possible. Stockpiles of refined products would be relatively expensive, but they could be dispersed and located near consumers, thereby reducing transportation requirements in an emergency.

Whatever forms of stockpiling were adopted might be supplemented by emergency production capacity within the United States. The National Petroleum Council has estimated that in 1974 existing major U.S. oil fields could temporarily increase production by 279,000 barrels a day after a delay of 90 days and by 307,000 barrels a day after 180 days. These levels, however, would require flaring of gas and could not be maintained much beyond 180 days.[12]

The U.S. government also could develop oil fields (in addition to the existing naval petroleum reserves) but not permit production from them except in an emergency. An important advantage of such shut-in capacity over stockpiling is that it reduces the need to decide exactly how long a supply interruption to hedge against. This is so because the shut-in wells could produce for a longer time than the likely duration of any emergency. For most shut-in capacity, a substantial reserve transportation system would have to be maintained. Production equipment, and possibly also skilled personnel, would also have to be kept in reserve, and this clearly would be expensive.

Attempting to Defuse the Arab Oil Weapon

A stable settlement of the Arab-Israeli dispute would by definition eliminate the grievance that has caused the Arabs to use their "oil weapon" in an effort to change U.S. policy toward Israel. The wisdom of doing everything possible to achieve a settlement is therefore obvious. What is less clear is whether there is any way short of a settlement to eliminate or significantly reduce the risk of a renewal of the embargo and oil supply restrictions.

Several important circumstances favor the U.S. position in the Middle East. The present leaders of Saudi Arabia and several of the smaller oil-exporting states appear to place a high value on the U.S. connection. They see the United States as a stabilizing influence in the area, a source of modern industrial technology and advanced armament, a valuable and growing market for oil, and one of the best places in which to invest some of their rapidly mounting oil revenues. Up to a point, even the close relations of the United States with Iran work to the U.S. advantage in the Arab states of the gulf, since they would not want to see Iran become the only nation in the area to enjoy the political, economic, and military advantages of a friendly association with the United States.

The United States should continue to build on the desire of many Arab leaders for a closer relationship with the United States by encouraging Arab investment in the United States, facilitating the transfer of U.S. industrial technology to interested Arab countries, making a special effort to increase U.S. exports to those countries, and providing them with military assistance. At the same time, the U.S. vulnerability to oil embargoes should be reduced by energy conservation measures, increased stockpiling, and possibly also by the creation of additional shut-in production capacity.

None of these courses of action is without problems. For example, how much can the U.S. government, operating within a free market, really do to strengthen economic ties between the United States and the Arab countries, thereby giving them a greater stake in the stability of all aspects of their relations with the United States? For the time being, the Arab countries appear to derive some reassurance from the involvement of the U.S. government in a variety of economic arrangements. They may, however, come to realize that they can obtain any nonmilitary products or technology they want by dealing directly with private U.S. firms, thereby avoiding even implied commitments to the U.S. government.

Military assistance to the Arab states bordering on the Persian Gulf poses especially difficult problems. Triggering an arms race between Iran and its Arab neighbors is a serious danger that has received little public attention. The most difficult problems of all arise, however, in connection with the Arab-Israeli dispute. How can the United States continue to supply arms to Israel (as it probably must if a stable balance of forces is to be maintained) and still convince the Arabs that it understands their position in the dispute and is not committed in every particular to the Israeli position? Similarly, how can the United States press the Israelis to make concessions in the interest of a settlement without undermining the sense of security that is needed if concessions are to be politically feasible?

Despite the successful negotiation of agreements separating Arab and Israeli forces on the Egyptian and Syrian fronts, optimism concerning the achievement of a stable settlement is not warranted. Continued effort, however, is justified. Persistent, even-handed diplomacy may well produce at least partial results and take some of the heat out of the Arab-Israeli confrontation. Moreover, so long as there is progress or even the hope of progress toward a settlement, the Arabs will be less disposed to turn once more to their oil weapon.

With the passage of time, the United States may, under the policies discussed above, hope to bring about a fundamental change in the nature of its relations with the Arab oil-exporting countries and certain other key Arab countries, as well. The Arabs may come to see the United States less as the ally of Israel and more as an increasingly important trading partner, an attractive place to invest excess oil revenues, and a source of advanced industrial and military equipment and technology. All of these reasons for maintaining good relations with the United States would weigh against any future decision by the Arab oil-exporting countries to impose a new embargo. The possibility that they might take such a step would be further reduced if the United States had been playing an active and impartial role in seeking to bring about a settlement—even if that settlement was not yet at hand.

A MULTILATERAL INTERNATIONAL OIL POLICY

Over the longer run, a multilateral international oil policy for the United States would rest on the common interest of both the importing and exporting nations in a healthy, stable oil industry serving a smoothly functioning world economy. For the present, however, and probably some time to come, such a policy would of necessity rest primarily on the interest of the oil-importing countries in adequate, uninterrupted supplies of oil at reasonable prices.

The Limits of Cooperation

As a practical matter, cooperation among oil importers means for most purposes cooperation among the leading industrialized nations of Western Europe and North America, plus Japan. Even among these nations, however, possibilities for effective cooperation are limited. A joint approach to the problem of peace in the Middle East has strong appeal in the United States, but most Europeans, Japanese, and Canadians see dangers and few benefits in close association with U.S. policy in that part of the world. Similarly, a common effort by major oil-importing countries to improve relations with the oil-exporting countries by helping them diversify their economies and by facilitating the investment of their excess oil revenues makes excellent sense in theory, but is not likely to come about in practice. These and other measures, including military assistance, that would be designed to forge stronger ties with the oil-exporting nations are much more likely to be undertaken on an uncoordinated national basis than to be incorporated in a broad multilateral program.

Prospects for cooperation among the major oil-importing countries in negotiations with OPEC are also poor. In its most ambitious form, such cooperation would involve a common negotiating front, a kind of buyers' cartel. A less ambitious variant would involve coordinated diplomatic support and "guidance" of the international companies from behind the scenes, with the companies continuing to handle direct relations with the host governments. An even more modest variant would be limited to intergovernmental consultations on problems associated with the production and marketing of oil.

Setting up a buyers' cartel is probably neither feasible nor desirable. The Japanese appear to have little or no interest in any such scheme. And achieving a coordinated position between the United States and the European Community, or even within the Community itself, would be a formidable task. Even if something resembling a buyers' cartel could be formed, its value would be questionable. The major importers acting collectively would be no more willing to risk an interruption of their oil supplies than they have in the past acting separately.

Consultation on negotiating problems represents no significant departure from present practice in the OECD Oil Committee. So long as consultations are conducted quietly and in a non-provocative manner, they can do little harm and may be moderately useful. The practical question therefore concerns the intermediate form of coordination. Should the industrialized nations try behind the scenes to concert their policies toward OPEC? Again, serious questions of both feasibility and desirability must be faced. Japanese participation in even this less visible form of consumer cooperation would not be likely. Some of the Europeans, most notably the French, would probably prefer to continue to play their own game. The companies' reactions to increased government intervention would be mixed, and their receptivity to guidance uncertain.

If, despite all obstacles, behind-the-scenes coordination could be made to work, the results would at best be modest. Diplomacy, unsupported by the will to apply real leverage, could affect the terms of oil supply only marginally. Coordination would limit the diplomacy of the oil-importing nations to the level of the least resolute of their number. Cooperation among the major importing countries in dealing with OPEC is therefore unlikely to advance beyond informal consultation, and the likely benefits of closer coordination would not be large.

Despite the limits on cooperation among the industrialized countries, opportunities for mutually beneficial joint action do exist in four major policy areas: (1) meeting oil supply emergencies; (2) relieving the energy-related problems of the developing countries; (3) increasing and diversifying energy supplies and conserving energy; and (4) stabilizing oil prices and production levels.

Cooperation in Emergencies

During the 1956 Suez crisis, which closed the canal and disrupted the flow of Middle Eastern oil to Western Europe, the European members of the Organisation for European Economic Co-operation (OEEC), the predecessor of OECD, established an emergency system under which its oil committee effectively allocated the reduced supply of oil. Ninety percent of the available oil was distributed in proportion to each country's recent imports and ten percent was used to meet special situations. The United States did not belong to the OEEC and was therefore not included in the system, but it provided essential support by calling on reserve oil-production capacity to help meet European needs.

In 1967, in the wake of the six day Arab-Israeli war, several Arab producers instigated an embargo directed primarily against the United States and the United Kingdom. Again, oil movements were disrupted, and the OECD Oil Committee instructed the companies to adjust their tanker operations in such a way as to avoid short falls. This action was taken with considerable reluctance on the part of some European governments which feared that any joint action would offend the Arabs. The embargo, however, proved to be both short-lived and ineffective and invoking the emergency allocation system was not necessary.

In 1972-73, efforts were made to include the United States, Japan, and other non-European members of OECD in the emergency allocation system. These efforts failed because of the insistence of the United States that the system include only oil moving in international commerce and the insistence of Japan, with considerable European support, that all the oil supplies of the participants be allocated.

When war broke out again in the Middle East in October 1973 and the Arabs imposed both oil supply reductions and a selective embargo, the emergency allocation system of OECD's European members still existed on paper, but the governments concerned proved unable to agree on putting the system into operation. By dividing the oil-importing countries into three cat-egories—friendly nations, which were to receive normal supplies, hostile nations, which were totally embargoed and neutral nations, which were left to scramble for their shares of a reduced total supply—the Arabs effectively sowed dissension within the European Community.

It would be foolhardy to assume that the Arabs will never again use oil for political purposes. The industrialized nations must treat a renewed embargo as a very real possibility and must consider how best to prepare for it.

The experience of 1973-74 reinforces the view that increased stockpiles are part of the answer. If fairly large stocks of oil had not existed throughout the industrialized world, adjusting to the Arab supply restrictions would have been even more difficult than it actually was. Before the outbreak of hostilities in October 1973, European Community policy called for increasing oil stocks from the then existing level of 65 days' consumption to 90 days by 1975. Japan had set a goal of 60 days' stocks by 1975.[13] The United States should also set realistic stockpiling goals for itself. The goals of all of the industrialized countries might then be made matters of OECD policy, thereby giving them the color of mutual obligations.

All of this is relatively noncontroversial. Creating an OECD-wide emergency allocation system would be more difficult. In this regard, the lesson of the 1973-74 embargo is not clear. No one will ever know whether the existence of such a system in 1973 would have deterred the Arabs from applying their selective embargo or caused them to lift it sooner. They might have been impressed by the damage-limiting capacity of the allocation system and by the political unity of the industrialized countries implicit in its establishment. But they might equally have concluded that the system merely made it more certain that by reducing exports to all industrialized countries they could exert heavy pressure upon their principal target, the United States. It can in fact be argued that in 1973-74 the industrialized countries accidentally reacted in the best possible way: their political disunity undercut the Arab rationale for hitting at the United States through its allies, and the oil companies evened out the effects of the embargo on Western Europe and Japan in much the same way as a government-backed allocation system would have done.

The strongest argument in favor of an emergency allocation system is not economic, but political. One of the most damaging consequences of the 1973-74 embargo was its divisive effect within Western Europe and between the United States and both Western Europe and Japan. Since cooperation among these three great industrialized areas is essential to the solution of so many international problems going far beyond those associated with oil, some means must be found to keep another oil embargo from further weakening the ability of the industrialized nations to work together.

Efforts should be continued to arrive at understandings among the major oil-importing nations concerning the actions to be taken in a future oil supply emergency. These actions should be of two kinds: measures to reduce the consumption of energy and a system for allocating available oil supplies. Consumption cuts should be deeper in some countries than in others, depending on the proportion of energy devoted to relatively nonessential purposes. All oil available to participating nations—both domestically produced oil and oil moving in international commerce—should in principle be included in the allocation system. However, in order to make the scheme acceptable to the U.S., British, and Canadian publics, something less than total sharing of domestic production may have to be accepted.*

*The Energy Coordinating Group formed by all of the thirteen participants in the Washington Energy Conference (February 1974) except France had by September 1974 reached general agreement on a contingency plan to meet future oil supply emergencies. The plan—which requires approval by the governments concerned—reportedly provides for stockpiling, restraints on consumption, and pooling of oil supplies, including at least some domestic production.[14]

In joining in this kind of emergency system, the United States would of course be giving up part of the advantage conferred by its large domestic energy sources. In return, it would receive improved prospects for a broadly cooperative relationship among the industrialized nations, including improved prospects for gaining support for a multilateral approach to other energy-related problems.

Cooperative Approaches to the Energy-Related Problems of the Developing Countries

Rising oil prices and the fear of future shortages are creating both problems and opportunities for the United States in its relations with those less developed countries (LDCs) that depend on imported oil. The common interest of *all* oil-importing countries in uninterrupted fuel supplies at moderate prices works against the threatened polarization of the world along North-South lines. The interests of the United States would clearly be served by strengthening and building on this common interest. At the same time, the increased cost of imported oil is a serious drag on the development of the poorer nations, and some of them are threatened by a disastrous curtailment of their ability to maintain essential imports. On humanitarian grounds and in the interest of international political stability and the healthy development of the world economy, the United States should do what it can to soften the impact of high oil prices on the LDCs.

For the immediate future, the LDCs hardest hit by high oil prices need emergency assistance. The United States should take the lead in mobilizing such assistance from all available sources--the international lending agencies, the excess revenues of oil-exporting countries, and the national aid programs of the industrialized countries. One possibility open to the United States would be to use existing authority under Public Law 480 to provide grain to LDCs on concessional terms (long-term, low-interest loans). The United States might propose that other major grain-exporting countries do the same and that the oil-exporting countries also finance part of their oil shipments to some of the LDCs on concessional terms.

Beyond the immediate emergency, the oil-importing LDCs will need help in adjusting to the sudden increase in their oil-import bills. They will, with good reason, want to develop other, primarily domestic, sources of energy. The United States can and should assist this effort by helping LDCs explore for oil and gas in geologically promising areas, including offshore waters, and by helping to finance economically sound energy projects.

The flow of development assistance to the LDCs should also be increased to counter the adverse impact of high oil-import costs on the economic growth of the LDCs. Some of this assistance should come from the United States and other industrialized countries. A new source of such assistance has, however, been created by the same high oil prices that gave rise to the problem. This source is of course the excess revenues of some of the oil-exporting countries.

Viewed broadly in terms of global economic development, the recently increased revenues of the oil-exporting countries are an unexpected accumulation of capital with great potentialities for good. The problem for the international community is how to channel this capital to where it is most needed. Understandably, the oil exporters will give priority to their own development needs, but those needs cannot absorb the total revenues prospectively available to several major oil exporters.

One way in which exporters with large excess revenues could channel resources to less fortunate LDCs would be to sell them oil at preferential prices. Administrative difficulties would be considerable, however, and the negotiations complex, particularly in deciding which countries might qualify for low prices.* But whether the U.S. government could do much to promote such arrangements is questionable. A more feasible and possibly a more desirable approach to the same end would be to induce the oil exporters to invest some of their excess revenues in the oil-importing LDCs. Except for limited subsidies from rich to poor Arab countries, the oil exporters cannot be expected to provide grant assistance or to make substantial loans on concessional terms. On the contrary, the problem is to make investment in the oil-importing LDCs more attractive in terms of both security and rate of return.

One possible approach to this end would be to devise special financial instruments through which the surplus oil revenues of the exporting countries could be channeled into concessional development lending. It is, after all, ironic that surplus funds from these countries should be used principally for investments in the capital-rich industrial countries. One way of changing this would be to use the World Bank and possibly the regional lending institutions as financial intermediaries. For example, the World Bank might issue a special long-term debt instrument whose exchange value in terms of Special Drawing Rights (SDRs) would be guaranteed by the

*Such an approach, however, might be attempted as part of a package in which the financially strong oil exporters invested in oil refineries and marketing facilities in oil-importing LDCs and, for a time at least, provided the oil at prices significantly below the world market.

industrial countries, directly or indirectly, and which therefore could carry a lower rate of interest than other prime debt instruments. A World Bank bond of this nature could be floated anywhere because of its attractive exchange-value guarantee; or its sale could be limited, at least initially, to Saudi Arabia and other oil-exporting countries with financial surpluses to invest.

Another possiblity would be to provide for special activations of SDRs for sale to oil-exporting countries whenever they wish to hold them in place of short-term financial balances; the proceeds would be used for development assistance—specifically for supplementary financing of the World Bank's International Development Association (IDA). None of the usual arguments against a link between SDR activation and development assistance would seem to apply in this case. The amount issued would be determined by the decisions of surplus oil-producing countries to hold a portion of their monetary reserves in SDRs rather than in another form. In this respect their preference for SDRs would constitute a legitimate independent requirement for international liquidity. And this form of link would not contribute to inflationary pressures since the SDRs in effect would be in payment for an equivalent export of oil.*

Still another possiblity would be for the World Bank to encourage the formation of joint ventures for the purpose of investing oil revenues in oil-importing LDCs. Partners in such ventures might be private firms based in the industrialized countries and public corporations formed by governments of oil-exporting countries. The former would supply technical skills and some capital; the latter would contribute principally capital. The Bank might also in some cases be a partner and provide supplementary financing as well as organizational skills.

All of these procedures—issuance of special World Bank bonds, a special allocation of SDRs linked to development assistance, and World Bank sponsorship of joint ventures—would provide a means of offsetting the producing countries' surplus receipts from oil sales by exports of goods and services from industrial to developing countries. And both would contribute to improving relations between industrial and developing countries through strengthening multilateral forms of cooperation between them.

*Since these SDRs would have to be treated as equivalent to "accepted" rather than newly allocated SDRs, they would carry a rate of interest. This could be paid (a) by keeping in the International Monetary Fund a portion of the interest-bearing assets for which they had been exchanged and using the interest earnings from those assets for this purpose; or (b) from the general revenues of the International Monetary Fund; or (c) by a special subscription from the industrial countries. In any event, the interest rate would be relatively low and the servicing burden small.

Cooperative Efforts to Avoid a Chronically Tight International Oil Market

Chapter 13 pointed out that, if the oil-exporting countries wish to keep supplies tight and prices high, they will probably have to slow down the rate of increase in oil production. The oil-importing countries could make this strategy more difficult to carry out—or even cause it to fail—by adopting effective oil conservation measures and by increasing energy production capabilities outside the control of the major oil-exporting countries. As such efforts took effect, the oil exporters would have to sacrifice more and more potential output to keep oil supply from exceeding demand at prevailing prices.

A formal agreement among the principal oil-importing countries to conserve energy may not be practicable, except in times of energy shortage. Much can be achieved, however, by the force of example and by an improved system for exchanging information on ways to save energy. A more promising area for sustained cooperation among importers lies in the expansion and diversification of energy production capabilities around the world.

Agreements to exchange information on technological advances in energy production could hasten and broaden their practical application, although serious problems of private proprietary rights to new technology would have to be resolved. Joint energy research and development projects, combining the resources and talents of several nations, would also appear to be in the common interest. Most such projects, however, could not be expected to have much effect on the world supply of energy, and thus on the international oil market, for many years.

More rapid, but by no means immediate, results might be obtained by large-scale joint investments in new energy production facilities. A number of such energy development projects may in fact be more suited to international cooperation than to unilateral national initiatives. A group of nations could more easily mobilize the necessary capital, provide any necessary market guarantees, and set rational investment priorities than could any nation acting alone.

In some instances, joint action would also have political advantages. Thus, Canada, a member of OECD, might be more receptive to a proposal by OECD or a group of its members to share in developing the Albertan tar sands than it would to a unilateral U.S. proposal. Similarly, Venezuela might prefer international development of its tar belt over a U.S. effort, although (since

Venezuela is not a member) OECD sponsorship would be less appropriate in this case than some *ad hoc* arrangement.

Joint efforts (U.S.-European and U.S.-Japanese) to assist in the development of Soviet and Chinese oil (and gas) resources might also have advantages over uncoordinated national efforts. For example, looking at possible Soviet gas projects from the point of view of the combined energy requirements of the industrialized non-Communist nations could result in dropping projects to ship liquefied natural gas from the Soviet Union to North America in favor of transporting larger quantities of Soviet gas to Western Europe by pipeline. Moreover, U.S. involvement in Soviet energy projects might reduce Japanese and West European fears of becoming dependent on the Soviet Union for a significant amount of their energy.

Other areas for possible joint action include the exploitation of U.S. coal and shale oil deposits, the intensified exploration of promising offshore areas for oil and gas, and the construction of new uranium enrichment facilities to keep pace with the fuel requirements of a rapidly expanding international nuclear energy industry.

In February 1974, the foreign ministers of the United States, Japan, and the countries belonging to the European Community met in Washington to discuss their common energy problems. At the end of the conference, they issued a joint communiqué in which all of them, except the French foreign minister, agreed on the need for a cooperative international effort in various areas, including "the acceleration of development of additional energy sources so as to diversify energy supplies."[15] A coordinating group was formed to study the practical steps needed to carry out this and other recommendations of the conference.

Agreement in principle among governments is of course only the first step. The execution of projects would be largely in private hands, and most of the capital required would be from private sources. The governments directly concerned with each project would provide diplomatic support, advice on priorities, and any needed investment guarantees and supplementary financing. In the case of high-cost energy sources, governmental action would probably also be needed to assure investors of the continued availability of a market at adequate prices.

After efforts such as those outlined above had been initiated, but before they yield large new supplies of energy, they could be expected to affect the marketing strategy of OPEC and its members. The oil exporters might conclude that by keeping prices high they would hasten the development of additional sources of energy and undermine their position over the longer run.

Cooperation among Both Importers and Exporters to Stabilize Prices and Production Levels

Even without special efforts by oil importers to expand and diversify energy sources, the oil-exporting countries may encounter increasing difficulty in restricting production to levels consistent with the volume that can be sold at continued high prices. The exporters may then become interested in negotiating an agreement with importers that would stabilize the market.

If and when such a change in the attitudes of the exporting countries does appear, the importing countries may be tempted to sit back and enjoy the benefits of falling prices. Certainly, they could not be expected to consider seriously joining in an effort to support the exceedingly high prices dictated by the exporters at the height of the 1973-74 Arab oil embargo. On the other hand, the importers should not let pass a rare opportunity to stabilize the international oil industry to the long run benefit of all parties concerned.

Successful negotiation of an international oil agreement would assure exporters of an expanding market at predictable prices. Importers would gain assurance of reliable supplies of an essential commodity and would have less reason to fear sudden, disruptive increases in prices, such as they experienced in 1973-74. The central feature of an international oil agreement would presumably be the periodic setting of a specific range within which prices would be permitted to move. When prices exceeded the upper limit, producers would be required to increase output; when prices fell below the lower end of the range, they would be permitted to restrict production. Understandings would be needed concerning how much each producer was to increase or decrease production when prices moved outside the agreed range. Agreement on these and other details might be reached through a system of weighted voting under which major exporters or importers of oil would exert greatest influence. Approval of a majority of both producers and consumers would probably be necessary.

Under the best of circumstances, negotiating an international oil agreement would not be easy. The effort should not in fact be made until a coincidence of uncertainties about future supply and demand causes both importers and exporters to welcome the predictability that an international agreement could provide. As a first step toward an international oil agreement, however, serious consideration should be given to the creation of a comprehensive international oil (or energy) organization that would include both exporters and importers. The institutional framework for negotiating an oil agreement would then be in place, if and when favorable circumstances arose. In the interim, the new organization would provide a useful forum for discussing international oil problems. The oil-importing LDCs would particularly benefit from the creation of a comprehensive international oil organization. The oil-exporting LDCs have OPEC and the industrialized oil-importing countries have OECD, but the oil-importing LDCs have no organized forum in which to further their interests.

ASSESSMENT OF POLICY ALTERNATIVES

A unilateral international oil policy is clearly the least desirable of the three alternatives discussed above. Such a policy would accept increasing dependence on imported oil and at the same time gamble on market forces bringing prices down and on a stable settlement being achieved in the Middle East. Stockpiling and other measures would provide only partial protection in emergencies. Special oil supply arrangements with selected exporting countries would be costly, would provide little or no protection against politically motivated embargoes, and could place the United States in an adversary relationship with its West European and Japanese allies. The scramble for "secure" oil supplies would make continued high prices more likely for all and further damage the oil-importing developing countries.

A strong case can be made for a policy of self-sufficiency. If successful, such a policy would assure the United States of reliable supplies of energy. The vulnerability of the United States to the Arab oil weapon would be greatly reduced, but not entirely eliminated, since the United States would suffer with the rest of the world if oil supply cutbacks brought on a depression. By holding down U.S. demand for foreign oil, the problems of other oil-importing countries would be somewhat eased.

On the other hand, the costs of self-sufficiency would probably be high in both financial and environmental terms. Results would be uncertain and slow to appear. Also, working toward self-sufficiency for the United States would provide little immediate relief for the oil-related problems of the world's poorer nations, nor would it do anything about the basic problem of stabilizing the international oil industry. Possibly most serious, if seeking self-sufficiency caused the United States to abandon interest in the energy problems of other nations, prospects for solving those problems would be diminished and all would eventually suffer, including the United States itself. The United States might opt out of the international oil market, but it would still be dependent on the world economic system, of which energy industries and markets form a critical sector.

It does not follow that a purely multilateral international oil policy would be best. Of the three alternatives, a multilateral policy promises the most. But by its very nature, the success of such a policy depends upon the cooperation of other nations, which might not be forthcoming. Also, like the policy of self-sufficiency, some of its results would be slow in coming. And unlike the self-sufficiency policy, a policy emphasizing international cooperation would not by itself reduce U.S. vulnerability to the Arab oil weapon, but might actually increase that vulnerability.

For at least the next few years, the best international oil policy for the United States would combine elements of the three strategies discussed in this chapter. The main thrust should be to seek the maximum feasible international cooperation—in preparing for emergencies, in alleviating the problems of the oil-importing LDCs, and in increasing and diversifying energy sources. At the same time, as a matter of simple prudence, dependence on imported oil should be limited by promoting conservation and developing domestic energy resources. The ability of the country to ride out supply interruptions should be improved by increased stockpiling and possibly by the creation of shut-in production capacity. Efforts to achieve a stable Middle Eastern settlement should be continued in the hope both of removing a serious threat to world peace and reducing the danger of a renewed Arab oil embargo.

As these various efforts yield results, the United States might find it possible to ease off somewhat in its efforts to achieve self-sufficiency. In most likely future circumstances, the risks involved in depending on foreign sources for 20 or 25 percent of the nation's oil requirements would be quite acceptable. Even if total requirements reached the 26 million barrels a day that was predicted before the October 1973 war—and that now seems unlikely—the United States would then have to import only between 5 and 7 million barrels of oil a day in that year. Particularly if a longer term oil supply arrangement was made with Venezuela, over 4 million barrels a day might be obtained in the Western Hemisphere. But even without such an arrangement, something approaching half of total oil imports could probably be bought in this hemisphere. Even if all of the remainder came from Arab countries, which would be unlikely, the vulnerability of this small proportion of total U.S. oil supplies to Arab embargo could be easily tolerated.

At some point, a combination of market forces and cooperative international efforts might create conditions favorable to the negotiation of an international oil agreement. Such an outcome should be a major objective of U.S. oil policy. Over the long run, an agreement that meets the needs of both producers and consumers would provide the best assurance that the oil industry will contribute to a smoothly functioning world economy.

NOTES FOR CHAPTER 19

1. Case III, in *U.S. Energy Outlook*, a summary report of the National Petroleum Council's Committee on U.S. Energy Outlook (Washington, D.C., 1972).

2. *New York Times*, November 8, 1973, p. 32.

3. *Oil and Gas Journal*, March 4, 1974, p. 21.

4. *New York Times*, December 13, 1973, p. 2.

5. Ministry of Energy, Mines and Resources, *An Energy Policy for Canada*, in 2 vols. (Ottawa, 1973), p. II: 127.

6. *World Energy Demands and the Middle East*, report of the Twenty-sixth Annual Conference of the Middle East Institute, Washington, D.C., September 29-30, 1972, in 2 parts (Washington, D.C., 1972), p. I: 99.

7. *New York Times*, April 14, 1974, p. 1.

8. *New York Times*, June 9, 1974, p. 1.

9. Article I of this Agreement states: "The Government of Iran is determined to resist aggression. In case of aggression against Iran, the Government of the United States of America, will take such appropriate action, including the use of armed forces, as may be mutually agreed upon and as is envisaged in the Joint Resolution to Promote Peace and Stability in the Middle East in order to assist the Government of Iran at its request." U.S. House of Representatives, Foreign Affairs Committee Print, *Collective Defense Treaties*, 91st Cong., 1st sess. (Washington, D.C.: Government Printing Office, 1969), p. 197.

10. See M. A. Adelman, "Is the Oil Shortage Real?" in *Foreign Policy*, Winter 1972-73, p. 69.

11. *Oil and Gas Journal*, December 25, 1972, pp. 66-68, and July 30, 1973, pp. 78-79.

12. *Oil and Gas Journal*, July 30, 1973, pp. 78-79.

13. Sam H. Schurr, Paul T. Homan, and associates, *Middle Eastern Oil and the Western World* (New York: American Elsevier, 1971), p. 80; "Japan Envisages Holding of 60 Days' Oil Stock by March 1975," *Japan Petroleum Weekly* (Tokyo), July 24, 1972, p. 1; "Directive du Conseil du 20 décembere 1968 faisant obligations aus Etats members de la C. E. F. de maintenir un niveau minimum de stocks de pétroliers" (68/414/CEF), *Journal Officiel des Communautés Européennes* (December 23, 1968), pp. 14-16.

14. *New York Times*, September 30, 1974, p. 57.

15. Text of communiqué issued by nations attending issued by nations attending Washington Energy Conference, *New York Times*, February 14, 1974, p. 26.

Chapter Twenty

International Nuclear Energy Policy

The policy of the United States with respect to the rapid spread of nuclear energy throughout the world should be made with three broad considerations in mind: the world's ever-increasing energy requirements, the security problems inherent in nuclear technology, and the desirability of increasing U.S. exports. These considerations are to some extent conflicting, and attempting to reconcile them will be a continuing problem.

The consequences of the diversion of nuclear fuel to military or criminal uses could be so serious that first priority must clearly be given to minimizing the security dangers accompanying the expanded use of nuclear energy. Second priority must go to the development of nuclear energy as one alternative to expensive and politically vulnerable petroleum. Least weight should be given to the promotion of nuclear-related exports. The likely size of such exports is in fact not large, so the sacrifice involved in not pushing them when to do so would run counter to security considerations would not be serious.*

Although U.S. predominance in nuclear energy matters may be on the wane, the nation's technological capabilities and worldwide industrial, as well as governmental, involvement in the nuclear field will enable U.S. policy makers to retain a major voice in shaping international nuclear energy policies. It is clear, however, that the United States cannot exert much influence over current LWR programs and policies in Japan and the industrialized countries

---

*U.S. exports of reactors, nuclear fuel, and other nuclear-related items and services in 1985 are expected to be only a very small fraction of total U.S. exports in that year. (See Appendix Table B-3.)

of Western Europe. Commercial patterns in those countries are well established, and tight policy coordination among governments in this field may not be a practicable alternative.

Emerging technologies in the reactor field and in uranium enrichment, on the other hand, may still be amenable to international policy coordination in ways that could both enhance economic benefits and reduce potentially undesirable security consequences. Similarly, the United States and other nations might profitably devote attention to the important economic and security issues that are about to be raised by the spread of nuclear energy systems to the developing nations.

Solutions must also be found to a number of safeguards problems. Existing deficiencies in the international safeguards system must be cured and the system adapted to emerging new technologies. Special precautions must be taken against the danger of nuclear theft. Possibly most important, adequate funding must be provided to enable the International Atomic Energy Agency to cope with a rapid expansion in the number of facilities and in the quantity of nuclear material that must be safeguarded.

U.S. POLICY OPTIONS CONCERNING ADVANCED NUCLEAR SYSTEMS

The High Temperature Gas Reactor

The introduction of the high temperature gas reactor, which uses highly enriched uranium for fuel, means that weapons-grade uranium will be produced for commercial use and will be in transit throughout the nuclear fuel cycle. The possibility of diversion of highly enriched uranium constitutes a new and extremely serious international security problem that must be added to the existing problem of plutonium diversions from conventional reactors.

Basically, HTGR technology is in the hands of one U.S. firm—General Atomic—and its licensees abroad, although there are a few small foreign indigenous efforts under way as well. The HTGR is expected to go into commercial operation in the United States in the late 1970s. The prototype is working well and several U.S. utilities have placed orders for the reactor. The technology for the HTGR is still classified by the U.S. government and the fuel for the prototype is being enriched at one of the government's gaseous diffusion plants. The opportunity exists for the United States to establish strong unilateral safeguards on the transfer of HTGR technology, particularly with respect to fuel fabrication. The United States should also push for strong international safeguards on the export of HTGR technology and fuel so that safeguards guidelines will have been established by the time other nations have developed the HTGR independently.

The United States could work for these kinds of international safeguards either through the International Atomic Energy Agency or through the Exporters Committee, an informal group established outside the IAEA framework "to develop arrangements that will ensure that all major exporting countries interpret their [NPT] obligations in a uniform way and do not try to get commercial advantage by cutting corners on safeguards."[1] The committee consists of government representatives from all non-Communist suppliers that have ratified or intend to ratify the Non-Proliferation Treaty. While the committee thus far has focused on establishing guidelines for safeguards on exports of nuclear materials and equipment to non-NPT members, it might usefully also become involved in establishing controls on all international transfers of nuclear technology.

Breeder Reactors

The development of the breeder reactor is the subject of extensive and growing controversy among nuclear industrialists, governmental agencies, environmentalists, economists, and scientists. The controversy centers on economic issues, environmental and safety issues, and potential international security problems. The capacity of the breeder to produce more fissionable material than it uses as fuel is both the source of its appeal and the cause of great concern. The breeder, if developed successfully, will make more efficient use of uranium ore than does the light water reactor; however, the larger amounts of plutonium that the breeder will produce could be used, essentially without change, to make nuclear bombs.

The evaluation of all aspects of the breeder controversy is beyond the scope of this book.[2] The focus here is on the major foreign policy issue associated with the breeder—i.e., the safeguards implications. From a safeguards standpoint, the case for the breeder essentially is that, even without this reactor, there will be increasing amounts of plutonium produced by light water and other reactors. More importantly, the plutonium produced by light water reactors, which is now stored, will probably begin to be recycled in the 1970s as fuel not only for breeders but also for LWRs. With plutonium recycling, the LWR will raise safeguards problems similar to those of the breeder in that plutonium will be moving through all phases of the fuel cycle and will therefore be subject to diversion or theft at each of the various fuel facilities and in transit. (Without plutonium recycling, the LWR is generally regarded as much less of a safeguards problem than advanced reactors, because plutonium is available only at the reprocessing plant and in transit to and from the reprocessing plant.) Thus, breeder proponents contend that development of this reactor will not create new safeguards problems.

The safeguards case against the breeder is that the vast stockpiles of plutonium created throughout the world by the breeder will indeed constitute a far more serious threat than would be the case if nuclear power was generated largely by LWRs, even with plutonium recycling. The tremendous increase in plutonium with the breeder, it is argued, will be so great that the problem will not be merely one of degree.

Since the breeder is still in the developmental stage, with commercialization unlikely before the 1990s, if then, it is difficult to analyze the dimensions of the plutonium problem with precision. Rough estimates indicate, however, that plutonium produced by the breeders will be on the order of three times that produced by LWRs.* The breeder will further exacerbate the safeguards problem because the plutonium that it produces will be of a much higher quality and easier to work with (from a weapons-potential stand-point) than that produced by LWRs.

Despite the continuing controversy, the international breeder race is so intense and the resources that governments and private industries have already poured into it are so great that the possibilities for halting development of the breeder are poor. Moreover, some European breeder programs appear to be ahead of the U.S. program, and the USSR has a demonstration plant in operation. Nevertheless, the United States should begin to explore with other industrialized nations the common problems raised by the possible future commercial development of breeder reactors. The IAEA has become one forum for such consultation. Another possible forum might be the Exporters Committee. Beyond this, a more substantial effort must be made to deal with breeders as part of discussions on the long-range energy problems of the advanced industrialized states. Given the cumbersome nature and slow pace of international agencies and conferences, however, multilateral consul-tation on security problems associated with the breeder will yield results—at best—for the medium or long run.

*The comparison was made between three 1000-megawatt breeders and three 1000-megawatt LWRs—one of which was on plutonium recycle; it takes approximately two conventional LWRs to provide enough fuel for one LWR on plutonium recycle.[3]

During the present stage of intensely competitive and nationalistic breeder programs, which offers little hope for forging international control or cooperation, the United States could most effectively influence events by slowing down its breeder program in order to reevaluate the merits of breeder reactors. From the safeguards point of view, the justification for this policy change does not rest exclusively on the problems associated with the large quantities of plutonium that will be produced by breeders. Rather, the basic problem from the safeguards standpoint is that heavy concentration on development of the breeder virtually locks the world in to nuclear power and its attendant and persistent threat to world security for many years beyond 2000. This is reflected in the high priority assigned to the breeder in government research and development funding both in the United States and in other industrialized countries. During the 1973 fiscal year, approximately 42 percent (about $260 million) of a total U.S. energy research and development budget of $622 million was allocated to the breeder.* In the 1974 fiscal year, an estimated $323 million was spent by the U.S. government alone on breeder development, and the administration has proposed raising expenditures for this purpose to $400 million in fiscal year 1975. In future years, large annual federal research funding will undoubtedly be necessary if this program is continued.

Major commitments to the development of other energy sources (as well as a commitment to the LMFBR) are of course possible. Indeed, the energy research budget for fiscal 1975 contains increased funding for research on alternative sources of energy. Nonetheless, as long as the breeder is considered "the answer" instead of one of several possible answers to future energy supplies, other potential energy sources are likely to receive lower priority and smaller funding. Since research and development programs tend to develop a life of their own, a substantial reduction in federal funding for breeder development would indeed constitute a major policy shift. A change in energy research and development priorities that provided for substantial research into other energy sources would permit more informed comparisons among several alternatives. Under present policies, the United States and the rest of the world have no way of knowing whether nuclear power is the best answer—or the only answer—to future energy needs. As discussed in

---

*By way of comparison, the United States spent about $5 million on the HTGR in the 1973 fiscal year.

Chapter 16, the world will not begin to run out of present proved reserves of uranium before the 1990s, and there is reason to believe that there is a great deal of undiscovered recoverable uranium on earth. In short, there is time to look into other energy sources and to make comparisons. A substantial slowdown rather than a termination of the breeder program is recommended, so that a return to the breeder is a viable alternative if other energy sources do not, after a strong research and development effort, appear acceptable.

A deliberate change in research and development emphasis in the energy field in the United States could have considerable impact abroad in that it would reduce the pressure on other industrialized countries to match the U.S. breeder effort. Such a policy could strengthen the hand of foreign experts who are questioning the priority and pace of foreign programs. Even if a slowdown in the domestic breeder program did not affect foreign programs, U.S. reactor manufacturers should not be seriously harmed commercially, as they almost certainly have little to lose in terms of foreign markets in the breeder race. Western Europe and Japan can be expected to supply their own markets, just as they are beginning to do in the light water reactor field. If breeders do go into commercial operation abroad, the U.S. government could, following the SST precedent, protect the domestic reactor market by pro-hibiting the importation of breeders on environmental and other grounds.

In Western Europe and Japan, concern over the long-term availability of imported uranium gives impetus to breeder reactor programs. As part of an international policy to bring breeders under greater control, therefore, the United States should seek ways of helping to ensure that world uranium resources are fully explored and made reliably available to other user nations.

Although uranium reserves are distributed among polit-ically diverse countries in various regions of the world, the number of countries with substantial known reserves is small. (See Appendix A, Table A-9.) Meetings of the major natural uranium exporters in 1972 and 1973 stimulated fears (in Japan, for example) that a "uranium OPEC" might emerge.[4] Even though the United States has large reserves of uranium ore itself, it should work against such a development. Ideally, a broad multilateral framework should be created that would deal with resource allocation, prices, and stability of supply. In the absence of such a broad framework, the United States can influence the world uranium market by its own export capabilities and policies. As the world market price for uranium rises above the artificially maintained U.S. price, the United States could enter the world market as a competitive supplier.

THE FUTURE U.S. ROLE IN ENRICHED URANIUM SUPPLY

For both economic and security reasons, some form of international cooperation or coordination in the field of uranium enrichment is imperative. The economic problem is to ensure adequate supplies of enriched uranium for the increasing number of nuclear power reactors, but to avoid the economic waste of building excess enrichment facilities. The security problem is posed principally by the new centrifuge technology for uranium enrichment.

Past U.S. plans for tight international control of enriched uranium equipment or service must be discarded as unrealistic in light of strong resistance by virtually all of the major industrialized countries. The United States should instead seek to influence future worldwide enriched uranium supply patterns through a systematic strategy that takes account of the close relationship between domestic decisions in this field and U.S. international nuclear policies. Decisions on research and development priorities, industrial access and private ownership of new enrichment facilities, and the terms of U.S. supply contracts must be made with foreign policy consequences and cooperative possibilities in mind. The success of such a strategy will depend on the United States maintaining a strong position as a supplier of enriched uranium.

In particular, in order to retain a share of the future enriched uranium market, the United States should avoid unilateral price increases and hardening of contract terms. This does not argue against obtaining advance commitments from foreign and domestic clients that will permit future capacity requirements to be sensibly planned and financed, nor does it suggest that U.S. facilities should not be operated on a commercially sound basis. But it does suggest that the AEC should ensure that privately run plants will adequately serve foreign needs and that foreign users will be consulted as a matter of course before final decisions are made regarding changes in prices or contract terms.

In early 1974, the United States took the lead in establishing an *ad hoc* group on uranium enrichment composed of representatives from the United States, Western Europe, and Japan.[5] The committee was established to look into possibilities for broad international cooperation in enrichment. In proposing formation of the group, the United States was especially interested in multilateral sharing of enrichment technology. There is still a serious foreign interest in building gaseous diffusion plants. For example, in late 1973 France announced its intention to build an enrichment plant based on French gaseous diffusion technology.[6] A loosening of U.S.

government controls over the sharing of its gaseous diffusion technology might interest various European nations and Japan in investing in facilities using this tested technology as a means of obtaining assured supplies of enriched uranium.

The new advisory committee would also be a valuable forum for sharing centrifuge technology. The United States should continue to strengthen its own position in this field and seek to bring centrifuge technology under some form of international policy control similar to the AEC's industrial access program. The influence of the United States, with its traditionally vigorous support of strong international safeguards, could be of great importance in shaping international centrifuge policies. It would be extremely useful if the *ad hoc* group became a standing committee and evolved into a body that would seek common approaches to the development, commercialization, and export of centrifuges.

THE INTERNATIONAL SAFEGUARDS SYSTEM AND U.S. POLICIES

The chief ingredients of an effective international safeguards system include the participation of all countries with commercial nuclear power programs, the availability of sufficient financial resources to operate the system, and the development and utilization of adequate safeguards procedures and techniques.

The Problem of Adherence to the Non-Proliferation Treaty

Achieving universal adherence to the NPT appears all but impossible. Years of diplomatic effort by the United States and other supporters of the treaty have reduced the group of holdouts to a relatively small number of countries, but some of them—including key countries such as France, China, India, Egypt, and Israel—appear impervious to ordinary persuasion. The question arises whether more powerful leverage should be applied and, specifically, whether the United States and other existing or potential suppliers should prohibit or restrict exports of reactors, nuclear fuel, and nuclear technology to nations that have not adhered to the NPT.

France would pose the principal obstacle to such a strategy. As a supplier nation itself, France would be less vulnerable than other holdouts to the denial of nuclear materials and technology. Also, France could not be expected to join an international effort to force others to do what it has refused to do, and the lack of French participation could undermine the whole effort, since other suppliers would hesitate to abandon lucrative nuclear export markets to French firms.

The political costs of trying to force adherence to the NPT could be high, and the prospects of success would be uncertain. Some of the non-NPT nations would respond, not by signing the treaty, but by developing autonomous nuclear fuel cycles and indigenous capabilities to manufacture reactors.

A more practical way to deal with the problem posed by the refusal of some nations to adhere to the NPT is to improve the effectiveness and increase the uniformity of the bilateral safeguards imposed by various supplier nations without reference to the NPT. The Exporters Committee provides a possible means of coordinating such an effort. Ideally, France might be persuaded to join the committee or at least to cooperate in requiring that all exports of nuclear materials and equipment be subjected to effective safeguards. But even without France, the other major suppliers could exert a strong restraint on the spread of unsafeguarded nuclear facilities.

Resources of the IAEA

Adequate financing is fundamental to an effective safeguards system. Even the most refined techniques will be of little use if the resources required for their application are lacking.

The IAEA is supported by voluntary contributions of member states. The total budget for 1972 was approximately $20 million, of which perhaps $2 million was devoted to safeguards activities. Annual financial requirements for safeguards activities are expected to rise to $5 million by 1975 and to reach at least $10 million by 1985.[7] Within the next few years, the number of IAEA inspectors is likely to increase from 50 to about 150, and by 1985 the inspection corps could easily reach approximately 400 professionals. These manpower estimates are conservative in that they are based on minimum rather than optimum inspection levels.

The IAEA bears the costs of its safeguards activities and reimburses states for any expenses incurred at its request, but the states bear all normal reporting and records expenses, as called for under the NPT. Some non-nuclear-weapons states, especially developing nations, have objected to the IAEA's using its resources for international safeguards tasks at the expense of other agency activities that they believe to be more important. Most of the major industrialized states are considerably less interested than the United States in the safeguards system, and it is not clear that they will be willing to increase their contributions significantly.

The U.S. contribution to the IAEA is limited by Congress to 25 percent of total agency funding from all countries. If the total budget rises from $20 million to, say, $30 million by 1985, (with most of the increase allocated to safeguards activities), the maximum U.S. contribution would be about $7.5 million. If other developed countries cannot be induced to increase their contributions, the United States will be faced with the choice of either raising the statutory limitation on the U.S. share, living with a greatly weakened international system, or seeking alternative sources of funding.

One alternative that might be explored is that of levying taxes on the major nuclear facilities of all IAEA member countries—including those located in the nuclear weapons states—on grounds that the cost of an effective safeguards program is an integral part of operating nuclear power systems. Obviously such a tax would be opposed by nuclear industry interests. Nonetheless, it should be possible to design the tax so that no company or country would be disproportionately affected.

The major emphasis in the last several years on persuading key countries to become parties to the NPT has overshadowed the crucial question of IAEA funding. With the anticipated rapid growth of nuclear power facilities throughout the world, and the accompanying growth in manpower requirements for the IAEA's inspection function, the thorny and potentially divisive financial issue will inevitably become much more prominent than heretofore.

Safeguards Techniques and Procedures

The precise terms of the safeguards agreements negotiated between the IAEA and individual nations and the specific procedures and techniques employed by the IAEA will do much to influence the overall effectiveness of international safeguards. Generalizations about safeguards mean little; details are crucial; and more than in most areas of foreign policy, U.S. policy on safeguards must take adequate account of technical problems.

A serious safeguards problem stems from the fact that certain material losses or uncertainties in nuclear inventories—designated as "material unaccounted for" or MUF—occur during the normal operation of commercial facilities. The obvious danger is that deliberate diversion of nuclear materials for military purposes could be concealed within the range of expected uncertainty. The MUF in a chemical reprocessing plant, for instance, could be as high as 2 or 3 percent of the total plant capacity—that is, 2 or 3 kilograms of plutonium for each 100 kilograms processed—a substantial uncertainty given the small amounts of plutonium needed for weapons fabrication. Experience in operating commercial nuclear facilities has provided data on expected MUF levels and on the causes for such uncertainties. While nuclear industries have strong economic and safety motivations to reduce the expected loss rate, MUF can never be totally eliminated. The IAEA faces the problem of setting acceptable MUF limits that are neither excessively narrow, which could place unreasonable burdens on industry and might increase the false-alarm rate, nor unnecessarily wide, which could facilitate diversion and undermine the value of international safeguards.

One approach taken by the IAEA to deal with the uncertainties in nuclear material measurements has been to emphasize "level of accountability"—a technique that establishes a defined amount of nuclear material "the loss or diversion of which is to be detected at a given level of confidence with a given probability of detection and within a given period of time."[8] This approach recognizes that the diversion of certain materials, such as plutonium and highly enriched uranium, poses a far greater threat than the diversion of natural or slightly enriched uranium. Consequently, it aims for relatively high assurance of detecting the loss of "critical amounts" of these materials—that is amounts sufficient to manufacture a nuclear bomb—while permitting higher tolerance levels in monitoring amounts of natural or slightly enriched uranium. The United States and other major NPT participants should support and help refine this "critical amounts" concept.

As important as this concept is, the single most important factor determining the overall effectiveness of the international safeguards system will probably become the precise number and nature of independent IAEA inspections applied to particular facilities.

The IAEA inspection rate is negotiated between the agency and a given state. The approved NPT safeguards concept specifies that the inspection effort, often termed the minimum routine inspection effort or MRIE, should be determined by applying a number of criteria, including the accessibility of nuclear material in particular facilities, the characteristics of a nation's overall fuel cycle, the development of new safeguards techniques, and perhaps most importantly, the overall effectiveness of a state's domestic safeguards system. Accordingly, given a relatively reliable national safeguards system and a reasonable set of assumptions regarding plant characteristics and verification techniques, the IAEA's inspection effort could drop by as much as 50 percent below the MRIE level for such crucial facilities as reprocessing plants, reactors, and plutonium storage sites. The IAEA's calculations have demonstrated that such reductions in inspection effort could, under appropriate conditions, still satisfy adequate safeguards.[9]

The possibility of reducing the frequency or intensity of international inspections could stimulate nations to establish more reliable domestic safeguards systems. On the other hand, many experts feels that the MRIE formulas already reflect concessions and that any further reduction would seriously weaken the international safeguards system. One U.S. observer, for example, recently stated that the "maximum routine inspection effort permitted under the NPT safeguards is . . . about the minimum required for a credible system."[10] Of particular concern is the possibility that the negotiation of less stringent inspection provisions with states having relatively effective domestic safeguard systems, as in the case of the European nations, will become the norm for all states. In such circumstances, the continuous inspection needed to monitor fuel reprocessing and other crucial commercial nuclear functions could be compromised and dangers of diversion would arise.

The original IAEA safeguards agreement developed prior to the NPT called for the right of access at all times by inspectors who were to be ·in residence in conversion, fabrication, or reprocessing plants during the entire period that safeguarded materials were being processed.[11] It would be unfortunate if the international safeguards system devised to support the NPT results in a substantially less effective safeguards system by excessive compromises in the face of national and commercial pressures.

A powerful strategy available to the agency is its ability to allocate extra inspection effort to suspicious facilities without exceeding the total inspection level for the category or facility involved. In negotiating safeguards agreements with the IAEA, however, some nations may seek to institute fixed numbers of inspections for particular facilities as subsidiary arrangements are reached. Such an arrangement would degrade the safeguards system, and the IAEA should insist upon retaining its flexibility to allocate annual inspection efforts on a nonproportional basis among facilities in designated categories. The United States should strongly support this position.

The effectiveness of international and domestic safeguards is not dependent simply upon inspection frequency and access. It is influenced also by the availability of physical devices and analytical techniques that can reduce measurement uncertainties without unduly increasing access to commercial processes or facilities. In certain cases, the use of sophisticated instruments could increase the cost of inspections, but such a price would be worth paying if inspections could be made easier, more reliable, and less intrusive. New techniques are presently being developed and tested by the IAEA and by the United States and other major nuclear nations, often through cooperative programs or under IAEA contracts. Consistent with its responsibilities under the NPT, the agency is also continuing to improve its capability to collate, store, retrieve, and analyze the growing volume of data produced by national safeguards systems.

A somewhat different technical approach to improving the efficiency and dependability of safeguards is the design of new facilities or the modification of existing ones. Such design features include greater use of physical barriers to isolate areas within plants, the establishment of secure storage areas, and the development of flow patterns that permit reliable sampling. Under present guidelines, the agency has no authority to force a state to redesign its facilities to make inspection easier, although it can suggest or recommend modification of a plant or process if adequate safeguarding cannot be accomplished. All nations would appear to have a self-interest in modifying plant designs or procedures to the extent practicable, simply as a means of reducing the intensity of visits by IAEA authorities. It would be useful, therefore, if specific design criteria were included in the model facility agreements to be published by the IAEA. National governments could then incorporate these features in licensing agreements for commercial nuclear facilities. Perhaps incentives could be offered for companies to utilize improved safeguards designs and procedures by minimizing the intensity of domestic safeguards as appropriate or by introducing financial penalties for domestic facilities refusing to implement reasonable design modifications.

The Centrifuge Problem

The safeguarding of centrifuge facilities deserves special consideration by the IAEA and by potential manufacturing countries because of the weapons proliferation dangers these facilities pose. One obstacle to effective centrifuge safeguards is the unwillingness of nations to permit IAEA inspectors to examine centrifuge units because to do so could compromise commercially sensitive design information.[12] Another inspection problem posed by the centrifuge is the fact that the small size of centrifuge units and their simplicity of operation facilitate clandestine activities.

Designing centrifuge facilities in a way that would facilitate inspections without exposing commercial secrets is one possibility that should be explored. The presence of resident IAEA inspectors also might help deter illegal operations and might lead to detection of suspicious activity. Unfortunately, continuous inspection would not be permitted, if—as now appears likely—centrifuge facilities are put in the same inspection category as enrichment plants, which are subject to low inspection requirements. Because of the critical importance of safeguarding centrifuge facilities and the difficulty of doing so, the United States should press for inclusion of these facilities in the same category as reprocessing plants and fabrication facilities using highly enriched uranium.

Multilateral Nuclear Export Controls

The Non-Proliferation Treaty has a direct bearing on international nuclear transfers and export policies. Under the no-transfer, no-acquisition clauses of Articles I and II, parties to the treaty cannot transfer nuclear weapons to non-nuclear-weapons states or assist such states in manufacturing such weapons, and the latter are obliged not to receive nuclear weapons or weapons manufacturing assistance. In addition, under Article III, all parties are obliged not to provide nuclear material or equipment to any non-nuclear-weapons state unless IAEA safeguards are applied.

The Exporters Committee has concentrated on developing common guidelines for imposing safeguards on exports to nations remaining outside the treaty. Exports of fissionable materials raise few problems since the need for safeguards is clear. The difficult question is how to define the types of equipment that should automatically trigger a safeguards requirement, taking into account the fact that many components and devices useful in nuclear programs are currently widely available and have many non-nuclear applications. The United States has consistently pressed for stronger controls on technology and nuclear-related equipment than other supplier nations have been willing to accept. Nevertheless, progress has been made in devising a "trigger list" of critical or unique items that all supplier nations should require be covered by IAEA safeguards when exported.

The Exporters Committee is an active and useful group, but at present it has no official standing. If arrangements can be made for French cooperation and Soviet support, the committee ultimately should become associated with the IAEA. The committee could then provide a vital institutional framework for dealing with a wide range of export policies and problems. The United States should not only continue to play a leadership role within the committee, but should also work toward its gaining official status either within the IAEA or as an affiliate.

The Exporters Committee could well become the appropriate international forum for developing common guidelines for centrifuge exports. A possible approach to this problem would be to set a lower limit on the size of centrifuge plants to be exported; to maintain agreed security classifications on centrifuge developments; and to develop an international registry of technology exports. It would also be important to develop policies to prevent centrifuge assistance from being used as bait for reactor sales. The bargaining position of the United States in pressing for strict controls over centrifuge exports would be strengthened by the other suppliers' realization that the United States could, if it wished, use its strong commercial and technological position to make it very difficult for them to compete for export sales of either centrifuges or enriched uranium.

As the number of countries ratifying the NPT increases, the Exporters Committee might properly turn its attention to the trade in other nuclear materials and technology among NPT members. Article IV of the NPT, however, provides that no party to the treaty will be deprived of the benefits of nuclear technology for peaceful uses. The committee should therefore not attempt to restrict the flow of technology among parties to the NPT. It could, however, establish a registry of trade among treaty members in nuclear equipment, fuels, and technology. This registry would provide the IAEA with useful information on what was moving where and would facilitate the application of safeguards. The establishment of the World Nuclear Fuel Market (WNFM) in January 1974 could make it easier for the Exporters Committee to keep an up-to-date registry of nuclear trade. The possibility of cooperation between the Committee and the WNFM for this purpose should be actively explored.

NUCLEAR ENERGY AND THE LESS DEVELOPED COUNTRIES

Nuclear power will continue to be unsuitable for many LDCs well into the 1990s, primarily because of the small size of their projected electric power grids. At the same time, nuclear power could provide an alternative to high-priced fossil fuels for the 15 to 20 large LDCs, a number of which already have reactors in operation or on order.

The spread of nuclear power into the developing countries will inevitably make the establishment and operation of an effective international safeguards system more difficult. This is not because the LDCs are any less reliable as a group than the industrialized countries. Most LDCs have in fact signed the Non-Proliferation

Treaty.* The problem is simply that the need to extend the international safeguards system to more and more countries will further strain the capabilities of the IAEA. The answer, however, is to provide the IAEA with the resources that it needs—not to try to restrict the introduction and growth of nuclear power in developing countries. To attempt to do so would be to ask the LDCs to forego a potential economic benefit that the industrialized countries have been unwilling to deny themselves.

The choice of the most efficient mix of energy sources is a difficult one and the considerations that must be weighed vary from country to country. The LDCs need technical assistance in planning their electric power programs, including help in evaluating the overall feasibility of alternative energy sources. The United States should support the expansion of the IAEA's recent efforts to provide this kind of technical assistance and should encourage the IAEA to continue to analyze a given country's future power system from the standpoint of the country's total development interests.

The United States should also encourage U.S. reactor manufacturers—by subsidizing research and development programs if necessary—to develop a standardized, light water reactor in the 200 to 300 megawatt range, a size that is considered appropriate for the initial nuclear power programs of most LDCs. A Canadian heavy water reactor in this megawatt range has already been developed and apparently operates with reasonable efficiency. Development of a small light water reactor would provide the LDCs with a choice between the two predominant reactor types and would improve the long-term competitive position of U.S. manufacturers in the developing countries.

THE NUCLEAR THEFT PROBLEM AND THE NEED FOR SAFEGUARDS

The threat of commercial nuclear power to international security is often seen only in terms of the "diversion" problem—i.e., the possibility that governments that do not now have nuclear weapons might divert nuclear fuels from commercial facilities to the manufacture of such weapons. Another important security problem is the danger of nuclear theft, blackmail, and hijacking.[13] Despite the clear international security risks posed by such actions, the international community is not moving vigorously enough to deal with them.

---

*All of the LDCs—with the important exceptions of India, Pakistan, Argentina, Brazil and Egypt—have signed the treaty. Israel, which is usually regarded as a developed country, also has not signed.

Under the existing IAEA structure, domestic safeguards systems are responsible for the physical protection of nuclear material within each country. In principle, domestic safeguards should be more stringent than standard IAEA safeguards. In actuality, the state of national safeguards systems throughout the world is far from ideal. It would clearly be desirable for the IAEA to include physical security as a prerequisite for concluding the safeguards agreements called for by the NPT.

The industrialized nations should coordinate nuclear intelligence and enforcement programs in order to facilitate the recovery of stolen material. Procedures for dealing with hijackings might be drawn up, along with an international system of rewards for information leading to the apprehension of persons involved in nuclear theft. The IAEA could play an active role in these efforts and could contribute directly by publishing standards for international nuclear security forces and by training personnel to deal with these problems. The agency could also become a clearinghouse for information on such matters.

A more sweeping response to the dangers of theft would be to impose IAEA control over plutonium stockpiles. These facilities could be operated as international "banks" where depositers would store their plutonium, draw registered amounts for use as fuel, and deposit additional material. Such a scheme would greatly reduce the possibilities for nuclear theft, but it would not deal with theft from reactors, from nuclear fuel or reprocessing facilities, or during transportation. Admittedly, the likelihood that a plutonium "bank" would be adopted is remote. However, other measures to lessen the dangers of theft from reactors, from nuclear fuel and reprocessing facilities, and during transport are available and practicable and their implementation should receive high priority.

*   *   *   *   *   *   *

In sum, the anticipated growth in the use of nuclear power as a commercial source of energy and the development of new nuclear technologies will greatly increase the size and difficulty of the task of preventing the diversion or theft of nuclear fuels. The safeguarding of commercial nuclear energy facilities will require a stronger policy commitment and the expenditure of more resources and effort than most governments of the world have thus far accepted.

The NPT is scheduled for review by treaty members in 1975. This will provide an excellent opportunity for international consideration of the safeguards problems discussed in this chapter and in Chapter 17.

NOTES FOR CHAPTER 20

1. Sigvard Eklund, IAEA director-general, quoted in *Nuclear Industry*, November-December 1972, pp. 63-64.

2. For an analysis of the economic and environmental issues, see Thomas B. Cochran, *The Liquid Metal Fast Breeder Reactor: An Economic and Environmental Critique* (Baltimore: Johns Hopkins University Press, 1974), p. 228.

3. Ibid., pp. 205-206.

4. *Nuclear Engineering International*, March 1973, p. 153.

5. *The Department of State Newsletter* (Washington, D.C.), March 1974, p. 36.

6. *The Economist*, December 1, 1973, p. 67.

7. These estimates are based on a number of sources, including Stockholm International Peace Research Institute (SIPRI), *The Near-Nuclear Countries and the NPT* (New York: Praeger Publishers, Inc., 1972); and Douglas E. George and Ralph F. Lamb, "International Safeguards," in *Civil Nuclear Power and International Security*, ed. Mason Willrich (New York: Praeger Publishers, Inc., 1970). The key assumptions made are to exclude major IAEA safeguards in the United States or the United Kingdom and to use inspector salaries as the primary cost parameter.

8. W. Gmelin, et al, "A Technical Basis for Safeguards," in *Peaceful Uses of Atomic Energy*, in 15 vols., Proceedings of the Fourth International Conference, United Nations and the International Atomic Energy Agency (New York, 1972), vol. 9, p. 508.

9. Ibid., pp. 503-10.

10. Mason Willrich, "Evaluation of International Safeguards and Their Impact on Nuclear Industry," paper presented at International Conference of the Atomic Industrial Forum and the American Nuclear Society, November 14, 1972. (See *Nuclear Industry*, November-December 1972, pp. 53-54, for summary of paper.)

11. George and Lamb, "International Safeguards," p. 44. The full text is given in Appendix E of Mason Willrich, *Non-Proliferation Treaty: Framework for Control* (Charlottesville, Va.:Michie Company, 1969).

12. See Victor Gilinsky, "The Military Potential of Civil Nuclear Power," in *Civil Nuclear Power and International Security*, ed. Willrich, pp. 14-40, and *Bulletin of the Atomic Scientists*, June 1970, p. 43, for discussions of the weapons proliferation implications of centrifuges.

13. For a thorough treatment of these problems from the standpoint of U.S. safeguards policies, see Mason Willrich and Theodore B. Taylor, *Nuclear Theft: Risks and Safeguards* (Cambridge, Mass.: Ballinger Publishing Company, 1974).

Chapter Twenty-one

Difficult Choices

Chapters 19 and 20 explored U.S. energy policy alternatives in some detail and arrived at a number of recommendations. At some risk of both repetition and oversimplification, it may be useful in this brief concluding chapter to lay out more starkly the difficult choices confronting the United States in the field of international energy policy.

IMMEDIATE PROBLEMS

The most pressing energy-related problems facing the United States and the other oil-importing countries are adjusting to the recent huge increase in oil prices and preparing for the possibility of a renewal of the Arab supply restrictions of 1973-74. The United States has thus far chosen to seek solutions to these problems through international cooperation, but it could have adopted a go-it-alone policy and as of mid-1974 that choice had by no means been entirely foreclosed.

The United States could still elect to take advantage of the fact that a disproportionate share of the swollen revenues of the oil-exporting countries will be spent or invested here and leave the other oil-importing countries to finance their oil-induced balance of payments deficits as best they can. The United States has, however, recognized its dependence on a healthy world economy and has initiated or supported a number of constructive international efforts to deal with the problem of adjusting to high oil prices. Whether these efforts will be sufficient, however, remains to be seen. Certainly, more must be done if the hardest hit developing countries are to be saved from disaster.

The United States could also, if it wished, try to protect itself from future oil supply emergencies by entering into special long-term supply agreements with selected oil-exporting countries. Here, too, the United States has recognized that taking care of its own energy needs would not be enough; it would not escape unharmed if restrictions in oil exports damaged the economies of its major trading partners. And in any case, special bilateral deals for oil could set off a scramble among the industrialized nations that could weaken both their general political relations and the multilateral trading system upon which they all depend.

The United States also faces a difficult choice with respect to sharing available oil supplies in a future oil supply emergency. At present, the United States could ride out any likely emergency with only a moderate degree of public inconvenience and a relatively minor disruption of normal economic activity. But once more, the indirect vulnerability of the United States to reduced economic activity in other major industrialized countries has caused the U.S. government to favor some kind of sharing system in future emergencies.

A secondary issue, however, is more difficult than the basic choice of whether or not to share: should domestic U.S. production be included in the oil supplies that are to be shared? At the Washington Energy Conference in February 1974, U.S. Secretary of State Kissinger gave an affirmative answer. But how far the U.S. public and Congress would be willing to go in the event of an actual emergency is much less clear.

In the field of nuclear energy, the most pressing international problem is probably establishing effective safeguards over the new centrifuge technology for enriching uranium. Devising such safeguards poses technical problems, but obtaining agreement on them may be even more difficult. The European backers of the URENCO centrifuge project might well regard a proposal to apply special safeguards to centrifuges as an effort to preserve the dominant position of the U.S. gaseous diffusion method of enriching uranium.

PRIORITIES IN THE MIDDLE EAST

The need to reconcile conflicting political and economic priorities in the Middle East could pose expecially difficult choices, both immediately and over the longer run.

In the autumn and winter of 1973-74, the more militant Arab leaders may have hoped to force the United States to choose between oil and the security of Israel. More moderate leaders, however, chose not to push the issue that far. They presumably realized that to do so would risk severely damaging the world economy on which they depend for their own development and could permanently rule out the good relations with the United States that they desire for economic and security reasons.

For the time being, a showdown has therefore been avoided, but the basic problem remains. Deciding between the security of Israel and the economic well-being of Western Europe, Japan, and indirectly the United States itself is one hard choice that the United States must by all means try to avoid having to make. This is all the more so because every crisis in the Middle East involves the risk of a dangerous confrontation between the United States and the Soviet Union. Some encouragement may be derived from the success achieved in separating the opposing forces, but whether further progress toward a settlement can be made is at best an open question in mid-1974.

If a stable settlement is achieved, the fundamental conflict of U.S. priorities will presumably be quite manageable. If not, the question for which there is not yet an answer is whether the United States can gain the kind of political and economic relationship with the leading Arab states that would survive a breakdown in diplomatic efforts to settle the Arab-Israeli dispute. Economic cooperation and military assistance can perhaps go some distance toward creating such a relationship. But the United States can scarcely hope to escape a series of difficult decisions concerning how far and how hard Israel should be pushed to accede to Arab demands.

LONG-TERM ENERGY POLICY

The most fundamental and difficult decision that the United Sates must make with respect to its long-term energy policy is choosing the proper degree of energy self-sufficiency. Complete self-sufficiency has three major attractions: it would eliminate the direct vulnerability of the United States to political manipulation of its oil supplies; it would remove the determination of U.S. crude oil prices from the hands of a group of foreign governments; and it would help the other oil-importing countries by taking U.S. demand out of the international oil market.

The true value of the first two of these arguments may, however, easily be overstated. The United States, as has already been pointed out, would still be indirectly vulnerable to a restriction of oil shipments to its trading partners. And since self-sufficiency could be achieved only by developing high-cost domestic sources of energy, it might bring higher rather than lower energy prices.

Complete self-sufficiency would also have political costs. Buying oil abroad can promote better relations with the oil-exporting countries. Also, as an oil-importing country, the United States is in a better position to use its influence and leadership to advance measures, such as an international oil agreement, to stabilize an industry that plays a vital role in the world economy of which the United States must remain a part, whether it is self-sufficient in energy or not.

The most prudent long-term energy policy appears to be to achieve and maintain the degree of energy self-sufficiency that would permit the United States to adjust without serious hardship to the direct impact of any likely oil supply interruption. Holding oil imports to between 20 and 25 percent of total oil requirements would be a reasonable goal.

The United States also faces long-run choices with respect to nuclear energy. It is too late to reconsider the wisdom of developing this form of energy at all, and the world may well need to increase its reliance on it greatly in the next century if other alternatives are not developed before oil and other hydrocarbon resources are exhausted or can be produced only at constantly increasing costs. Nuclear energy should not, however, be developed without proper regard for the serious security problems that have been discussed in this study, or for the equally serious safety and environmental problems that have been analyzed in other studies.

The most difficult foreign policy decision that the United States must make with respect to nuclear energy concerns the fast breeder reactors that are under development in the United States and several other industrialized countries. The large amounts of plutonium that will be produced by the fast breeders if they are put into commercial use will greatly complicate the task of preventing diversion or theft of fissile material present at various stages of the nuclear fuel cycle. Moreover, concentration of research and development efforts on the breeder to the relative neglect of other possibilities, such as controlled nuclear fusion, could in time amount to a decision to prolong the era of heavy reliance on power from nuclear fission, with all of its security, safety, and environmental problems. On the other hand, large sums of money have already been spent on the breeder, and its advocates argue that breeder technology is needed to conserve the world's uranium resources.

To proceed with rapid development of the breeder before the security and safety problems that it poses have been solved would be foolhardy—especially since it is entirely possible that other new energy technologies will become feasible before even currently known reserves of relatively low-cost uranium approach exhaustion. But stopping present breeder development programs in the United States and elsewhere is almost certainly impossible, given their momentum and the degree to which government leaders have committed their prestige to this particular solution to the energy problems of the future. The only way to resolve this dilemma may be to curtail the U.S. breeder program significantly, but not terminate it. The option of putting the breeder into commercial use would thereby be maintained, but no irreversible commitment to this course of action would be made.

CONCLUSIONS

This study's review of the major uncertainties and difficult choices that confront the makers of U.S. foreign policy in the field of energy leads to several broad conclusions.

First, the nature and importance of the uncertainties give special value to a flexible approach. To the extent possible, alternative options should be preserved and all bets should be hedged. No irreversible gambles should be made on particular sources of energy supply or on the stability of the energy-related policies of other nations.

Second, while the problems of international energy policy are both difficult and persistent, they are manageable. Few, if any, perfect solutions can be expected, but the United States should be able to avoid both extreme risks and intolerable political or economic costs.

And, finally, most of the energy-related problems faced by the United States abroad are best attacked in cooperation with other nations. This principle applies to problems as different as increasing and diversifying energy supplies, safeguarding the nuclear power industry, and dealing with the effects of high oil prices on the poorer countries of the world. But since the needed degree of cooperation may not be forthcoming in every instance, the United States must be prepared, when necessary, to act by itself. The best international energy policy is one that seeks the maximum cooperation with others but that will not leave the United States helpless to preserve essential U.S. interests if such cooperation is lacking.

Appendix A

World Patterns of Energy Production, Consumption, Trade and Reserves

PRODUCTION AND CONSUMPTION

The regional and major country breakdowns of world energy production, consumption, trade flows, and reserves shown in the accompanying tables indicate the heavy dependence of two of the major energy-consuming regions—Western Europe and Japan—on energy imports. Three other principal industrialized regions—North America, the USSR, and Eastern Europe—emerge as major producing and consuming areas; in 1971 none of the three was heavily involved in world energy trade.

Total world production of energy increased by approximately 70 percent from 22.1 billion barrels of oil equivalent to 35.6 billion barrels between 1962 and 1971. (See Table A-1.) During the decade, significant changes occurred both in the components of energy produced and consumed and in the major producing and consuming areas. Tables A-1 through A-7 primarily pertain to conventional fossil fuels. Production, consumption, trade, and reserve figures relevant to nuclear energy are shown in Tables A-8 through A-11.

Coal production increased by only 16 percent during the ten-year period; thus, in 1962 coal accounted for nearly one-half of world energy production and consumption, but by 1971 its share had dropped to approximately one-third. Production of petroleum and natural gas more than doubled during the same period. In the

Middle East, production of oil nearly tripled; North American, Latin American, and Western European production increased; and Africa (including North Africa) by 1971 had emerged as a major oil-producing area. Natural gas production increased significantly in North America, Western Europe, and Latin America, but remained at low levels in the oil-producing countries of Africa and the Middle East.*

Total energy production in North America increased by some 47 percent, but in Western Europe it remained fairly steady over the decade. The distribution of production among the various energy sources changed considerably in Western Europe—natural gas production increased dramatically and coal production declined. In North America output of all major energy sources increased substantially.

In North America and in Western Europe, total energy consumption rose by roughly 53 percent, while in Japan the increase was nearly 160 percent over the ten-year period. Total energy consumption in the less developed regions in the 1960s increased at a more rapid rate than in most developed countries. In the Middle East, Far East, and Latin America, total consumption more or less doubled.

INTERREGIONAL TRADE IN ENERGY

Table A-2 shows major interregional flows in solid and liquid fuels in 1971. In that year, interregional trade in solid fuels was 566 million barrels of oil equivalent; trade in liquid fuels was about eighteen times as great and accounted for over 10 billion barrels of oil equivalent. Note that intraregional trade in energy is not included, so that, for example, the significant trade in coal among West European countries and within Eastern Europe does not appear in the figures. For the purposes of Table A-2 and A-3, the United States, Canada, and Japan are considered as separate regions. China is included among the Communist countries rather than as part of the Far East. In fact, no trade between Communist countries is shown.

In Table A-3, imports are the combined total of imports for domestic consumption and imports entering bonded warehouses or foreign free trade zones. Exports include national exports and reexports. Bunkers were not intentionally included in export figures, although some bunkers unavoidably slipped in. Liquid fuels include

*Proved reserves of natural gas in the Middle East and Africa are relatively high (see Table A-6). Production in these areas is low, however, because the costs of shipping natural gas have limited exports.

Table A–1. World Production, Trade, and Consumption of Energy, by Regions and Selected Countries, 1962 and 1971 (in thousand barrels of oil equivalent)

		Production								Consumption			
		Total Energy	Coal & Lignite	Crude Petroleum	Natural Gas	Hydro & Nuclear Electricity	Net Imports	Bunkers	Total Energy	Solid Fuels	Liquid Fuels	Natural & Imported Gas	Hydro, Nuclear & Imported Electricity
WORLD	1962	22,101,675	10,143,025	7,897,056	3,587,148	474,447	-185,053	650,534	21,646,020	10,166,741	7,433,849	3,570,557	474,874
	1971	35,576,406	11,738,587	15,537,650	7,482,000	818,163	-243,976	1,079,147	34,772,272	11,696,266	14,843,438	7,414,803	817,766
NORTH AMERICA	1962	7,507,065	1,981,663	2,638,714	2,716,090	170,598	612,226	110,270	8,113,141	1,868,419	3,366,829	2,707,642	170,251
	1971	11,041,101	2,550,607	3,581,831	4,618,951	289,713	1,094,268	131,276	12,363,477	2,333,047	5,225,713	4,545,118	289,600
Canada	1962	491,622	38,999	223,528	165,351	63,744	51,945	17,900	538,544	90,479	283,509	101,602	62,955
	1971	1,097,982	76,195	458,910	461,830	101,048	-113,440	23,726	986,973	119,354	470,983	297,881	98,755
United States	1962	7,015,315	1,942,536	2,415,186	2,550,739	106,854	558,193	90,993	7,573,661	1,777,720	3,082,605	2,606,041	107,295
	1971	9,943,041	2,474,333	3,122,922	4,157,121	188,665	1,203,602	105,399	11,404,466	2,213,536	4,752,853	4,247,237	190,840
LATIN AMERICA[a]	1962	1,618,534	38,009	1,414,601	142,027	23,897	-858,122	94,462	785,764	48,726	579,405	134,412	24,221
	1971	2,072,073	51,372	1,691,264	277,722	51,715	-761,362	127,037	1,349,209	72,755	939,595	276,032	51,827
Mexico	1962	176,356	5,424	109,211	58,349	3,371	-22,094	598	163,893	5,694	108,707	50,798	3,695
	1971	290,060	9,849	152,400	118,918	8,894	10,432	568	316,349	12,235	177,880	117,228	9,006
Venezuela	1962	1,101,633	132	1,067,235	33,869	397	-1,015,211	20,345	112,411	696	77,449	33,869	397
	1971	1,251,960	201	1,188,618	61,123	2,024	-1,158,654	21,046	131,752	1,671	66,934	61,123	2,024
WEST EUROPE	1962	2,804,456	2,438,926	112,406	102,234	150,891	1,774,025	203,918	4,458,177	2,642,746	1,562,537	101,690	151,204
	1971	2,868,244	1,844,409	119,996	663,382	240,458	4,388,215	355,848	6,799,485	2,010,725	3,865,301	683,844	239,615
France	1962	334,920	265,031	16,778	30,939	22,173	273,675	17,562	601,294	350,169	198,200	31,213	21,712
	1971	265,688	169,859	13,975	46,653	35,202	770,652	35,319	987,228	245,774	626,686	80,438	34,329
West Germany	1962	909,278	849,920	43,164	8,448	7,747	189,111	22,908	1,087,462	772,740	297,381	7,718	9,624
	1971	860,004	697,966	47,314	102,557	12,167	791,056	33,986	1,569,313	597,379	811,156	144,584	16,195
United Kingdom	1962	989,197	982,960	813	779	4,645	339,835	39,210	1,295,839	937,576	352,785	779	4,699
	1971	862,469	720,623	1,313	121,393	19,139	702,802	46,643	1,502,737	706,707	650,054	126,778	19,198
USSR & EAST EUROPE	1962	5,247,523	3,345,617	1,283,624	569,939	48,343	-359,699	n.a.	4,905,184	3,236,337	1,050,379	569,983	48,486
	1971	8,247,415	4,005,187	2,522,422	1,631,646	88,161	-484,909	n.a.	7,792,054	3,849,386	2,178,315	1,675,746	88,607

Table A-1. (continued)

		Production						Consumption				
	Total Energy	Coal & Lignite	Crude Petroleum	Natural Gas	Hydro & Nuclear Electricity	Net Imports	Bunkers	Total Energy	Solid Fuels	Liquid Fuels	Natural & Imported Gas	Hydro, Nuclear & Imported Electricity
Czechoslovakia												
1962	348,086	337,316	1,127	7,801	1,842	24,030	n.a.	366,295	336,571	20,070	7,771	1,882
1971	400,791	389,927	1,235	7,982	1,646	87,558	n.a.	470,434	384,861	61,583	18,718	5,272
East Germany												
1962	376,609	375,698	191	348	372	56,296	n.a.	435,395	418,656	15,930	544	265
1971	412,061	390,535	1,911	18,620	995	113,249	n.a.	527,034	438,859	68,032	18,870	1,274
Poland												
1962	560,609	553,362	1,421	5,351	475	-72,672	n.a.	485,879	458,135	20,100	7,183	461
1971	802,645	763,645	2,690	35,133	1,176	-92,537	n.a.	701,842	599,657	56,218	44,835	1,132
USSR												
1962	3,634,061	1,923,740	1,186,373	479,882	44,066	-370,058	n.a.	3,287,533	1,841,141	924,542	477,926	43,924
1971	6,126,470	2,258,410	2,401,965	1,386,225	79,870	-735,377	n.a.	5,447,614	2,161,282	1,801,157	1,409,598	75,578
MIDDLE EAST												
1962	2,009,510	23,304	1,965,400	19,874	931	-1,776,794	101,709	171,461	24,020	126,636	19,874	931
1971	5,397,781	33,349	5,198,880	161,827	3,724	-4,879,141	170,231	440,882	33,658	278,683	124,818	3,724
Iran												
1962	427,138	1,005	419,205	6,865	64	-379,373	18,596	37,162	1,005	29,229	6,865	64
1971	1,554,716	2,940	1,448,220	101,915	1,642	-1,385,059	42,263	149,749	2,940	80,257	64,910	1,642
Iraq												
1962	317,407	...	313,198	4,209	...	-297,361	270	18,321	5	14,107	4,209	...
1971	539,323	...	533,644	5,679	...	-506,792	657	30,458	5	24,774	5,679	...
Kuwait												
1962	595,816	...	587,167	8,649	...	-557,395	27,136	22,481	...	13,833	8,649	...
1971	966,946	...	937,576	29,371	...	-942,177	39,749	41,547	...	12,177	29,371	...
Saudi Arabia												
1962	482,606	...	482,528	78	...	-462,634	17,238	8,477	...	8,399	78	...
1971	1,442,712	...	1,423,132	19,580	...	-1,353,199	62,269	37,578	...	17,998	19,580	...
United Arab Emirates												
1962	5,076	...	5,076	...	...	-4,998	...	83	...	83	...	...
1971	329,525	...	329,525	...	...	-327,849	809	862	...	862	...	...
FAR EAST[b]												
1962	2,433,678	2,092,354	233,608	34,511	73,206	514,627	99,793	2,857,293	2,127,164	622,442	34,481	73,206
1971	3,833,084	2,940,853	659,437	104,684	28,111	1,842,429	229,688	5,407,856	3,086,559	2,095,622	97,564	128,111
Mainland China[c]												
1962	1,381,163	1,322,206	43,473	245	15,239	19,370	n.a.	1,398,279	1,315,567	67,228	245	15,239
1971	2,297,968	2,096,984	163,072	7,120	30,792	43,022	n.a.	2,341,965	2,095,323	208,730	7,120	30,792

Table A-1. (continued)

		Production						Consumption				
		Total Energy	Coal & Lignite	Crude Petroleum	Natural Gas	Hydro & Nuclear Electricity	Net Imports	Bunkers	Total Energy	Solid Fuels	Liquid Fuels	Natural & Imported Gas	Hydro, Nuclear & Imported Electricity
India	1962	314,967	301,056	6,860		7,051	48,466	3,572	358,391	294,887	56,453	...	7,051
	1971	416,069	349,698	45,766	3,655	16,949	90,910	5,037	503,294	355,686	127,003	3,655	16,949
Japan	1962	325,037	271,999	4,910	8,923	39,205	379,157	49,314	650,779	326,766	275,885	8,923	39,205
	1971	242,418	164,032	4,817	17,557	56,012	1,643,935	136,901	1,675,334	391,995	1,200,490	26,837	56,012
Indonesia	1962	166,429	2,313	146,025	17,655	436	-112,886	3,322	52,954	2,401	32,463	17,655	436
	1971	305,540	970	281,490	22,295	784	-212,812	2,891	76,518	1,044	52,396	22,295	784
Australia	1962	147,005	143,310	...	15	3,680	60,486	9,859	207,054	126,547	76,812	15	3,680
	1971	369,308	253,266	93,766	14,607	7,669	-31,130	17,870	339,751	164,689	152,787	14,607	7,669
AFRICAd	1962	480,911	223,151	248,704	2,475	6,581	-91,316	39,244	354,005	219,329	125,621	2,475	6,581
	1971	2,116,707	312,811	1,763,819	23,794	16,283	-1,443,476	59,462	598,310	310,136	260,210	11,682	16,283
Algeria	1962	134,579	260	131,859	2,303	157	-120,418	2,318	12,069	848	8,756	2,303	162
	1971	260,087	74	240,468	19,350	196	-223,391	2,867	36,084	1,524	24,103	10,261	196
Libya	1962	55,934	...	55,934	...	...	-1,891	676	1,842	186	1,656	...	...
	1971	843,364	...	843,364	3,033	...	-842,575	421	5,581	...	5,566	15	...
Nigeria	1962	24,373	3,107	21,197	...	69	-16,268	211	8,056	2,950	5,037	...	69
	1971	489,147	951	486,501	730	965	-456,832	294	16,337	1,019	13,622	730	965
South Africa	1962	202,297	202,277	...	...	20	21,217	3,660	221,264	194,437	26,808	...	20
	1971	288,262	288,198	...	...	64	94,844	26,019	351,355	278,942	72,349	...	64
UAR (Egypt)	1962	30,561	...	29,841	...	720	13,289	6,262	38,259	1,338	36,201	...	720
	1971	97,147	...	93,869	...	3,278	-46,344	1,656	47,525	2,602	41,645	...	3,278

aIncludes Caribbean America.
bIncludes Oceania.
cIncludes North Korea, North Vietnam, and Mongolia.
dIncludes North Africa.

Source: United Nations, Department of Economic and Social Affairs, *World Energy Supplies, 1961-1970* and *1968-1971,* Statistical Papers, Series J, No. 15 and No. 16 (New York).

Note: In column 6, "Net Imports of Total Energy," a minus indicates that the country or region is a net exporter.

Table A-2. Estimated Major Interregional Trade in Solid and Liquid Fuels, by Region of Origin and Destination, 1971 (in thousand barrels of oil equivalent)

Importing Region	World	Canada	United States	Western Europe	Caribbean[a]	Other Latin America	Middle East	North Africa	Sub-Saharan Africa	Far East[b]	Japan	Communist Countries[c]
World:												
Solid Fuels	566,072.5	35,819.0	260,582.0	6,272.0	0	0	0	2,523.5	5,439.0	100,009.0	0	155,428.0
Liquid Fuels	10,159,921.0	299,409.5	82,709.0	101,214.5	1,333,637.0	221,153.5	5,489,271.5	1,353,237.5	605,498.5	271,815.5	28,433.5	373,541.0
Total	10,725,993.5	335,228.5	343,291.0	107,486.5	1,333,637.0	221,153.5	5,489,271.5	1,355,761.0	610,937.5	371,824.5	28,433.5	528,969.0
Canada:												
Solid Fuels	82,393.5		81,732.0	661.5	0	0	0	0	0	0	0	0
Liquid Fuels	263,384.0		10,110.5	2,190.0	159,833.5	7,665.0	58,217.5	4,927.5	20,440.0	0	0	0
Total	345,777.5	···	91,842.5	2,851.5	159,833.5	7,665.0	58,217.5	4,927.5	20,440.0	0	0	0
United States:												
Solid Fuels	1,102.5	955.5		73.5	0	0	0	0	73.5	0	0	0
Liquid Fuels	1,435,143.5	296,891.0		54,166.0	617,543.5	202,356.0	146,876.0	32,959.5	37,704.5	43,362.0	803.0	2,482.0
Total	1,436,246.0	297,846.5	···	54,239.5	617,543.5	202,356.0	146,876.0	32,959.5	37,778.0	43,362.0	803.0	2,482.0
Western Europe:												
Solid Fuels	224,003.5	1,935.5	75,019.0		0	0	0	122.5	4,410.0	15,508.5	0	127,008.0
Liquid Fuels	4,964,511.0	365.0	23,506.0		192,391.5	7,482.5	2,843,824.5	1,147,742.5	411,318.5	547.5	0	337,333.0
Total	5,188,514.5	2,300.5	98,525.0	···	192,391.5	7,482.5	2,843,824.5	1,147,865.0	415,728.5	16,056.0	0	464,341.0
Caribbean:												
Solid Fuels	3,699.5	0	2,205.0	122.5		0	0	0	0	759.5	0	612.5
Liquid Fuels	130,597.0	0	15,366.5	146.0		0	38,033.0	26,170.5	45,442.5	5,365.5	73.0	0
Total	134,296.5	0	17,571.5	268.5	···	0	38,033.0	26,170.5	45,442.5	6,125.0	73.0	612.5
Other Latin America:												
Solid Fuels	13,940.5	0	12,593.0	318.5	0		0	0	0	318.5	0	710.5
Liquid Fuels	691,127.5	401.5	11,789.5	4,416.5	345,947.0		150,635.5	101,506.5	76,285.0	109.5	36.5	0
Total	705,068.0	401.5	24,382.5	4,735.0	345,947.0	···	150,635.5	101,506.5	76,285.0	428.0	36.5	710.5
Middle East:												
Solid Fuels	0	0	0	0	0	0		0	0	0	0	0
Liquid Fuels	4,088.0	0	0	2,993.0	0	0		···	0	0	0	1,095.0
Total	4,088.0	0	0	2,993.0	0	0	···	···	0	0	0	1,095.0
North Africa:												
Solid Fuels	2,548.0	0	0	0	0	0	0		0	0	0	0
Liquid Fuels	51,392.0	0	0	17,447.0	0	0	16,388.5		0	0	0	0
Total	53,940.0	0	0	17,447.0	0	0	16,388.5	···	0	0	0	0

Table A-2. (continued)

Exporting Region → Importing Region ↓	World	Canada	United States	Western Europe	Caribbean[a]	Other Latin America	Middle East	North Africa	Sub-Saharan Africa	Far East[b]	Japan	Communist Countries[c]
Sub-Saharan Africa:												
Solid Fuels	4,753.0	0	220.5	1,641.5	0	0	0	2,205.0		0	0	686.0
Liquid Fuels	206,334.5	0	1,679.0	11,424.5	8,687.0	3,650.0	176,076.0	2,591.5	…	584.0	0	1,642.5
Total	211,087.5	0	1,899.5	13,066.0	8,687.0	3,650.0	176,076.0	4,796.5		584.0	0	2,328.5
Far East:												
Solid Fuels	1,788.5	0	49.0	220.5	0	0	0	0	98.0	0	0	1,421.0
Liquid Fuels	736,898.5	0	6,205.0	2,153.5	5,402.0	0	688,317.0	3,650.0	0		27,521.0	3,650.0
Total	738,687.0	0	6,254.0	2,374.0	5,402.0	0	688,317.0	3,650.0	98.0		27,521.0	5,071.0
Japan:												
Solid Fuels	227,899.0	32,928.0	88,200.0	0	0	0	0	196.0	857.5	83,275.5	…	22,442.0
Liquid Fuels	1,609,942.0	1,752.0	14,052.5	1,533.0	3,540.5	0	1,335,717.5	7,409.5	14,308.0	221,847.0	…	9,782.0
Total	1,837,841.0	34,680.0	102,252.5	1,533.0	3,540.5	0	1,335,717.5	7,605.5	15,165.5	305,122.5		32,224.0
Communist Countries:												
Solid Fuels	3,944.5	0	563.5	3,234.0	0	0	0	0	0	0	0	
Liquid Fuels	66,503.0	0	0	4,745.0	292.0	0	35,186.0	26,280.0	0	147.0	0	
Total	70,447.5	0	563.5	7,979.0	292.0	0	35,186.0	26,280.0	0	147.0	0	…

Sources: *Solid Fuels* - - United Nations, Department of Economic and Social Affairs, *World Energy Supplies, 1968-1971,* Statistical Papers, Series J, No. 16 (New York, 1973), Table 4, pp. 42-47. *Liquid fuels* - - U.S. Department of Interior, Office of Oil and Gas, *1971 Petroleum Supply and Demand in the Non-Communist World* (Washington, D.C., May 1973), p. 19.

a. Includes Mexico.

b. Includes Australia and New Zealand; excludes Japan.

c. Includes mainland China and other Asian Communist countries; also includes USSR and Eastern Europe.

Table A–3.   Exports and Imports of Solid and Liquid Fuels: Interregional Flows by Destination and Origin, 1971

A. Percentage Distribution of Exports to Regions Listed at Left

Destination \ Origin	World	Canada	USA	Western Europe	Caribbean	Other Latin America	Middle East	North Africa	Sub-Saharan Africa	Far East	Japan	Communist Countries
Canada	3.22	...	26.75	2.65	11.98	3.47	1.06	0.36	3.35	0	0	0
United States	13.39	88.85	...	50.46	46.31	91.50	2.68	2.43	6.18	11.66	2.82	0.47
Western Europe	48.37	0.69	28.70	...	14.43	3.38	51.81	84.67	68.05	4.32	0	87.78
Caribbean	1.25	0	5.12	0.25	...	0	0.69	1.93	7.44	1.65	0.26	0.12
Other Latin America	6.57	0.12	7.10	4.41	25.94	...	2.74	7.49	12.49	0.12	0.13	0.13
Middle East	0.04	0	0	2.78	0	0	...	0	0	0	0	0.21
North Africa	0.50	0	0	16.23	0	0	0.30	...	0	0	0	3.80
Sub-Saharan Africa	1.97	0	0.55	12.16	0.65	1.65	3.21	0.35	...	0.16	0	0.44
Far East	6.89	0	1.82	2.21	0.41	0	12.54	0.27	0.02	...	96.79	0.96
Japan	17.13	10.35	29.79	1.43	0.27	0	24.33	0.56	2.48	82.06	...	6.09
Communist Countries	0.66	0	0.16	7.42	0.02	0	0.64	1.94	0	0.04	0	...

B. Percentage Distribution of Imports

Destination \ Origin	World	Canada	USA	Western Europe	Caribbean	Other Latin America	Middle East	North Africa	Sub-Saharan Africa	Far East	Japan	Communist Countries
Canada	3.13	...	20.74	0.04	0	0.06	0	0	0	0	1.89	0
United States	3.20	26.56	...	1.90	13.08	3.46	0	0	0.90	0.85	5.56	0.80
Western Europe	1.00	0.82	3.78	...	0.20	0.67	73.21	32.35	6.20	0.32	0.08	11.33
Caribbean	12.43	46.22	43.00	3.71	...	49.07	0	0	4.12	0.73	0.19	0.41
Other Latin America	2.06	2.22	14.09	0.14	0	...	0	0	1.73	0	0	0
Middle East	51.18	16.84	10.23	54.81	28.32	21.36	...	30.38	83.41	93.18	72.68	49.95
North Africa	12.64	1.43	2.29	22.12	19.49	14.40	0	...	2.27	0.49	0.41	37.30
Sub-Saharan Africa	5.70	5.91	2.63	8.01	33.84	10.82	0	0	...	0.01	0.83	0
Far East	3.47	0	3.02	0.31	4.56	0.06	0	0	0.28	...	16.60	0.21
Japan	0.27	0	0.06	0	0.05	0.01	0	0	0	3.73	...	0
Communist Countries	4.93	0	0.17	8.95	0.46	0.10	26.79	37.27	1.10	0.69	1.75	...

Source: Table A-2.
Note: Excludes natural gas.
Percentages may not add up to 100.00 because of rounding.
Exports = Imports because bunker fuels are omitted.

Table A–4. U.S. Imports of Crude Oil and Products, 1973 (thousand barrels)

Origin	Crude Oil			Refined Products			Total Imports		
	Per Day	Total for Year	Percent of Total	Per Day	Total for Year	Percent of Total	Per Day	Total for Year	Percent of Total
Canada	1,001.0	365,370	30.9	311.9	113,854	10.6	1,312.9	479,224	21.2
Caribbean (includes Mexico)	1.3	489[a]	0.1[b]	1,196.2	436,601	40.4	1,197.5	437,090	19.3
Other Latin America	455.8	166,379	14.0	994.1	362,838	33.6	1,449.9	529,217	23.4
Western Europe	· ·	· · ·	· ·	254.9	93,046	8.6	254.9	93,046	4.1
North Africa	285.1	104,041	8.8	45.9	16,755	1.6	331.0	120,796	5.3
Sub-Saharan Africa	497.1	181,440	15.3	13.4	4,900	0.5	510.5	186,340	8.2
Middle East	802.7	292,988	24.7	53.5	19,539	1.8	856.2	312,527	13.8
Japan	· ·	· · ·	· ·	8.7	3,171	0.3	8.7	3,171	0.2
Far East	200.8	73,289	6.2	45.6	16,635	1.5	246.4	89,924	4.0
USSR & Eastern Europe	· ·	· · ·	· ·	33.4	12,188	1.1	33.4	12,188	0.5
TOTAL	3,243.8	1,183,996	100.0	2,957.6	1,079,527	100.0	6,201.4	2,263,523	100.0

Source: U.S. Department of the Interior, Bureau of Mines, "Crude Petroleum, Petroleum Products, and Natural-Gas-Liquids," *Mineral Industry Surveys* (Washington, D.C., January 1974), p. 34, Table 30.

[a]Entirely from Mexico.

[b]Actually 0.04 percent of total.

Table A–5.   World Movement of Natural Gas, 1971 (in million cubic meters)

Destination \ Origin	World	Algeria[a]	Canada	United States	Mexico	Iran	Afghanistan	Brunei	Netherlands[b]	Romania	USSR
World	60,780	1,390	25,570	2,280	660	5,670	2,510	50	17,430	200	4,560
Canada	410	...	—	410	...	...	...	...	...	...	...
United States	26,310	80	25,570	—	660	...	...	...	...	...	...
Caribbean America[c]	450	...	...	450	—	...	...	...	...	...	...
Sarawak	50	...	...	...	...	...	...	50	...	...	...
Japan	1,420	...	...	1,420[a]	...	...	...	...	...	...	...
Belgium-Luxembourg	6,300	...	...	...	...	...	...	...	6,300	...	...
France	5,170	480	...	...	...	...	...	...	4,690	...	...
West Germany	6,410	...	...	...	...	...	...	...	6,410	...	...
Austria	1,430	...	...	...	...	...	...	...	...	...	1,430
United Kingdom	830	830	...	...	...	...	...	...	...	...	...
Spain	370	...	...	...	...	...	...	...	...	...	...
Czechoslovakia	1,640	...	...	...	...	...	...	...	...	...	1,640
Hungary	200	...	...	...	...	...	...	...	...	200	...
Poland	1,490	...	...	...	...	...	...	...	...	...	1,490
USSR	8,180	...	...	...	...	5,670	2,510	...	...	...	...

Source: United Nations, Department of Economic and Social Affairs, *World Energy Supplies, 1968-1971, Statistical Papers, Series J, No. 16* (New York, 1973), Table 16, pp. 132-35.

[a]Liquefied natural gas.

[b]From Netherlands, 17,400; West Germany, 20; France, 10.

[c]Entirely to Mexico.

crude oil and refined products. Solid fuels include hard coal and lignite made into briquettes and coke. (Natural gas trade, which accounts for a very small fraction of total world energy movements, is shown in Table A-5.)

Table A-3 shows both the percentage of total world exports of solid and liquid fuels destined for each of the major regions and the proportion of total world imports emanating from each region. Approximately 13 percent of total world exports went to the United States in 1971, while imports from the United States accounted for some 3 percent of total world imports. (Note that Table A-1 shows that the United States had an energy trade deficit of 1.2 billion barrels of oil equivalent in 1971.) Exports to the United States came chiefly from Latin America and Canada. About one-half of total world exports went to Western Europe, whose principal suppliers were the Middle East and North Africa. Japan, supplied primarily by the Middle East, the Far East (Indonesia and Australia), and the United States, received the second largest share of total world exports—17 percent. Participation by Communist countries in foreign trade in energy was relatively slight. While 5 percent of total world imports came from those countries, they received less than 1 percent of world exports.

1973 U.S. Oil Imports

Because U.S. oil imports have risen substantially since 1970 (the most recent year for which worldwide figures are available), U.S. oil imports by region of origin are shown in Table A-4 as a supplement to Tables A-2 and A-3. By the end of 1973, U.S. imports of crude oil and products amounted to roughly 35 percent of total domestic oil consumption. In addition to the marked increase in oil imports in both absolute and relative terms in recent years, the supply pattern of U.S. imports has changed markedly just since 1970. The Caribbean provided over one-half of U.S. oil imports in 1970; by 1973 the proportion had dropped to about one-third. The percentage of total imports from the Middle East and from Africa doubled.

Natural Gas Flows

Table A-5 shows world trade in natural gas for 1971. Shipments of liquefied natural gas (LNG) had just begun (from Algeria to Western Europe), and almost all of the trade was by pipeline. Nearly one-half of world movement of natural gas involved flows from Canada and Mexico to the United States. About another one-fourth involved flows from the Netherlands to Western Europe.

Table A-6.   Proved World Reserves of Oil and Gas by Region and
Selected Countries, January 1974

Country	Oil (thousand bbl.)	Gas (million cu. mtr.)
Asia, of which	35,635,040	3,801,700
Indonesia	10,500,000	424,929
Mainland China	20,000,000	566,572
Australia	2,300,000	1,067,989
Middle East, of which	350,162,500	11,708,924
Abu Dhabi	21,500,000	354,108
Iran	60,000,000	7,648,725
Iraq	31,500,000	623,229
Kuwait	64,000,000	920,680
Neutral Zone	17,500,000	226,629
Saudi Arabia	132,000,000	1,441,926
USSR and Eastern Europe	83,000,000	20,266,289
USSR	80,000,000	20,000,000
Western Europe, of which	15,990,500	5,490,000
Netherlands	251,000	2,606,232
Norway	4,000,000	651,558
United Kingdom	10,000,000	1,416,431
Africa, of which	67,303,750	5,317,847
Algeria	7,640,000	3,001,275
Libya	25,500,000	764,873
Nigeria	20,000,000	1,133,144
Western Hemisphere, of which	75,764,669	11,017,847
Canada	9,424,170	1,424,901
United States	34,700,249	7,005,949
Ecuador	5,675,000	141,643
Venezuela	14,000,000	1,189,802
Total World	627,856,459	57,602,606

Source: *Oil and Gas Journal,* December 31, 1973, pp. 86-87.

The USSR exported relatively small quantities to Western Europe
and also received natural gas from Afghanistan and Iran.

RESERVES

With the shift from coal to oil and natural gas as the primary sources
of energy and the intensive emphasis on exploration, estimates of
proved oil and gas reserves are the object of great interest and change
frequently from year to year. The estimates, by and large, are made
and published by the major oil companies rather than by the
governments involved. The concept of proved reserves for oil and gas
is too restrictive to be useful for long-range analysis of world energy
resources because reserves become "proved" only in response to

commercial demand. At the same time, estimates of unproved resources vary tremendously. Thus, for present purposes data will be presented in terms of proved reserves (see Table A-6).

Coal is a different matter altogether. The figures for world coal reserves in Table A-7 are estimates of total original coal resources. ("Original" estimates include coal that has already been recovered.) The same figures for coal reserves are often used for years—a primary reason is that production typically is only a very small faction of what is available in the ground. Thus, the coal reserve estimates presented here are more comprehensive than the reserve figures for oil and gas, which can be assumed to be even moderately useful for at most about ten years.

Oil and Gas

There appears to be considerable latitude in what constitutes "proved" crude oil reserves. Definitions of "proved" reserves vary from one country to another and among oil companies. A frequently cited example of the definitional variations is that the figure for proved reserves for a Middle Eastern country in any given year may be roughly twice as high as would be the case if the oil fields were located in North America. Moreover, reserve figures are subject to frequent revision because the ultimate size of newly discovered reservoirs (whether in old or new fields) is seldom known in the year of discovery. Similar difficulties are associated with estimates of natural gas.

Table A-7. Estimated World Original Coal Resources, by Region (in billion metric tons)

	Coal Resources Determined by Mapping & Exploration	Probable Additional Resources (Unmapped & Unexplored)	Estimated Total
North America	1,560.0	2,612.2	4,172.2
(of which United States)	(1,434.0)	(1,490.2)	(2,924.2)
Europe	562.4	190.4	752.8
Oceania	54.4	63.5	117.9
USSR	5,895.5	2,721.0	8,616.5
Latin America	18.1	9.1	27.2
Asia	453.5	907.0	1,360.5
Africa	72.6	145.1	217.7
Total	8,616.5	6,648.3	15,264.8

Source: Donald A. Brobst and Walden P. Pratt, eds., *United States Mineral Resources*, Geological Survey Professional Paper 20 (Washington, D.C.: United States Government Printing Office, 1973), p. 137 and p. 140.

A more or less standard operating definition of proved crude oil reserves for the United States and Canada was spelled out in 1971 in a joint publication of the American Gas Association, the American Petroleum Institute, and the Canadian Petroleum Association:

Reservoirs are considered proved if economic producibility is supported by either actual production or conclusive formation tests. The area . . . considered proved includes: (1) that portion delineated by drilling and defined by gas-oil or oil-water contacts . . . (2) the immediately adjoining portions not yet drilled but which can be reasonably judged as economically productive on the basis of available geological and engineering data . . . .

Reserves of crude oil which can be produced economically through the application of improved recovery techniques . . . are included in the "proved" classification when successful testing by a pilot project . . . provides support for the engineering analysis on which the project . . . was based.[1]

Within the context of the preceding caveats concerning the validity and comparability of estimates of gas and oil reserves, as of January 1974 total world reserves of crude oil were estimated at just under 628 billion barrels (see Table A-6). The leading holders of reserves were Saudi Arabia, the USSR, Kuwait, Iran, the United States, Iraq, Libya, and Abu Dhabi. World estimates of proved natural gas reserves were 57,603 billion cubic meters (see Table A-6). The leading countries were the USSR, Iran, the United States, Algeria, Saudi Arabia, Canada, the United Kingdom, and Nigeria.

Coal

Table A-7, which provides estimates of world coal reserves by region, indicates that more than one-half of the world's coal resources are in the USSR. The United States ranks second, but its coal resource base is much smaller than that of the USSR. Most of the reserves in Asia are located in China.

The recoverability rate of coal reserves under present technological conditions is generally regarded as about 50 percent. Thus, the estimates shown in Table A-7 would be reduced by half, so that proved world coal resources, given present technology, would be about 4.3 billion metric tons. This means that an enormous quantity of coal is available to meet future world energy requirements. Darmstadter notes:

If energy consumption from all fuels were to grow at the end of the present century at the annual 5 percent rate . . . cumulative energy requirements to the end of the century . . . might amount to 400 billion tons of coal equivalent. Not only could the estimated 4.3 billion tons of estimated recoverable coal resources meet this entire growth of energy demand, but in the year 2000, at then prevailing rates of total energy consumption, enough coal would be left in the ground to meet the entire bill for a century and a half beyond.[2]

Uranium

Uranium and thorium are the two natural resources essential for the manufacture of nuclear fuel. Table A-8 shows production of electricity (the only energy use for which nuclear energy is now interchangeable with fossil fuels) by nuclear power plants in major countries in 1971. While nuclear energy presently accounts for about only 2 percent of total world energy production, it is generally forecast to account for a much more significant proportion by 1985.[3]

Table A- 8. Production of Electricity by Nuclear Power Plants in Major Producing Countries, 1971 (in million killowatt-hours)

Country	Quantity
Canada	3,988
United States	37,899
Total North America	41,887
Japan	8,010
India	1,793
Total Asia	9,803
France	8,743
West Germany	5,812
Italy	3,365
Switzerland	2,700
United Kingdom	26,937
Other Western Europe	3,018
Total Western Europe	50,575
USSR	4,300
Eastern Europe[a]	404
Total USSR and Eastern Europe	4,704

Source: United Nations, Department of Economic and Social Affairs, *World Energy Supplies 1968-1971*, Statistical Papers, Series J, No. 16 (New York, 1973), Table 20, pp. 164-75.

[a]Entirely from East Germany.

Table A-9 shows world reserves by major producing countries in the non-Communist world of the uranium oxide (U_3O_8)

used in the nuclear fuel cycle. Estimates are shown for reserves at a price up to $10 a pound and also for reserves in the $10 to $15 price range. The United States, South Africa, Canada, and Australia possess the bulk of reserves in both the "reasonably assured resources" category and in the "estimated additional resources" category at under $10 per pound. These same countries dominate the scene in the $10 to $15 price range, except that Sweden becomes important in this higher priced category.

With a few exceptions, U_3O_8 is produced primarily in the countries that hold the largest deposits. (See Table A-10.) Australian production, which was relatively low in 1969 and 1970, was discontinued in 1971. France, whose resources are relatively modest, has been a major producer and, because of its reserve position, has invested in production in two former colonies, Gabon and Niger.

Table A-11, which shows U_3O_8 imports of major consuming nations, reflects the dependence of most of the major consumers (with the exception of the United States) on uranium resources outside their respective national borders. France appears as both an exporter and importer; its lack of self-sufficiency in uranium is reflected in its efforts to assure continuity of supplies by investments in Africa. The United States, which is self-sufficient in uranium, has had an embargo on uranium imports since 1966 and, thus, does not appear in the table. Because the U.S. domestic price for U_3O_8 for years was above the world price, the United States has exported very limited quantities of the ore concentrate; U.S. exports are so small they do not appear in a table showing major flows.

Table A-9. Uranium Resources by Major Countries, 1973
(1000s of short tons)

	Price Range up to $10/lb U_3O_8 [a]		Price Range $10-15/lb U_3O_8	
	Reasonably Assured Resources	*Estimated Additional Resources*	*Reasonably Assured Resources*	*Estimated Additional Resources*
Argentina	12.0	18.0	10.0	30.0
Australia	92.0	102.0	38.3	38.0
Brazil	. . .	3.3	0.9	. . .
Canada	241.0	247.0	158.0	284.0
Central African Republic	10.5	10.5	. . .	. . .
Denmark (Greenland)	7.0	13.0	. . .	. . .
Finland	. . .	. . .	1.7	. . .
France	47.5	31.5	26.0	32.5
Gabon	26.0	6.5	. . .	6.5
India	. . .	. . .	3.0	1.0
Italy	1.6	. . .	. . .	. . .
Japan	3.6	. . .	5.4	. . .
Mexico	1.3	. . .	1.2	. . .
Niger	52.0	26.0	13.0	13.0
Portugal (Europe)	9.3	7.7	1.3	13.0
(Angola)	. . .	. . .	. . .	17.0
South Africa	263.0	10.4	80.6	33.8
Spain	11.0	. . .	10.0	. . .
Sweden	. . .	. . .	351.0	52.0
Turkey	2.8	. . .	0.6	. . .
United States	337.0	700.0	183.0	300.0
Yugoslavia	7.8	13.0	. . .	. . .
Zaire	2.3	2.2	. . .	. . .
TOTAL	1,127.7	1,191.1	884.0	820.8

Source: Organisation for Economic Co-operation and Development, *Uranium Resources Production and Demand*, a joint report by the OECD Nuclear Energy Agency and the International Atomic Energy Agency (Paris, August 1973), p. 14.

[a]Dollar value of March 1973: $1 = 0.829; EMA u/a* = 0.829 SDR (Special Drawing Rights). This dollar value corresponds to $42.22 per fine ounce of gold.

*European Monetary Agreement units of account.

Table A- 10.   World Uranium Production by Major Countries
(short tons U_3O_8)

	1969	1970	1971	1972
Argentina	55	60	60	33
Australia	330	330	. . .	. . .
Canada	4,450	4,580	4,980	5,200
France	1,530	1,630	1,630	1,800
Gabon	650	520	700	270
Germany	. . .	. . .	. . .	. . .
Japan	. . .	. . .	. . .	20
Mexico	40	. . .	. . .	. . .
Niger	. . .	. . .	560	1,130
Portugal	122	. . .	105	105
South Africa	4,000	4,117	4,186	4,000
Spain	72	66	78	78
Sweden	38	18	10	9
United States	11,600	12,900	12,300	12,900
Total	22,880	24,200	24,610	25,550

Source: Organisation for Economic Co-operation and Development, *Uranium Resources Production and Demand,* a joint report by the OECD Nuclear Energy Agency and the International Atomic Energy Agency (Paris, August 1973), p. 16.

Table A–11. Major Free World Flows of Natural Uranium Ores and Concentrates by Importing Countries for Years Specified (in short tons)

Importing Countries		Gabon	South Africa	France	Portugal	Niger	Canada	Unspecified	Total
Japan[a]			6,000	5,000[b]			32,000	17,000	60,000
West Germany	1968		11	233[c]					294
	1969			283					283
	1970		19	299					318
	1971			260					260
France	1968	542[d]			731[e]			261	1,534
	1969	526							526
	1970	647							647
	1971	646				538[f]			1,184
United Kingdom	1970		850				850		1,700
	1971		1,090[g]				1,090		2,180
	1972		1,100				1,100		2,200

Sources: Japan - Embassy of Japan, Washington, D.C.; West Germany - U.S. Bureau of Mines; France - U.S. Bureau of Mines; United Kingdom - Estimates for Canadian imports are derived from *Statistics Canada, 1972.*

NOTES FOR APPENDIX A

1. *Reserves of Crude Oil, Natural Gas Liquids, and Natural Gas in the United States and Canada and United States Productive Capacity as of December 31, 1971.* Vol. 26, May 1971.

2. Joel Darmstadter and associates, *Energy in the World Economy* (Baltimore: Johns Hopkins University Press, 1971), p. 47.

3. Thorium reserves are not included in the accompanying tables. Thorium reserves have been explored only slightly, because very little thorium is now used in nuclear fuel production, and its use is not expected to increase much in the foreseeable future.

Appendix B

Nuclear Energy Statistics

Table B-1. Transfer Patterns in the World Reactor Market by Country, March 1973 (number of reactors and generating capacity)

Country	100% U.S. Built	80% U.S., 20% Abroad	50% U.S., 50% Abroad	50-100% Indigenous	Non-U.S. Import	TOTAL
Western Europe:						
U.K.				22*		22
				4235		4235
France	1			6*		7
	266			2285		2551
West	1			4+1*		6
Germany	237			1730/100		2067
Spain	2				1*	3
	593				480	1073
Italy	1	1			1*	3
	150	247			210	607
Sweden				2		2
				1200		1200
Switz.			3			3
			1006			1006
Other					1	1
W. Europe					450	450
Asia:						
Japan	4		1	1	1*	7
	1862		460	500	166*	2988
Other	2				3*	5
Asia	380				925	1305
Australia						
Africa						
Canada				6*		6
				2490		2490
Latin America						
Subtotal	11	1	4	42	7	65
	3488	247	1466	12540	2231	19972
East				1		1
Europe				150		150
U.S.S.R.				6		6
				1745		1745
TOTAL	11	1	4	49	7	72
	3488	247	1466	14435	2231	21867

Table B-1 (continued)

UNDER CONSTRUCTION

	100% U.S. Built	80% U.S., 20% Abroad	50% U.S., 50% Abroad	50-100% Indigenous	Non-U.S. Import	TOTAL
Western Europe:				8*		8
U.K.				5090		5090
			1			1
France			850			850
				6		6
West				7118		7118
Germany		6				6
Spain		5412				5412
		1				1
Italy		783				783
			1	2		3
Sweden			900	1160		2060
			1			1
Switz.			850			850
				3	1	4
Other W. Europe				2430	440	2870
Asia:	1		1	6		8
Japan	1122		559	4174		5855
	1	2			2*	5
Other Asia	564	1208			800	2572
Australia						
Africa						
Canada						
					1	1
Latin America					318	318
	2	9	4	25	4	44
Subtotal	1686	7403	3159	19972	1558	33778
					4	4
East Europe					2640	2640
U.S.S.R.				12		12
				9640		9640
	2	9	4	37	8	60
TOTAL	1686	7403	3159	29612	4198	46058

Table B-1 (continued)

	100% U.S. Built	80% U.S., 20% Abroad	50% U.S., 50% Abroad	50-100% Indigenous	Non-U.S. Import	TOTAL
Western Europe				2*		2
U.K.				1320		1320
			1	2		3
France			850	1850		2700
				5		5
West Germany Spain				4924		4924
Italy		2		2		4
Sweden		1480		1800		3280
				1		1
Switz.				850		850
					1	1
Other W. Europe					700	700
Asia:		3		3		6
Japan		3322		2315		5637
		5				4
Other Asia		2364				2364
Australia						
Africa				4*		4
Canada				3000		3000
		2				2
Latin America		1426				1426
	0	11	1	19	1	32
Subtotal	0	8592	850	16059	700	26201
				5		5
East Europe				1760		1760
U.S.S.R.				1		1
				1000		1000
	0	11	1	25	1	38
TOTAL	0	8592	850	18819	700	28961

Table B-1 (continued)

	Total Without Proposed	Proposed	TOTAL
Western Europe:	32	8*	40
U.K.	10645	8410	19055
	11		11
France	6101		6101
	17	15	32
West Germany	14109	13626	27735
	9	5	14
Spain	6485	5730	12215
	4	1	5
Italy	1390	750	2140
	9	7	16
Sweden	6540	5300	11840
	5	5	10
Switz.	2706	4300	7006
	6	8	14
Other W. Europe	4020	4240	8260
Asia:	21	25	46
Japan	14480	24039	38519
	14	5+2*	21
Other Asia	6241	2684+440	9365
		2	2
Australia		1000	1000
		2	2
Africa		1000	1000
	10	8	18
Canada	5490	6000	11490
	3	6	9
Latin America	1744	2800	4544
	141	97	238
Subtotal	79951	79319	159270
	10	8	18
East Europe	4550	5834	10384
	19	4	23
U.S.S.R.	12385	6500	18885
	170	109	279
TOTAL	96886	91653	188539

Sources: U.S. Atomic Energy Commission, *Nuclear Power 1973-2000* (Washington, D.C.: Government Printing Office, 1972); additional data provided by the Commission's Division of International Programs.

"Power Reactors '73," *Nuclear Engineering International*, April 1973.

Notes: The United States is not included because virtually all U.S. reactors are built domestically.

The top figure in each column shows the number of reactors of 100 MWe or more. The bottom figure shows the total MWe capacity of reactors in each category.

An asterisk (*) indicates reactor type other than light water reactor.

Table B-2.   Annual Enriched Uranium Requirements
(in thousands of kilograms of separative work units)

	1980	1985	1990
United States	15,300	30,000	52,800
Total Foreign	13,784	27,920	47,700
Western Europe	9,284	18,520	29,000
West Germany	3,500	6,000	8,500
U.K. & Netherlands	1,380	4,300	6,080
France	990	2,200	4,000
Other Western Europe	3,414	6,020	10,420
Japan	4,200	6,300	10,300
Others	300	3,100	8,400

Sources: U.S. Atomic Energy Commission, *Nuclear Power 1973-2000* (Washington, D.C.: Government Printing Office, 1972); additional data provided by the Commission's Division of International Programs.

Note: West Germany, the United Kingdom, and the Netherlands are members of the Tripartite Centrifuge Organization.

Table B-3.   U.S. Nuclear-Related Exports, Selected
Fiscal Years (in millions of U.S. dollars)

Exports	Cumulative 1959 thru 1970	1971	1972	1973	1980 ('73 prices)	1985 ('73 prices)
Light Water Reactors, Power Plants, and Equipment	1,000	700	800	1,000	1,400–1,900[a]	1,400–2,200[b]
Enrichment Services[c]	500	72	44	61	356	600[d]
Miscellaneous		57	24	23	neg.	neg.
Total	1,500	829	868	1,084	1,756–2,256	2,000–2,800

Sources: Cumulative 1959-70 figures were obtained from statement of Myron Kratzer in *Hearings* of the U.S. House of Representatives, Subcommittee on International Cooperation in Science and Space, 92nd Cong., lst sess. (May 1971); and U.S. Atomic Energy Commission, *Major Activities in the Atomic Energy Programs, January-December 1971* (Washington, D.C.: Government Printing Office, 1972); and U.S. Atomic Energy Commission, *Financial Report* (FY 1972) (Washington, D.C.: Government Printing Office, 1973). Prices for enrichment and miscellaneous services supplied by the U.S. Atomic Energy Commission, Division of International Programs.

Note: The estimates in this table do not represent absolute dollar flows for the U.S. balance of payments for 1980 and 1985. The actual dollar earnings in U.S. exports for 1980, for example, would amount to the sum of the payments by overseas purchasers for reactors ordered between 1976 and 1980. These data were not available from the Atomic Energy Commission; hence, estimates were made of the value of the market for 1980 and 1985 (i.e., the value of contracts for reactor sales for those two years).

Canada is excluded from this table on the assumption that Canada will continue to rely on heavy water reactors for domestic use and for export.

ᵃEstimate was calculated as follows. The size of the reactor market for 1980 was computed on the assumption that the growth in nuclear power in each country between 1985 and 1980 (see Table 16-2) would determine the size of the 1980 market, given the long lead time between placement of orders and plant operations. The value of the market was computed by multiplying the kilowatt growth by $400 per kilowatt for Japan and Western Europe (the present consensus estimate of plant construction costs in developed countries) and $500 per kilowatt for less developed countries. Of total construction costs, the Atomic Energy Commission estimates that the United States realizes about 25 percent per reactor sale as export earnings for developed nations and about 30 percent per reactor sale as export earnings for less developed nations. Thus,

Region	Total Value of Reactor Export Market for 1980 (million of U.S. dollars)		U.S. Export Earnings (Range)		
			Low	High	
Supplier countries	2,000	U.S. gets 0% of market	0	300	U.S. gets 15% of market
Non-supplier Western Europe	750	U.S. gets 50% of market	400	500	U.S. gets 66% of market
Other	1,500	U.S. gets 66% of market	1,000 —— 1,400	1,100 —— 1,900	U.S. gets 75% of market

Note: the United Kingdom is not considered a supplier because it does not presently produce light water reactors. For purposes of this table, the United Kingdom was considered as part of "non-supplier Western Europe" and 25% of the UK's total new nuclear capacity was expected to be provided by LWRs by 1980.

ᵇThe same procedure was followed as for 1980.. Construction costs per kilowatt hour were held constant.

Region	Total Value of Reactor Export Market for 1985 (millions of U.S. dollars)		U.S. Export Earnings (Range)		
			Low	High	
Supplier countries	3,000	U.S. gets 0% of market	0	15	U.S. gets 5% of market
Non-supplier Western Europe	750	U.S. gets 25% of market	200	300	U.S. gets 30% of market
Other	3,500 —— 7,250	U.S. gets 33% of market	1,200 —— 1,400	1,750 —— 2,200	U.S. gets 50% of market

Note: the United Kingdom is considered a supplier nation (LWRs) by 1985, and 75% of UK's increased nuclear capacity is assumed to be provided by light water reactors.

cPrices for separative work unit (SWU):

$26.00 per SWU through February 1971.
$28.70 per SWU from February 1971 through November 1971.
$32.00 per SWU from November 1971 through August 1973.
$36.00 per SWU thereafter.

dAppendix Table B-2 indicates non-U.S., non-Communist enrichment requirements for 1985 (28 million separative work units). Figure is derived by multiplying 28 million SWU's by $36 (1973 average price per SWU). The Atomic Energy Commission assumes that the United States will supply 60% of non-Communist foreign uranium requirements in 1985.

Index

About the Authors

Joseph A. Yager is a senior fellow at the Brookings Institution. He was born in 1916 in Owensville, Indiana and educated in the public schools of Toledo, Ohio and at the University of Michigan where he specialized in law and economics.

He served with the Office of Strategic Services in China during World War II. After the war, he joined the U.S. Department of State and stayed with that Department in a series of foreign and domestic assignments for more than twenty years. Before coming to Brookings, he was a deputy division director at the Institute for Defense Analyses.

Eleanor B. Steinberg, who was born in 1936 in Washington, D.C., is a research associate at the Brookings Institution. A graduate of Oberlin College, she also studied at the London School of Economics and holds a master's degree in Political Science from Yale University, where she was a Woodrow Wilson Fellow.

Mrs. Steinberg's main professional interest has been the economic, environmental, and governmental problems of the developing countries. She has been a consultant to the World Bank and has worked at the U.S. Department of Commerce and the Urban Institute.

She is co-author, with Edwin T. Haefele, of *Government Controls on Transport: An African Case* (Brookings, 1965) and has contributed several journal articles.